STUDY GUIDE

for

McFadden/Keeton's

Biology: An Exploration of Life

STUDY GUIDE
for
McFadden/Keeton's
Biology: An Exploration of Life

CAROL HARDY McFADDEN
Cornell University

LINDSAY GOODLOE
Cornell University

W · W · NORTON & COMPANY · NEW YORK · LONDON

Copyright © 1995 by W.W. Norton & Company, Inc.

All Rights Reserved

Printed in the United States of America

First Edition

Cover design by Linda Kosarin

Cover painting by Charles DuFresne: *Tropical Forest (1919)*,
Paris Museum of Modern Art from Art Resources of New York.
Cover photo © R. Frerck/Tony Stone Images

ISBN 0-393-95718-7

W.W. Norton & Company, Inc., 500 Fifth Avenue, New York, NY 10110
W.W. Norton & Company Ltd., 10 Coptic Street, London, WC1A 1PU

CONTENTS

To the Student — vii

Chapter 1 Introduction — 1

Part I The Chemical and Cellular Basis of Life

Chapter 2 Building Blocks of Matter: Atoms, Bonds, and Simple Molecules — 7
Chapter 3 The Molecules of Life — 14
Chapter 4 The Flow of Energy in Biochemical Reactions — 26
Chapter 5 At the Boundary of the Cell — 31
Chapter 6 Inside the Cell — 39
Chapter 7 Energy Transformations: Photosynthesis — 46
Chapter 8 Energy Transformations: Respiration — 56

Part II The Perpetuation of Life

Chapter 9 How Cells Reproduce — 67
Chapter 10 The Structure and Replication of DNA — 75
Chapter 11 Transcription and Translation — 79
Chapter 12 The Control of Gene Expression — 86
Chapter 13 Recombinant DNA Technology — 94
Chapter 14 Inheritance — 98

Part III The Biology of Organisms

Chapter 15 Multicellular Organization — 112
Chapter 16 Nutrient Procurement and Gas Exchange in Plants and Other Autotrophs — 116
Chapter 17 Nutrient Procurement in Heterotrophic Organisms — 122
Chapter 18 Gas Exchange in Animals — 130
Chapter 19 Internal Transport in Plants — 136
Chapter 20 Internal Transport in Animals — 145
Chapter 21 Defense of the Human Body: The Immune System — 153
Chapter 22 Regulation of Body Fluids — 160
Chapter 23 Chemical Control in Plants — 169
Chapter 24 Hormonal Control of Reproduction and Development in Flowering Plants — 173
Chapter 25 Chemical Control in Animals — 179
Chapter 26 Hormonal Control of Vertebrate Reproduction — 188
Chapter 27 Development of Multicellular Animals — 195
Chapter 28 Nervous Control — 206
Chapter 29 Sensory Perception and Processing — 219
Chapter 30 Effectors and Animal Locomotion — 228

Part IV The Biology of Populations and Communities

Chapter 31 Evolution: Adaptation 235
Chapter 32 Evolution: Speciation and Phylogeny 245
Chapter 33 Ecology 253
Chapter 34 Ecosystems and Biogeography 262
Chapter 35 Animal Behavior 271

Part V The Genesis and Diversity of Organisms

Chapter 36 The Origin and Evolution of Early Life 283
Chapter 37 Viruses and Bacteria 291
Chapter 38 The Protistan Kingdom 299
Chapter 39 The Plant Kingdom 306
Chapter 40 The Fungal Kingdom 314
Chapter 41 The Animal Kingdom: The Radiates and the Protostomes 318
Chapter 42 The Animal Kingdom: The Deuterostomes 327
Chapter 43 The Evolution of Primates 333

TO THE STUDENT

Research has shown that almost 90 percent of the material a student hears or reads is forgotten within a few months—a most discouraging fact! Fortunately, retention can be increased by working with the material and by explaining it to others. This *Study Guide* gives you the opportunity to do just that; it requires you to use the material you are studying. To answer many of the questions, you must synthesize, analyze, and interpret information. The *Study Guide* is designed to help you focus on the important concepts you need in order to learn the material presented in *Biology: An Exploration of Life*. To help you organize your thinking as you begin studying biology, a description of the parts of each study guide chapter and a section called "Central Themes in the Study of Life" are provided here.

CHAPTER PARTS

Each chapter of the *Study Guide* has ten parts: A General Guide to the Reading, Key Concepts, Crosslinking Concepts, Objectives, Key Terms, Summary, Concept Maps, Key Diagrams, Questions, and Answers.

A General Guide to the Reading You should begin your study of each chapter in the text by reading the general guide to the reading in the corresponding chapter of the *Study Guide*. This is a helpful guide to the text chapter, pointing out areas of difficulty, figures that are particularly useful, sections that are important for understanding subsequent chapters and other related material, and providing hints for learning. You will want to keep this part of the *Study Guide* handy as you do your reading in the text.

Key Concepts Each chapter contains a list of the important concepts, which provide a framework for facts that the chapter presents.

Crosslinking Concepts The *Study Guide* makes a special effort to help you integrate biological knowledge. The goal is to minimize memorization and increase retention of the subject matter. Crosslinking concepts link together concepts within a chapter, between chapters, and between disciplines.

Objectives Each chapter provides clear objectives. These objectives describe what you should be able to do when you have mastered the material. They are not all-inclusive; they are directed toward the chapter's major concepts. They suggest a plan for study, define the goals toward which you should aim, and help you organize the content of the chapter.

Once you have read the text chapter, using the general guide in the reading, take a look at the key concepts and objectives. Then go back over the text, trying to keep the key concepts in mind. At the end of each section, stop and try to relate what you have read to the corresponding concept; then review the objectives, and see if you could meet them. Next, reread any sections that are still unclear. The page references given for each concept will make it easy to find the relevant sections in the text.

Key Terms The *Study Guide* provides a list of key terms, normally those terms that are printed in bold in the text. You will want to study this list and make sure you are familiar with the terms. Again, page references make it easy to locate in the text any terms that are unfamiliar.

Summary The *Study Guide* offers a summary of each chapter in the text. Since many of the chapters are long, the summaries can help you master the material.

Concept Maps Concept maps help you to identify and remember the connections among the most important concepts in each chapter. We have attempted to include one or two maps for most chapters: usually an overview map and sometimes a more detailed map of particularly significant or difficult sections. We encourage you to map other sections on your own.

Key Diagrams A series of coloring exercises has been included to increase your retention of anatomical detail. Although coloring may at first seem to be a "kindergarten" activity, research has shown that the act of coloring actually enhances learning. Coloring the name and associated structures with the same color forces you to make a visual connection and frees you from rote memorization. When you do the exercises, first color in the outlined name, then the corresponding structure. The reference diagrams in the text are cited should you need guidance.

Questions In most chapters the questions are divided into three sections: "Testing recall," "Testing knowledge and understanding," and "For further thought." The "recall" questions require you to remember the basic facts and vocabulary presented in the text. The questions are varied: true or false, matching, fill-in-the-blank, and multiple choice. When you have finished reading the chapter in the text, do the questions in this section and check your answers; the results will help you identify areas that need further study. Page references to the text are given for each question, so you can look up the material you have missed. Once you answer the "recall" questions correctly, you can go on to the next section.

In general, the questions in "Testing knowledge and understanding" are more demanding. You will need not only to recall information, but also to analyze and to integrate it. Most of the questions in this section are multiple choice. You should choose the best, most specific answer from among the alternatives given. When you choose the answer you think is correct, be sure you understand why each one of the other possibilities is wrong. Once again, the page references next to the questions will help you find the relevant material in the text. All of the questions have been tested and evaluated; they are drawn from twenty years of examinations in the introductory biology course at Cornell University.

Answers Answers for the "recall" and "knowledge" questions are provided at the end of each chapter.

CENTRAL THEMES IN THE STUDY OF LIFE

To help you in the task of making connections among facts and of identifying the central organizing ideas of biology, we list below some concepts that have such broad applicability to life science that they should be pointed out at the beginning of your study. These are the unifying themes that serve as the solid foundation for biological knowledge. They can be counted on as fixed principles that will still apply when much of the specific information in any current textbook is obsolete. And you will discover that, like some themes in a musical composition, they will recur frequently in various contexts as you study biology during the coming year.

1 **Living organisms are composed of the same chemical and physical components as nonliving things, and all life processes obey the laws of chemistry and physics.** It follows from this statement that many complex biological processes are best understood by studying their simpler components, such as molecules. For instance, cycles of matter and energy flows that describe ecosystems, one of the broadest levels of biological study,

rest upon the laws of chemistry and physics. It also follows that vitalism—the old belief that life is driven by unique forces that defy explanation—is rejected by modern biologists.

2 **All living organisms must take in energy and materials to maintain their internal organization, because all things in the universe, living and nonliving, tend naturally toward a state of disorder. Living things use energy to overcome disorder.**

3 **The cell represents the lowest level of structure capable of performing all the functions of life.** Biological systems are characterized by numerous levels of organization, each more complex than the one below it. Thus organs make up the body of a complex multicellular organism, but are themselves made of tissues, which in turn consist of cells. Progressively broader levels include organisms, populations, communities, ecosystems, and biosphere. Within this hierarchy, the cell is the basic structural unit of life.

4 **The organization and function of biological systems at many levels are maintained by feedback control—mechanisms that maintain the constancy and control the activity of the system.** The classic example of feedback control is the thermostat: it keeps the room temperature constant by detecting when the temperature goes below or above its setting and turning the furnace on or off accordingly. A large array of analogous mechanisms enable cells and organisms to maintain constant internal conditions.

5 **At all levels from the molecular to the macroscopic, biological structure is closely related to function.** Whether we consider the structure of hemoglobin or of the kidney, we find that an understanding of its structure is essential to a full understanding of how it works. The converse is equally true.

6 **Living organisms are extremely diverse, but possess an underlying unity.** For example, many biochemical pathways found in bacteria are also present in human beings Many other aspects of cellular structure and function are universal among living organisms.

7 **All organisms are capable of self-reproduction based on a set of instructions known as the genome.** The genome in all cellular organisms is composed of DNA—a striking example of the unity in diversity mentioned above.

8 **Life as it exists today is the product of evolution: the change in the genomic composition of populations of organisms over time.** The course of evolutionary change is directed by natural selection. Evolution by means of natural selection is the most important unifying theme in biology. It explains the unity in diversity, the complementarity of structure and function, and the adaptations to varying niches exhibited by life.

ACKNOWLEDGMENTS

We are indebted to Kraig Adler, Carolyn Eberhard, Antonie Blackler, Paul R. Ecklund, Carl Hopkins, and Robert Turgeon for many of the questions used in the *Study Guide*. We also thank Joseph Calvo, from whom we learned much about writing objectives. We are also grateful to Jane Heinze-Fry, who wrote most of the crosslinking concepts and produced the concept maps. Particular appreciation is due the students in introductory biology at Cornell University for their criticisms, comments, and suggestions, which often led to improvements in the *Study Guide*.

We are especially grateful to the late William T. Keeton, who gave the senior author the opportunity to develop the first *Study Guide for Biological Science* and from whom we learned the pleasure of teaching introductory biology.

Finally, we thank all the people at W. W. Norton who have been so helpful to us in the preparation of this book, especially Stephen Mosberg, Rachel Warren, and Jessica Avery.

We hope the *Study Guide* will help you in your study of biology. Your comments and suggestions would be most welcome.

Ithaca, New York   Carol Hardy McFadden
March 1995   Lindsay Goodloe

Chapter 1

INTRODUCTION

A GENERAL GUIDE TO THE READING

Chapter 1 is designed to give you a preliminary idea of the way science works and of where the major organizing principles of biology have come from. The pages that follow in the *Study Guide* will outline these ideas and give you ample opportunity to test your recall and understanding of various details. As you read Chapter 1 in your text, you will find the following major topics presented.

1. The flow of information. One of the themes of the text is the flow of chemical information in the cell. Figure 1.2 (p. 2) is the first introduction to this topic; you may want to spend some time studying this diagram.

2. The flow of energy. Energy from the sun powers all living things on earth. Focus on energy flow from the sun to green plants to animals. (p. 1)

3. The scientific method. The scientific method allows many variations in experimental design (pp. 2–4). You will see the method at work in many experiments described throughout the book.

4. The rise of modern biological science. A brief summary of the development of scientific thought from the ancient Greeks up to Darwin is presented. The beginnings of modern astronomy, physics, and biology are discussed.

5. Darwin's theory of evolution. Understanding this theory (pp. 7–13) is of crucial importance, because the concept is the major organizing concept in biology and a unifying theme of the text. Pay special attention to the summary on page 13.

6. The six kingdoms of organisms. These kingdoms, and the major groups of organisms in each, are described briefly on pages 13 to 24. This introduction to the diversity of life is provided here to familiarize you with the major groups that will be mentioned in later chapters.

KEY CONCEPTS

1. Every living organism has a set of instructions that directs its metabolism, organization, and reproduction and is the raw material on which evolution acts (p. 2). This theme will be particularly significant for your study of genetics in Part II.

2. Energy moves from the sun through all living things. This theme will be significant for your study of how plant cells trap the sun's energy to make sugar (Chapter 7) and how cells harvest the energy stored in sugar (Chapter 8). Your study of ecology will trace the theme of energy flow throughout the biosphere (Chapter 34).

3. All science is concerned with the material universe; it seeks to discover facts about the material universe and to fit these facts into theories or laws that clarify the relationships among them (p. 2). Descriptions of classic experiments throughout the text continually illustrate this endeavor.

4. Organisms change over time, from generation to generation. Evolutionary change is determined by natural selection (pp. 11–13). The theory underlies the presentation of many topics throughout the text and will be

discussed at greater length with other evolutionary mechanisms and controversies in Chapters 31 and 32.

5   Living things are classified on the basis of the evolutionary relationships thought to exist among them. (pp. 13–14)

OBJECTIVES

1   Briefly summarize the flow of energy from the sun through living organisms. (p. 1)

2   Briefly summarize the flow of information through living organisms. (p. 2)

3   Discuss the scientific method and its applications and limitations. (pp. 2–4)

4   Briefly describe the contribution of Aristotle, Nicolaus Copernicus, and Isaac Newton to the development of scientific thought. (pp. 5–6)

5   Briefly describe the contribution of Harvey, Leeuwenhoek, Pasteur, and Lister to the development of modern biology. (pp. 6–7)

6   Contrast Jean Baptiste de Lamarck's explanation for the mechanism of evolutionary change with Charles Darwin's. (pp. 8–13)

7   Discuss the two parts of Darwin's theory of evolution and give the five basic assumptions on which it rests. Indicate the types of evidence that Darwin used in formulating his theory. (pp. 7–13)

8   Name the six kingdoms of organisms and give the distinguishing characteristics of each.

9   Give one important characteristic of each of the following groups of plants: green algae, mosses, and vascular plants.

10 Give one important characteristic and one example of each of the following groups of animals: cnidarians, flatworms, molluscs, annelids, arthropods, echinoderms, and chordates.

KEY TERMS

scientific method (p. 2)
hypothesis (p. 3)
theory (p. 3)
controlled experiment (p. 4)
natural selection (p. 7)
procaryote (p. 14)
eucaryote (p. 14)
Archaebacteria (p. 15)
Eubacteria (p. 15)
Protista (p. 16)
Plantae (p. 18)
vascular plant (p. 18)
seed plant (p. 19)
Fungi (p. 19)
Animalia (p. 20)
chordates (p. 24)
vertebrates (p. 24)

SUMMARY

Biology is the science of living things. All living organisms are chemically complex, are highly organized, utilize energy, undergo development, and reproduce. In addition, every living organism has a set of instructions resident in its genes that directs its metabolism, organization, and reproduction, and is the raw material on which natural selection acts.

All science is concerned with the material universe; it seeks to discover facts about the material universe and to fit these facts into theories or laws that will clarify their relationships.

Scientific method   The scientific method involves formulating a question and making repeated, careful observations in an attempt to answer it, and then using these observations to generate a hypothesis. Hypotheses are then tested, using controlled experiments. If the test results do not confirm a hypothesis, the hypothesis must be altered to conform with the evidence or rejected. If all evidence continues to support the new hypothesis, it may eventually become widely accepted and called a theory. Sometimes a leap of intuition will generate new explanations for anomalies in a theory. The insistence on testability of hypotheses imposes limitations on what science can do. Science cannot make value and moral judgments.

The rise of modern biological science   The physical sciences underwent explosive growth during the sixteenth and seventeenth centuries, but biological science did not enter its modern era until 1859, when Charles Darwin proposed his theory of evolution by natural selection. To this day the theory remains one of the most important unifying principles in all biology.

Darwin's theory   Darwin postulated that all organisms living today have descended by slow, gradual changes from ancient ancestors quite unlike themselves. The course of this evolutionary change is determined by natural selection. Darwin's theory depends on five basic assumptions:

1   Many more individuals are born in each generation than will survive and reproduce.

2   There is variation among individuals.

3 Individuals with certain characteristics have a better chance of surviving and reproducing than individuals with other characteristics. They therefore tend, on the average, to leave more offspring.

4 At least some of the characteristics resulting in differential survival and reproduction are heritable.

5 Long spans of time are available for slow, gradual change.

The Diversity of Life There are two kingdoms of unicellular procaryotic organisms: the *Eubacteria* and the *Archaebacteria*. *Procaryotic cells* lack the membrane-enclosed nucleus and other membranous structures present in the cells of other organisms, which are called *eucaryotic cells*. The kingdom *Protista* includes a wide variety of organisms, largely unicellular but many multicellular groups as well. Two major groups of Protista are the photosynthetic protists (algae), which synthesize their own food, and the protozoa, which cannot manufacture their own food and must ingest food already made. The red and brown algae are photosynthetic, multicellular protists whose bodies show little tissue differentiation. The organisms belonging to the kingdom *Plantae* live on land; all have cells with rigid cellulose cell walls and chloroplasts. Thus they are photosynthetic and can synthesize their own food. The members of this kingdom, the mosses, liverworts, and vascular plants, probably evolved from ancestral green algae. Organisms of the kingdom *Fungi* also have cell walls, but of different composition, and they lack chlorophyll and cannot manufacture their own food. They must obtain their nutrients already synthesized. Fungi depend entirely on absorption of nutrient molecules. Members of the kingdom *Animalia* differ from plants and fungi in their lack of rigid cell walls and in their mode of nutrition; animals ingest their food. Cnidarians, flatworms, molluscs, annelids, arthropods, echinoderms, and chordates will be frequently referred to throughout the book.

CONCEPT MAPS

Concept mapping is a tool for organizing, learning, and retaining complex concepts. It is most powerful when actively employed; reading someone else's maps can only take you so far. Your chances of success will greatly increase if you make the effort to create your own. Here are some suggestions for getting started.

How to Make Concept Maps

Evolution is the major unifying theme of biology. Following you will find three figures that show step-by-step construction of a concept map of "Evolution."

Step 1. Concepts Read the section or chapter carefully. List the concepts in the section you are trying to learn. Defined words in bold print are generally the most important concepts, though other terms also represent important concepts. Put the words on a piece of paper with the more general concepts at the top and more specific concepts at the bottom. Circle each concept. (Map 1.1)

Map 1.1 Evolution: key concepts

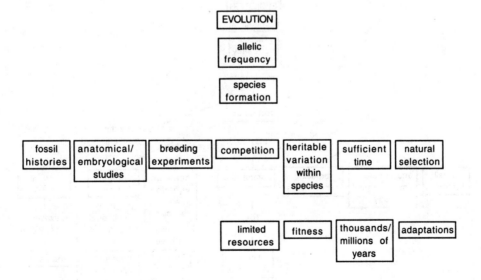

Step 2. Propositional linkages Draw lines between concepts that are related to each other. Write in a word or phrase that describes or proposes the relationship between the two concepts. (These connections are called propositional linkages.) (Map 1.2)

Step 3. Crosslinking Try to connect concepts in different parts of the map. (These connections are called crosslinks and are especially helpful in weaving the material together.) (Map 1.3)

Map 1.2 Evolution: key concepts with propositional linkages

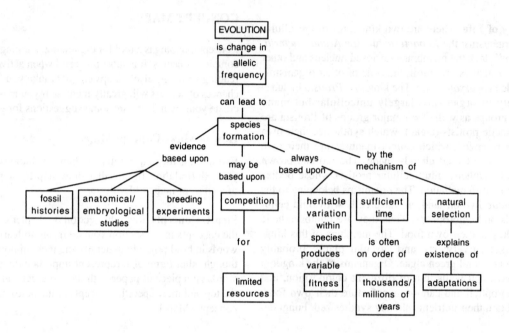

Map 1.3 Evolution: key concepts with propositional linkages and crosslinks

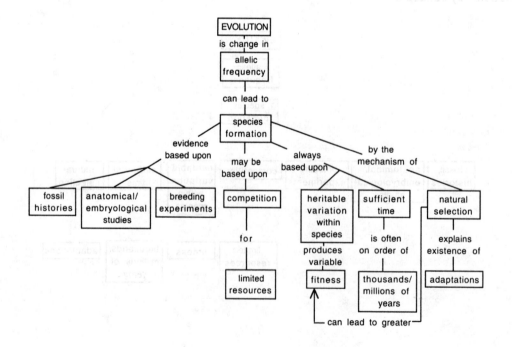

QUESTIONS

Choose the one best answer.

1. A general statement that is a tentative causal explanation for a group of observations is referred to as

   a theory.
   b law.
   c hypothesis.
   d prediction.
   e control. (p. 3)

2. Which one of the following biologists is *mismatched* with his work?

   a Jean Baptiste de Lamarck—proposed that organisms have a natural tendency toward perfection
   b Andreas Vesalius—traced the circulation of blood
   c Anthony van Leeuwenhoek—discovered microorganisms
   d Louis Pasteur—disproved the theory of spontaneous generation
   e Joseph Lister—demonstrated the effectiveness of antiseptics (pp. 6–9)

3. For consistency with the theory of natural selection as the agent of evolution, it is necessary to postulate that

   a in each generation, all individuals well adapted for their environment live longer than those not so well adapted.
   b the deaths of individuals occur completely at random with respect to the environment.
   c some of the reproductive success of individual organisms is heritable.
   d most deaths of individual organisms occur soon after birth.
   e more than half of the poorly adapted individuals must be eliminated by natural selection in each generation. (p. 13)

4. In formulating the theory of evolution, Charles Darwin collected and synthesized data on a variety of subjects. All the following were lines of evidence used by Darwin in formulating his theory *except*

   a the existence of fossils of extinct animals.
   b the similarities of living organisms.
   c creation of the earth in 4004 B.C.E.
   d gradual morphological changes in fossils over time.
   e structural changes in domestic animals and plants. (pp. 9–13)

Below are listed the six kingdoms of organisms used in the classification system presented in your textbook. Answer questions 5 to 10 with the appropriate letter or letters.

   a Animalia
   b Archaebacteria
   c Eubacteria
   d Fungi
   e Plantae
   f Protista

5. Which kingdoms have only unicellular organisms? (p. 14)

6. Which kingdoms have multicellular organisms? (p. 14)

7. Which kingdoms include organisms whose cells have cell walls? (p. 20)

8. Which kingdoms include organisms that ingest particulate food? (p. 21)

9. In which kingdom are nearly all the organisms photosynthetic? (p. 18)

10. In which kingdom are all the organisms totally dependent on absorption to obtain their high-energy nutrients? (p. 19)

Below are listed some of the important animal groups that you are expected to become familiar with. Match each of the characteristics in questions 11 to 17 with the appropriate group or groups of animals.

   a annelids
   b arthropods
   c chordates
   d cnidarians
   e echinoderms
   f flatworms
   g molluscs

11. digestive cavity with one opening (p. 21)

12. two distinct tissue layers (p. 21)

13. jointed legs and hard outer covering (p. 23)

14. radial symmetry (p. 21)

15. shells (p. 22)

16. internal hard skeleton (p. 24)

17. segmented worms (p. 23)

Referring to the list above, match each animal in questions 18 to 30 with the group to which it belongs.

18. grasshopper (p. 23)

19. clam (p. 22)

20. lobster (p. 23)

21. planarian (p. 22)

22. hydra (p. 21)

23. sea star (p. 24)

24. fish (p. 24)

25. crayfish (p. 23)

26. jellyfish (p. 21)

27. snake (p. 24)

28. frog (p. 24)

29. human being (p. 24)

30. earthworm (p. 23)

For questions 31 to 38, use the following key.

a	algae	*c*	mosses
b	fungi	*d*	vascular plants

31 Which group consists of the dominant land plants? (p. 19)

32 Which group obtains its nutrients only by absorption? (p. 19)

33 Which group includes angiosperms and gymnosperms? (p. 19)

34 Which group includes aquatic plants that show little tissue differentiation? (p. 18)

35 Which group shows the greatest tissue specialization? (p. 18)

36 Which group includes the flowering plants? (p. 19)

37 Which group lacks chlorophyll? (p. 19)

38 Which group possesses an effective transport system? (p. 19)

For further thought

Some people think science is absolute truth. They become frustrated when scientists do not always agree about what is true. Cite examples from the text where scientists disagreed about what was true. Explain the basis of the disagreement. Can you cite examples of disagreement in today's scientific community?

ANSWERS

1	*c*	11	*d, f*	21	*f*	30	*a*
2	*b*	12	*d*	22	*d*	31	*d*
3	*c*	13	*b*	23	*e*	32	*b*
4	*c*	14	*d, e*	24	*c*	33	*d*
5	*b, c*	15	*g*	25	*b*	34	*a*
6	*a, d, e, f*	16	*c*	26	*d*	35	*d*
7	*b, c, d, e, f*	17	*a*	27	*c*	36	*d*
8	*a, f*	18	*b*	28	*c*	37	*b*
9	*e*	19	*g*	29	*c*	38	*d*
10	*d*	20	*b*				

Chapter 2

BUILDING BLOCKS OF MATTER: ATOMS, BONDS, AND SIMPLE MOLECULES

A GENERAL GUIDE TO THE READING

Chemistry is basic to all biology. You will continually return to the concepts presented in this chapter and the next, so take the time to learn them now. The following topics are of central importance in Chapter 2.

1. Atomic structure. You will want to be familiar with the basic structure of an atom (pp. 30–36), since understanding atomic structure is fundamental to understanding the material on isotopes, energy levels in atoms, and formation of chemical bonds (pp. 36–42), as well as the material in organic chemistry (Chapter 3).

2. Chemical bonds. You will need to learn about ionic and covalent bonds. The concept of electronegativity and its relationship to polar and nonpolar bonds is the basis for understanding solubility and transport across cell membranes. (p. 36–42)

3. Biologically important weak bonds. Pay particular attention to ionic bonds, hydrogen bonds, and hydrophobic interactions (pp. 40–42). Understanding these weak bonds will help you understand many of the other biological concepts that follow, including the special properties of water, protein structure and function (cell membrane and enzyme activity), and DNA structure and function.

4. Properties of water. The sections describing water as a solvent and its special physical properties (pp. 42–48) show why it is fundamental to life. Grasp of the concepts presented in these sections is essential for understanding the relative solubilities of various organic molecules (pp. 42–44) and the structure of the cell membrane (Chapter 5).

KEY CONCEPTS

1. Living organisms are integral parts of the physical universe and must obey the fundamental laws of chemistry and physics. (pp. 29–34)

2. An atom's properties depend on the number of electrons in its outer shell. Atoms tend to complete their outer shells by reacting with other atoms. (pp. 35–36)

3. Covalent bonds are strong bonds; they can be broken apart only by relatively large amounts of energy. Hence covalently bonded molecules are stable. (pp. 36–38)

4. In a polar covalent bond, electrons are shared unequally among the constituent atoms; in a nonpolar covalent bond, they are shared equally. The type of covalent bond formed by an atom depends on its electronegativity. Highly, electronegative atoms such as nitrogen and oxygen tend to form polar covalent bonds, whereas bonds between carbon atoms and between carbon and hydrogen are essentially nonpolar. (pp. 37–38)

5. Weak chemical bonds play a crucial role in stabilizing the three-dimensional shape of many of the large molecules

found in living matter and in holding different molecules together. (pp. 40–42)

6. Water molecules are polar and have a strong tendency to form hydrogen bonds with one another; the consequent ordering of the molecules gives water many of its special properties, making it the medium of life. (pp. 42–48)

CROSSLINKING CONCEPTS

1. The polar structure of the water molecule is the key to its function. The hydrogen bonding among water molecules gives water such fundamental properties as surface tension and capillarity, and makes bodies of water effective buffers against extreme fluctuations in temperature.

OBJECTIVES

1. Describe the structure of an atom, using the terms proton, neutron, electron, mass number, atomic weight, and orbital. Indicate what is meant by electrons in an excited state and those in a ground state. (pp. 30–33)

2. Draw the electron dot structures of carbon, hydrogen, oxygen, and nitrogen. Indicate how many electrons are available for covalent bonding in each. (pp. 35–37)

3. Describe a covalent bond. Relate the structure of an atom (e.g., the number of electrons in its outer shell) to its chemical properties and to the number and type of chemical bond(s) it forms. (pp. 36–40)

4. Using diagrams such as Figures 2.8 (p. 37) and 2.9 (p. 38), explain the difference between a nonpolar covalent bonds and a polar covalent bonds. Then, in a molecule like the one below, indicate which of the bonds marked by arrows are polar covalent bonds and which are nonpolar covalent bonds. (pp. 37–38) Indicate why or why not a sharp distinction exists between ionic, polar covalent, and nonpolar covalent bonds. (p. 38)

$$\begin{array}{c} H \\ | \\ H-C-O-H^+ \\ | \\ H-C-H \\ | \\ H \end{array}$$

5. Using diagrams such as Figures 2.10 (p. 39) and 2.11 (p. 40), explain what an ion is and how it forms. Describe an ionic bond and explain how it differs from a covalent bond. Indicate why or why not, sodium chloride exists as a molecule when in solution and when in a solid. (p. 40)

6. Explain the crucial role of weak chemical bonds (such as ionic bonds, hydrogen bonds, and hydrophobic interactions) in the organization of living materials. (pp. 40–48)

7. Explain what is meant by pH. Specify the pH of the material within most living cells and indicate whether this is acidic or basic. Describe what happens if the pH changes appreciably from this value. Give the name used for compounds that help minimize changes in pH. (p. 45)

8. Describe the special physical properties of water. In doing so, draw two water molecules in a way that illustrates a hydrogen bond; specify the number of hydrogen bonds a single water molecule can form with other water molecules in ice; explain why water is a good solvent for ionic and polar compounds, but not for hydrophobic substances; and explain the basis for the high surface tension of water and for capillarity. (pp. 46–48)

9. Indicate why the inorganic molecules O_2 and CO_2 are basic to life, and name the principal source of each of these molecules. (p. 48)

KEY TERMS

atom (p. 30)
atomic nucleus (p. 30)
proton (p. 30)
neutron (p. 30)
atomic number (p. 30)
mass number (p. 30)
isotope (p. 30)
electron (p. 31)
orbital (p. 33)
ground state (p. 33)
excited state (p. 33)
Periodic law (p. 36)
valence electrons (p. 36)
chemical bond (p. 36)
covalent bond (p. 36)
molecule (p. 36)
nonpolar covalent bond (p. 37)
compound (p. 38)
electronegativity (p. 38)
polar covalent bond (p. 38)
polar molecule (p. 38)
electron acceptor (p. 38)
electron donor (p. 38)
ion (p. 38)
ionization (p. 38)

BUILDING BLOCKS OF MATTER: ATOMS, BONDS, AND SIMPLE MOLECULES • 9

electrostatic attraction (p. 39)
ionic bond (p. 39)
dissociate (p. 40)
hydrogen bond (p. 41)
hydrophobic interactions (p. 41)
hydrated (p. 43)
hydrophilic (p. 43)
acid (p. 45)
base (p. 45)
pH (p. 45)
buffer (p. 45)
surface tension (p. 46)
capillarity (p. 46)
greenhouse effect (p. 48)
carbon cycle (p. 49)

SUMMARY

The matter of the universe is composed of a limited number of substances called *elements*. Of the 92 nuturally occurring elements, only a few are important for life. Prominent among them are hydrogen, carbon, oxygen, nitrogen, phosphorus, and sulfur. An atom, the basic chemical unit of matter, are made up of a *nucleus*, which contains positively charged *protons* and uncharged *neutrons*, surrounded by a cloud of negatively charged *electrons*. The number of protons is equal to the number of electrons. The *atomic number* of an atom represents the number of protons; the *mass number* represents the total number of protons and neutrons. *Isotopes* of an element contain different numbers of neutrons. Some isotopes are radioactive; they give off various particles to reach a more stable form.

Electrons are in constant motion outside the nucleus and occupy discrete *energy levels*. The higher the energy of an electron, the greater its probable distance from the nucleus. The volume of space within which the electron would probably be found 90 percent of the time is its *orbital*. Electrons occupy discrete energy levels in an atom; the lowest energy level has one orbital (holding a maximum of two electrons) and the second energy level has a total of four orbitals (holding a maximum of eight electrons).

The chemical properties of an element are determined largely by the number of *valence electrons* in its outermost energy level. Most atoms are particularly stable when their outermost (valence) energy levels contain eight electrons. Atoms tend to complete their outer levels by reacting with other atoms and forming *chemical bonds*.

There are setvertal types of chemical bonds. *Covalent bonds*, in which electrons are shared between atoms, are strong; hence covalently bonded molecules are stable. When the electrons in a covalent bond are shared equally, the bond is *nonpolar covalent*; when the electrons are pulled closer to one atom than to the other, the charge is distributed unequally, and the bond is *polar covalent*.

An *ionic bond* is formed when one or more electrons is transferred from one atom to another. The atom that gives up an electron (*electron donor*) becomes a positively charged ion, while the atom that takes the electron (*electron acceptor*) becomes a negatively charged ion. The electrostatic attraction between the oppositely charged ions creates an ionic bond. Ionic bonds are strong in the solid state but weak in aqueous solutions.

Two other types of weak chemical bonds, *hydrogen bonds* and *hydrophobic interactions* are important in stabilizing the shape of many of the large, complex molecules found in living matter.

Molecules may be classified into two types: organic and inorganic. *Organic* compounds always contain the element carbon and have at least one carbon-hydrogen bond. All other compounds are said to be *inorganic*. Water, oxygen, and carbon dioxide are three inorganic molecules basic to life.

The water molecule is polar because of the polarity of bonds within the molecule and the V-shaped arrangement of its atoms. Each water molecule can form hydrogen bonds with four other water molecules, and thus permit an orderly arrangement of molecules and give water many special properties. Because of its polarity, water is an excellent solvent for ionic and polar substances. Electrically neutral and nonpolar substances are not soluble in water because they cannot interact with the polar water molecules.

Acidity and *alkalinity* are expressed in terms of *pH*, a measure of the contentration of hydrogen ions. Acids act as hydrogen-ion donors, bases as hydrogen-ion acceptors. Solutions with a pH less than 7 are acidic, whereas those with a pH greater than 7 are basic. Living matter is extremely sensitive to changes in pH.

Like water, molecular oxygen and carbon dioxide play important roles in life processes; they are basic to the chemistry of life.

CONCEPT MAPS

How to Interpret a Concept Map

Try to interpret Maps 2.1 (an overview map) and 2.2 (a detail map). Note the following characteristics, which should help you in interpreting maps throughout the book.

1 Concepts (words found in boxes) are words that were generally written in bold in the textbook.

2 Concepts at or near the top of the map (molecules, elements, atoms) are more general than those at the bottom of the map (protons, neutrons, electrons).

3 Concepts are connected by lines and words that describe the relationship between the concepts (e.g., atoms "form" molecules).

Map 2.1

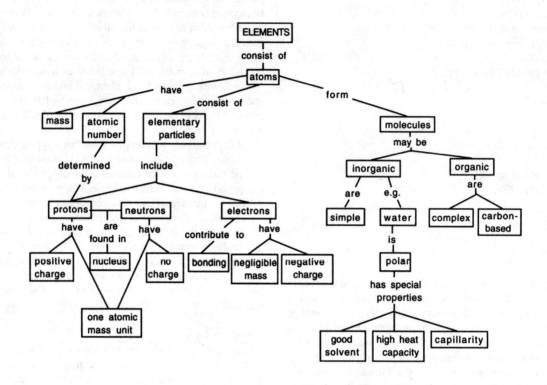

Map 2.2

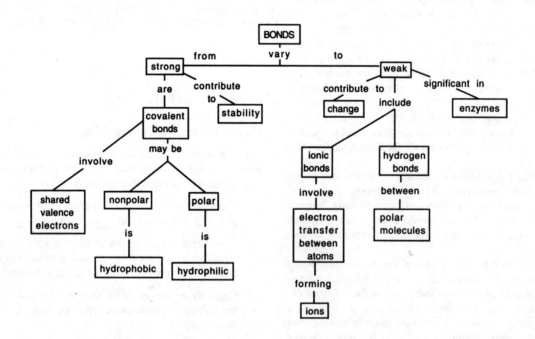

KEY DIAGRAMS

1. Below is a diagram of an atom that has four electrons in its outer energy level. When struck by a photon of light, an electron absorbs energy and moves to a higher energy level. The "excited" electron quickly reemits the absorbed energy and returns to its original energy level. In the diagram below, color each name and the corresponding structure(s) with the same color. (Fig. 2.4, p. 33)

ELECTRONS

EXCITED ELECTRON

PROTONS

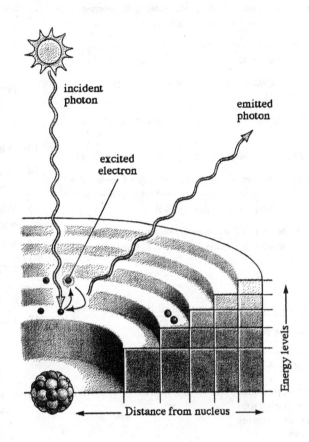

2. Below are diagrams of the four atoms that make up the bulk of living matter. The symbol, atomic number, and mass number (atomic mass) are given. For each, determine the number of electrons in both energy levels and draw in the electrons, using different colors for each energy level. (Fig. 2.7, p. 35)

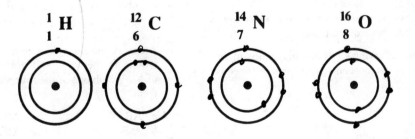

QUESTIONS

Testing recall

Decide whether the following statements are true or false, and correct the false statements.

1. The isotope ^{14}C is heavier than ^{12}C because its atomic nucleus contains two additional protons. (p. 30)

2. The number of electrons at the outermost energy level determines the chemical properties of an atom. (p. 35)

3. Ions are formed when a neutral atom loses or gains electrons. (p. 38)

4. In solution, ionic bonds are stronger than covalent bonds. (p. 40)

5. When a solution is acidic, its pH is greater than 7. (p. 45)

6. The orderly three-dimensional arrangement of molecules in pure water is due primarily to the formation of ionic bonds between water molecules. (p. 46)

7. The marked polarity of water molecules makes water an excellent solvent for many important classes of chemicals. (p. 43)

Testing knowledge and understanding

Choose the one best answer.

8. The chemical properties of an atom are primarily determined by the

   a number of protons.
   b number of electrons.
   c number of neutrons.
   d mass number.
   e number of isotopes. (p. 35)

9. A nonpolar covalent bond occurs

   a when one atom in a molecule has a greater affinity for electrons than another.
   b when a molecule's constituent atoms attract the electrons equally.
   c when an electron from one atom is completely transferred to another atom.
   d between atoms whose outer energy levels are complete.
   e when a molecule becomes ionized. (p. 37)

10. Which one of the following kinds of biologically important bonds requires the most energy to break in aqueous solution?

    a ionic bond
    b covalent bond
    c hydrophobic interaction
    d hydrogen bond (pp. 36–41)

11. When matter is broken down into units indivisible by ordinary chemical means, the resulting structure is the

    a proton.   d molecule.
    b atom.   e nucleus.
    c neutron.   (p. 30)

Questions 12 to 17 concern the following atoms.

   a $^{19}_{9}F$ d $^{40}_{18}Ar$
   b $^{12}_{6}C$ e $^{28}_{14}Si$
   c $^{40}_{20}Ca$

12. How many protons does fluorine (*a*) have?

    a 9   c 19
    b 10   d 26   (p. 30)

13. Which one of the above would be relatively inert? (p. 36)

14. Which one of the above would have two electrons in its outer orbital? (p. 35)

15. Which one of the above would be a strong electron acceptor? (p. 38)

16. Which two of the above have similar chemical properties?

    a *a* and *c*   d *b* and *e*
    b *b* and *c*   e *c* and *e*
    c *a* and *d*   (p. 35)

17. How many electrons would a Ca^{++} ion have?

    a 2   d 22
    b 18   e 40
    c 20   (p. 38)

18. In an electrically neutral atom, how many electrons would be at the outermost energy level or shell if the atom had an approximate atomic mass of 12 and an atomic number of 6?

    a 0   d 6
    b 2   e some other number
    c 4   (p. 35)

19. Which one of the following would be the strongest electron donor?

    a $^{12}_{6}C$ d $^{17}_{16}S$
    b $^{19}_{9}F$ e $^{39}_{19}K$
    c $^{14}_{7}N$ (p. 38)

20. Atoms A and B combine together readily. Atom A has an atomic number of 12 and atom B has an atomic number of 17. Which one of the following molecules would they form?

    a AB   d AB_2
    b A_2B e AB_3
    c A_2B_2 (p. 39)

21 For a hypothetical atom in which the number of protons is X, the number of neutrons is Y, and the number of electrons is Z, the atomic number would be

a X.
b Y.
c Z.
d the sum of X and Y.
e the sum of Y and Z.   (p. 30)

22 In an ionic bond

a electrons are shared.
b there is a mutual attraction between two electrically neutral atoms. No electron transfer is involved.
c there is a mutual attraction between two charged atoms. Electron transfer is involved.
d there is an unequal sharing of electrons.   (p. 38)

23

The bond indicated by the arrow is

a a hydrogen bond.
b a covalent bond.
c a polar bond.
d an ionic bond.
e both a and c.   (p. 37)

24 Solution A has a pH of 4 and solution B has a pH of 8. The hydrogen ion concentration of A is _____ times that of B.

a 10,000
b 1,000
c 4
d 0.001
e 0.0001   (p. 45)

25 Solution A has a pH of 8 and solution B has a pH of 2. Which one of the following statements is *correct*?

a Solution A is basic and solution B is acidic.
b Solution A is acidic and solution B is basic.
c Solution A has more hydrogen ions than solution B.
d Solution A is basic and solution B is neutral.
e Two of the above are correct.   (p. 45)

26 Below is a molecule composed of carbon, nitrogen, oxygen, and hydrogen. The hydrogen atoms have not been drawn in. What is the correct chemical formula?

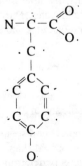

a $C_9H_6NO_3$
b $C_9H_9NO_3$
c $C_9H_{11}NO_3$
d $C_9H_{13}NO_3$
e None of the above are correct.   (p. 37)

27 The cohesiveness among water molecules is directly due to

a hydrogen bonds.
b polar covalent bonds.
c nonpolar covalent bonds.
d hydrophobic interactions.
e van der Waals interactions.   (p. 41)

Questions 28 to 29 refer to the following diagrams of water molecules.

28 The bond labeled *x*, from O to H, is a(n)

a polar covalent bond.
b nonpolar covalent bond.
c hydrogen bond.
d hydrophobic interaction.
e ionic bond   (p. 38)

29 When ice melts, breaks occur in the bond(s) marked

a x.   b y.   c both x and y.   (p. 47)

For further thought

1 In aqueous solutions, the product of the H^+ ion concentration times the hydroxide (OH^-) concentration is a constant number, 1×10^{14}. (That is, $[H^+] \times [OH^-] = 1 \times 10^{14}$, where [ ] is the symbol meaning "concentration of.") Suppose a solution has a pH of 5. What is the concentration of H^+ ions? Of OH^- ions?

2 Explain why the molecule O_2 is relatively insoluble in water whereas ammonia, NH_3, is highly soluble.

ANSWERS

Testing recall

1 false—two additional neutrons
2 true
3 true
4 false—ionic bonds are weaker
5 false—less than 7
6 false—hydrogen bonds
7 true

Testing knowledge and understanding

8 b	14 c	20 d	25 a
9 b	15 a	21 a	26 c
10 b	16 d	22 c	27 a
11 b	17 b	23 b	28 a
12 a	18 c	24 a	29 b
13 d	19 e		

Chapter 3

THE MOLECULES OF LIFE

A GENERAL GUIDE TO THE READING

This chapter, like the preceding one, presents several ideas that are central to an understanding of modern biology. Organic chemistry is difficult for many students, primarily because the molecules seem so complex. They really are not; organic molecules are composed of relatively few elements, and you will soon learn to recognize the key features of the various kinds of molecules. As you do your reading, remember that you do not have to memorize the structures shown in the figures; most are provided only as examples of particular classes of molecules. If you focus instead on what makes each class unique, you will find this chapter much easier and more meaningful. As you read Chapter 3 in our text, you will want to concentrate on the following topics.

1. Carbohydrates, lipids. Read carefully the sections on these compounds (pp. 52–62); the material they present will be referred to in your study of photosynthesis (Chapter 7), respiration (Chatper 8), nutrition (Chapter 17), and other topics. Glucose, in particular, will be referred to repeatedly.

2. Proteins. The discussion of proteins (pp. 64–71) is crucial; you need to finish this chapter with an understanding of the three-dimensional shape of proteins and the relationship between their shape and their function. You will find that changes in protein shape are involved in many cellular activities that you will be studying.

3. Nucleic acids (pp. 72–74) will be discussed in more detail in later chapters; at this point you need only a general understanding of what they are.

KEY CONCEPTS

1. Both living and nonliving matter are made up of the same fundamental particles. The only difference between living and nonliving things seems to be in the way these basic materials are organized. (See Chapter 2, p. 1.) The carbon atom is the key to the diversity of organic compounds.

2. All complex organic molecules are composed of many simpler building-block molecules, bonded together by condensation reactions; they can be broken down to their building-block molecules by hydrolysis. (pp. 55–65)

3. Carbohydrates are an important energy source. Some are also important as structural components of organisms, or as raw materials for building other classes of organic molecules. (pp. 52–58)

4. Lipids are generally nonpolar molecules and are therefore relatively insoluble in water. They are especially important as energy-storage molecules and as components of cell membranes. (pp. 59–61)

5. Proteins are much more complex and diverse than carbohydrates or lipids. They are the catalysts for virtually all chemical reactions in cells. Many others play important structural roles in organisms.

6. The amino acid content and sequence of a protein determine its three-dimensional shape. It is the R groups of amino acids that largely determine the structure and properties of proteins. Of particular importance is the fact that some R groups are hydrophilic, while others are hydrophobic. Weak bonds between these groups play a

crucial role in determining the three-dimensional structure of proteins. Alteration in the shape of a protein can lead to a change in its biological function. (pp. 64–71)

7 Nucleic acids encode the hereditary information that determines the structural and functional characteristics of living things. (pp. 72–74)

CROSSLINKING CONCEPTS

1 The concepts of atomic structure and chemical bonding from Chapter 2 govern synthesis of larger molecules from simple building blocks.

2 Weak bonds (Chapter 2) within parts of a polypeptide chain or between polypeptide chains give proteins their unique shape and their ability to change shape with a modest expenditure of energy. These structural qualities make proteins function well as enzymes.

3 The cell membrane, which you will study in Chapter 5, is composed of phospholipids and proteins. Their structure governs the function of the selectivity of the cell membrane.

4 Nucleic acids control our genetic inheritance. Their structure is key to the perpetuation of life (Part II).

OBJECTIVES

1 Identify the various functional groups listed in Table 3.1 (p. 53). Given an unknown organic molecule, recognize and name the various functional groups and indicate whether each group is charged, polar, or nonpolar and whether it is hydrophilic or hydrophobic. (p. 53)

2 Describe the structure of a typical monosaccharide such as glucose. Write out a condensation reaction between the two given glucose molecules, and explain what is meant by hydrolysis. (pp. 54–55)

3 Using Figure 3.10 (p. 59), point out the carboxyl(-acid) group and the hydrocarbon chain of a fatty acid. Explain the difference between a saturated fatty acid and an unsaturated fatty acid. For butter and corn oil, indicate whether each is considered a saturated or an unsaturated fat. Using a diagram such as Figure 3.15 (p. 61), explain how three fatty acids can react with glycerol to make a fat. (p. 61)

4 Using Figure 3.16 (p. 61), point out the polar and nonpolar portions of a phospholipid molecule. Specify which end of the molecule would be soluble in water. (pp. 60–61)

5 Given the structure of an amino acid, point out the carboxyl group, the amino group, and the R group (all attached to the same carbon atom). Be able to tell whether the R group is charged or polar, and therefore hydrophilic, or nonpolar and hydrophobic (p. 65).

6 Given the structure of two amino acids, write out a condensation reaction between the two (see Fig. 3.20, p. 65) and circle the resulting peptide bond. Explain how a polypeptide chain is formed and how it is broken down by hydrolysis. (p. 65)

7 Given a picture of a protein such as the one below, point out the number of polypeptide chains it contains, point out any regions of alpha helix and label the region of the protein that would be stabilized by ionic bonds, label the region stabilized by hydrogen bonds between R groups, and circle a disulfide bond and state whether the bond is strong or weak. (pp. 65–70)

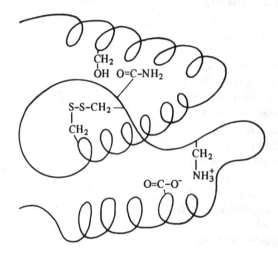

8 Differentiate among the various levels of protein structure—primary, secondary, tertiary, and quaternary. Specify the hightest level of structure shown in the protein in the diagram above. Specify the level(s) of structure shown by fibrous proteins, such as hair, and the levels shown by globular proteins. Explain why proteins are so sensitive to changes in temperature and pH. (pp. 66–70)

9 Using Figure 3.34 (p. 73), point out an individual nucleotide and the five-carbon sugar, the phosphate group, and the nitrogen-containing base of which it is made; a base from a nucleotide on one chain bonded to a base on the opposite side; and the hydrogen bonds between the bases. (pp. 72–73)

10 Identify examples of each of the four main classes of biologically important organic molecuels and the building-block units of which they are composed. (pp. 74–75)

KEY TERMS

hydrocarbon (p. 51)
isomer (p. 52)
functional group (p. 52)
hydroxyl (p. 53)
alcohol (p. 54)

carboxyl (p. 53)
amino (p. 54)
aldehyde (p. 53)
ketone (p. 53)
phosphate (p. 54)
monomer (p. 53)
monosaccharide (p. 53)
disaccharide (p. 55)
condensation reaction (p. 55)
dehydration reaction (p. 55)
hydrolysis (p. 55)
polysaccharide (p. 56)
polymer (p. 56)
lipid (p. 58)
fatty acid (p. 58)
triglyceride (p. 60)
saturated fat (p. 59)
unsaturated fat (p. 59)
phospholipid (p. 60)
steroid (p. 62)
amino acid (p. 64)
peptide bond (p. 65)
polypeptide (p. 65)
disulfide bond (p. 65)
primary structure (p. 66)
alpha helix (p. 66)
secondary structure (p. 66)
fibrous proteins (p. 66)
beta sheet (p. 66)
globular protein (p. 68)
tertiary structure (p. 69)
quaternary structure (p. 69)

denatured protein (p. 70)
conjugated protein (p. 71)
prosthetic group (p. 70)
nucleic acid (p. 72)
nucleotide (p. 72)
DNA (p. 72)
adenine (p. 72)
guanine (p. 72)
cytosine (p. 72)
thymine (p. 72)
RNA (p. 74)
uracil (p. 74)

SUMMARY

Carbon, with its covalent bonding capacity of four, commonly forms bonds with hydrogen, nitrogen, and other carbon atoms. The carbon atoms usually join in long chains or rings and form molecules that may be very complex. Four major classes of organic molecules found in living organisms are carbohydrates, fats, proteins, and nucleic acids.

Although these four classes of molecules differ in structure and function, they are similar in that all are complex molecules composed of many simpler "building-block" molecules bonded together. In each case these building-block molecules are combined by the removal of water molecules in condensation reactions. In such a reaction a hydrogen atom is removed from the end of one building-block molecule and a hydroxyl (—OH) group from the end of a second molecule. The two building-block molecules are now joined together, and a molecule of water has been formed:

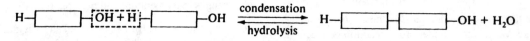

Condensation reactions are reversible; the complex organic molecules can be hydrolyzed into the simpler building-block molecules.

The basic building-block molecules of carbohydrates are simple sugars, or monosaccharides. When two simple sugars are bonded together, a disaccharide is formed. Many simple sugars bond together in long chains to form a polysaccharide. Starch, glycogen, and cellulose are examples of polysaccharides. The carbohydrates are a major structural component of plants and an important energy source for all organisms.

Lipids, the fats and fatlike substances, tend to be insoluble in water. Fats are composed of three fatty acids joined to glycerol by condensation reactions. Phospholipids are derived from the fats; they are important constituents of cell membranes.

The basic building-block molecules of the proteins are amino acids. Amino acids are bonded together to form a protein by condensation reactions. The resulting bond is the peptide bond and the chains produced are polypeptide chains. The primary structure of each protein is the sequence and type of amino acids making up the polypeptide chains, and the disulfide bonds that form between sulfur-containing amino acids. Because hydrogen bonds form between one amino acid and another, the chain assumes a stable regular shape known as the secondary structure typically an α-helix or ß-sheet. These regular molecules may in turn be folded into complicated globular shapes by weak attractions between the different R groups within the chain and thus form the tertiary structure of the protein. Some globular proteins are made up of two or more polypeptide chains held together by weak bonds; the way these chains fit together determines the quaternary stucture. Fibrous proteins have only primary and secondary structure, whereas globular proteins have tertiary and sometimes quaternary structure. Because the shape of a protein depends on weak bonds, it is easily altered by temperature or pH; this causes a change in its biological function.

The building-block unit of nucleic acids is the nucleotide, which is made up of a five-carbon sugar attached to a phosphate group and to a nitrogen-containing base. Nucleotide units are joined together to form a polymer through condensation reactions between the sugar of one nucleotide and the phosphate group of the next. There are four different nucleotides in each nucleic acid; it is the different sequences of the nucleotides that encode the hereditary information. The two types of nucleic acids, DNA and RNA, differ in the types of sugar, certain nitrogen-containing bases, the number of strands in the molecules, and their biological roles.

CONCEPT MAPS

Map 3.1 Complex organic molecules. Take particular note of the hierarchical pattern. From the top to the bottom of the map you see the name of the complex organic compound, the molecular composition, the building blocks of the compounds, the important functional groups of the building blocks, and the chemistry determined by the functional groups (in terms of water solubility).

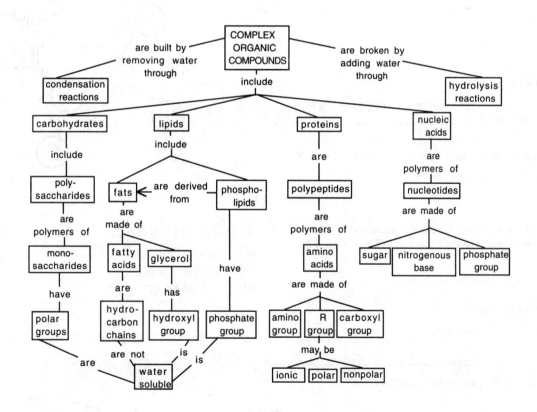

18 • CHAPTER 3

KEY DIAGRAMS

A section of an alpha-helical structure of a protein
Color each name and the appropriate structure(s) with the same color (Fig. 3.23, p. 67)

BONDS BETWEEN ATOMS:

COVALENT BONDS

HYDROGEN BONDS

COMPONENT ATOMS:

CARBON ATOMS

NITROGEN ATOMS

OXYGEN ATOMS

HYDROGEN ATOMS

R GROUPS

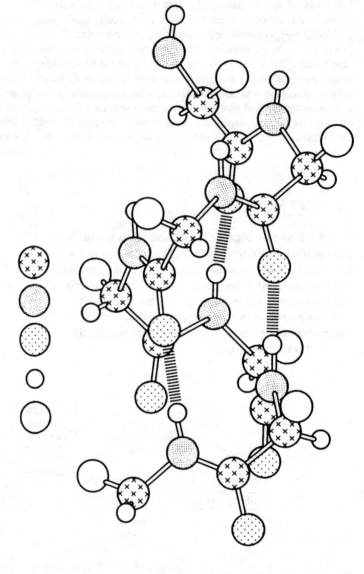

5-CARBON SUGAR

NITROGENOUS BASE

PHOSPHATE GROUP

HYDROXYL GROUP

Related questions

1. Which of the two types of bonds (covalent, hydrogen) are strong bonds? _____
   Weak bonds? _____

2. Which of the bonds help stabilize the alpha helix?
   _____

Structure of a nucleotide Color each name and the appropriate structure with the same color (Fig. 3.31, p. 72)

Related questions

1. If a second nucleotide were to be added to the above nucleotide, where would it be added?
   _____

2. What are the four nitogenous bases in DNA?
   _____

3. Which bases are purines? _____
   pyrimidines? _____

QUESTIONS

Testing recall

Decide whether the following statements are true or false, and correct the false statements.

1. Denaturation of a protein involves a loss of the protein's natural shape. (p. 70)
2. The hormone insulin is a steroid. (p. 66)
3. Phospholipids are molecules with hydrophilic and hydrophobic regions. (pp. 60–61)
4. Adding hydrogen to an unsaturated fat will make it more saturated. (p. 59)
5. Chemical reactions involving the combination of smaller building-block molecules with the removal of water molecules are hydrolysis reactions. (p. 55)
6. The bond between two adjacent amino acids in a protein molecule is a peptide bond. (p. 65)

Questions 7 to 10 refer to the examples of various building-block molecules shown in the diagrams on the right. Before answering the questions, see if you can place each molecule in the class to which it belongs. The following procedures will help you with the identification.

First, circle all the hydroxyl, or alcohol, groups, —OH. Next, cirlce all the carboxyl, or acid, groups.

$$-C\overset{O}{\underset{OH}{\diagup}} \quad or \quad -C\overset{O}{\underset{O^-}{\diagup}}$$

Then circle all the amino groups, —NH3+. Now that you have circled the major functional groups, use the information below to assign each one of the above building-block molecules to the class of molecules to which it belongs.

7. Carbohydrates have hydrogen and oxygen atoms present in the same proportions as in water; there are two hydrogen atoms for every one oxygen atom. Therefore, carbohydrates have many hydroxyl groups, and the grouping H—C—OH recurs frequently. Using this information, pick out any carbohydrates among the molecules above. (pp. 52–54)

8. Amino acids, as their name suggests, have both an amino and an acid group attached to the same carbon. Pick out any amino acids among the molecules above. (p. 64)

9. Fatty acids have a hydrocarbon chain and an acid group. Pick out any fatty acids among the molecules above. (p. 58)

10. Nucleotides are composed of three parts: a five-carbon sugar, a phosphate group, and a nitrogenous base. Pick out the nucleotide amont the molecules above, and circle the sugar, the phosphate, and the nitrogenous base. (p. 72)

Testing knowledge and understanding

11 Complete the following chemical reactions.
(pp. 55–65)

a)
$$H_2N-\underset{\underset{H}{|}}{\overset{\overset{H}{|}}{C}}-\overset{O}{\underset{OH}{\overset{\|}{C}}} \quad + \quad \underset{H}{\overset{H}{\underset{H}{\overset{|}{N}}}}-\underset{\underset{CH_2-C_6H_4-OH}{|}}{\overset{\overset{H}{|}}{C}}-\overset{O}{\underset{OH}{\overset{\|}{C}}} \xrightarrow{\text{condensation}}$$

b) (triglyceride with three hexanoate-like chains) + 3H₂O →ʰʸᵈʳᵒˡʸˢⁱˢ

c) (glucose) + (glucose) →ᶜᵒⁿᵈᵉⁿˢᵃᵗⁱᵒⁿ

For questions 12 to 17, use the structural formulas for some important organic molecules given below.

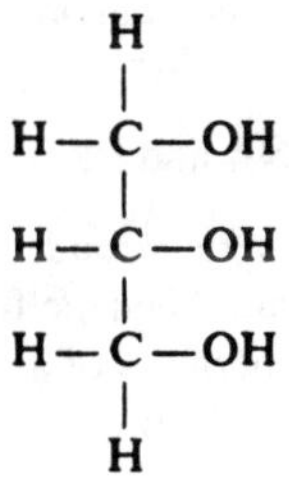

12 Which one of the molecules is a building-block molecule for proteins? (p. 64)

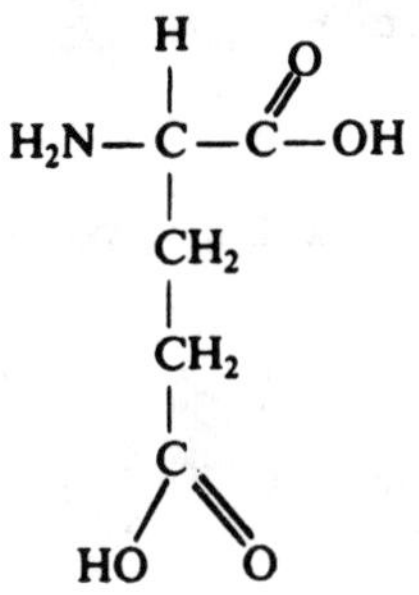

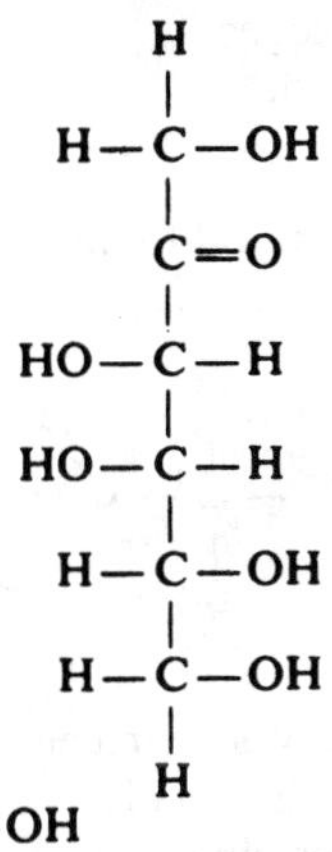

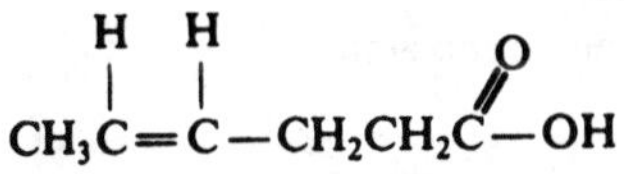

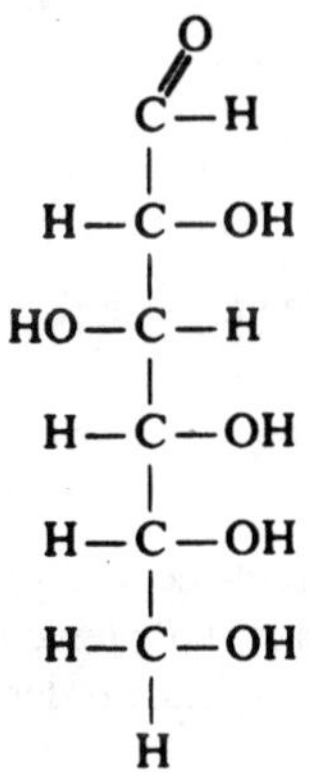

13 Which one of the above molecules is glucose? (p. 54)

14 Which one of the above molecules is a fatty acid? (p. 58)

15 Which one of the above molecules would be least soluble in water? (p. 58)

16 Molecules *b* and *d* are examples of

　a isomers.　　　　*d* hydrophobic substances.
　b isotopes.　　　　*e* disaccharides.
　c organic acids.　　　　　　　　(p. 54)

17 Which two of the molecules on the left could be the products of the hydrolysis of a fat?

　a *a* and *b*　　　*d* *c* and *e*
　b *b* and *c*　　　*e* *c* and *d*
　c *a* and *e*　　　　　　　(p. 61)

For questions 18 to 22, use the diagram below, which shows a polypeptide composed of four amino acids. Specific bonds are indicated by arrows with letters; functional groups are circled with broken lines and labeled with numbers. Use the letters and numbers on the diagram to answer the questions.

18 Which bond is a peptide bond? (Use letters on the diagram.) (p. 65)

19 Which functional groups or R groups are hydrophobic? (Use numbers on the diagram.) (p. 65)

20 Which functional groups or R groups are hydrophilic? (p. 65)

21 Which functional group is an amino group? (p. 54)

22 Which functional group is an acid? (p. 53)

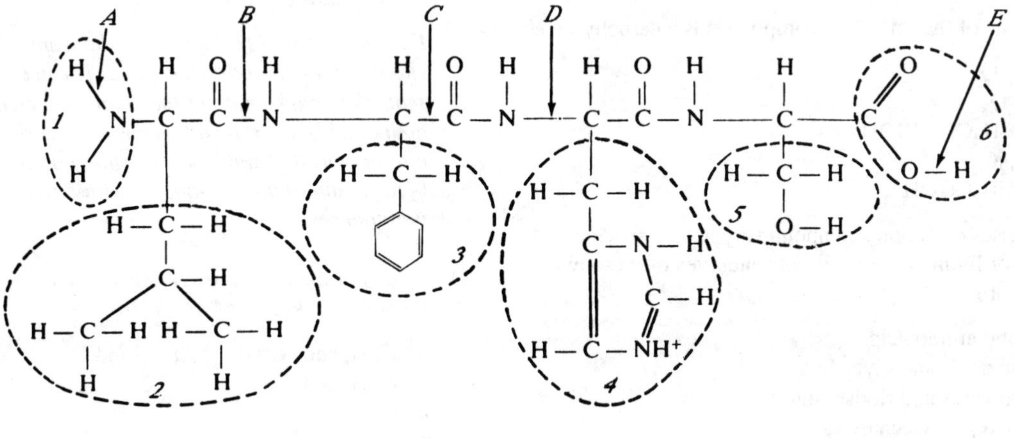

Choose the one best answer.

23 The carbon atom can form so many different chemical compounds because

   a its unstable nucleus easily gives up neutrons.
   b its outer energy level contains four electrons.
   c its electron shells are stable.
   d it can form both ionic and covalent bonds.
   e it tends to give up electrons to electron acceptors.
   (p. 51)

24 Below is the structural formula for a molecule composed of carbon, nitrogen, and hydrogen. The hydrogen atoms have not been drawn in. Which is the correct chemcial formula?

   a $C_5H_{11}N$ d $C_5H_{12}N$
   b C_5H_9N e C_5H_6N
   c $C_5H_{10}N$ (Chapter 2, p. 37; p. 51)

25 Compounds with the same atomic content but differing structures and properties are called

   a isotopes.
   b isomers.
   c ionic compounds.
   d polar covalent compounds.
   e nonpolar covalent compounds.   (p. 52)

26 All the following are polymers with many subunits *except*

   a fat.   d DNA
   b protein.   e glycogen.
   c starch.   (p. 59)

27 Which one of the following compounds is a carbohydrate?

   a $C_5H_{10}O_5$
   b $C_3H_8O_3$
   c $CH_3CH_2CH_2COOH$
   d $C_6H_{12}O_2$
   e H_2NCH_2COOH (p. 52)

28 Two classes or organic compounds typically provide energy for living systems. Representatives of these two classses are

   a fats and amino acids.
   b amino acids and glycogen.
   c amino acids and ribose sugars.
   d fats and polysaccharides.
   e nucleic acids and phospholipids.   (p. 74)

29 If two 5-carbon sugars are combined to form a disaccharide molecule with 10 carbons, how many hydrogen atoms will it have?

   a 10   d 20
   b 12   e 22
   c 18   (p. 55)

30 Plants commonly store carbohydrates for an energy source as

   a glycogen.   d sucrose.
   b starch.   e fat.
   c cellulose.   (p. 56)

31 The chemical formula below represents the product of a condensation reaction between

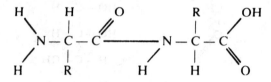

   a a fatty acid and an amino acid.
   b two fatty acids.
   c two amino acids.
   d an amino acid and an alcohol.
   e a fatty acid an an alcohol.   (p. 65)

32 Both DNA and RNA

   a are single-stranded molecules.
   b contain the same four nucleotide bases.
   c are polymers of amino acids.
   d have the same five-carbon sugar.
   e contain phosphate groups.   (p. 72)

33 The conformation of a protein molecule depends on several different types of bonds and group interactions. Which of these remain intact when a protein is denatured?

   a peptide bonds
   b ionic bonds
   c hydrogen bonds
   d hydrophobic interactions
   e none of the above   (p. 70)

For questions 34 to 38, use the diagram of a polypeptide chain at the top of the next page. The heavy line represents the "backbone" of the chain. Selected R groups of amino acids are shown together with various bonds and interactions that stabilize the folding of the chain. Each bond or interaction is labeled with a roman numeral in parentheses.

34 Which of the atoms below are bonded together linearly to form the backbone of the chain?

   a C, H, and N   d C and O
   b C, H, and O   e C, N, and O
   c C and N   (p. 65)

35 Which of the bonds between R groups is (are) ionic bond(s)?

   a I   d III and IV
   b III   e I and II
   c IV   (p. 39)

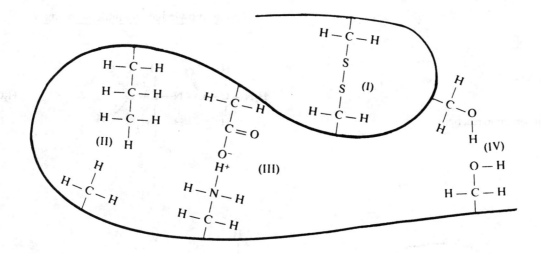

36 Which of the R-group interactions is (are) hydrophobic?

   a I
   b II
   c I and II
   d III
   e III and IV   (p. 65)

37 Which of the bonds between R groups is (are) hydrogen bond(s)?

   a II
   b III
   c IV
   d II and III
   e II and IV   (Chapter 2, p. 41)

38 Which of the bonds between R groups is (are) covalent?

   a I
   b II
   c III
   d IV
   e I and IV   (p. 65)

For questions 39 to 42, use the drawing on the right which represents a protein molecule.

39 The region labeled (1), taken alone, illustrates which level of protein structure?

   a primary
   b secondary
   c tertiary
   d quaternary   (p. 66)

40 In the drawing the bond labeled (2) is

   a a covalent bond.
   b a hydrogen bond.
   c an ionic bond.
   d a hydrophobic bond.   (p. 65)

41 The region of the protein molecule labeled (3) is

   a hydrophobic.
   b hydrophilic
   c nonpolar.
   d alpha-helical   (p. 65)

42 The highest level of protein structure shown in the drawing is

   a primary.
   b secondary.
   c tertiary.
   d quaternary.   (p. 69)

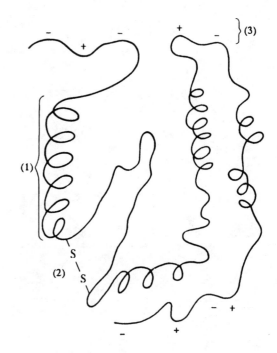

ANSWERS

Testing recall

1 true
2 false—protein
3 true
4 true
5 false—condensation reactions
6 true
7 *b*, *d*
8 *a*, *e*
9 *c*
10 *f*

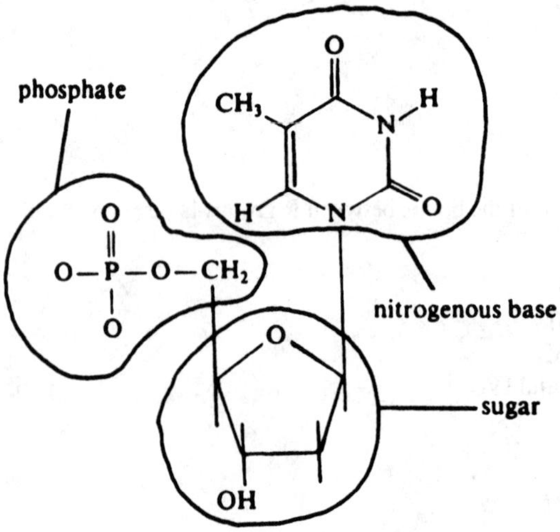
phosphate
nitrogenous base
sugar

Testing knowledge and understanding

11

a

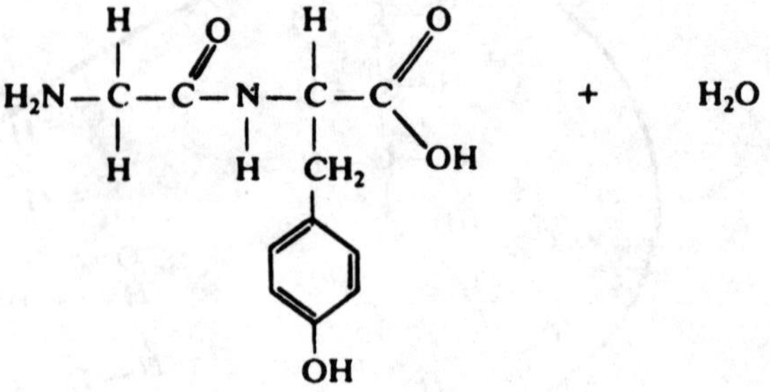

b

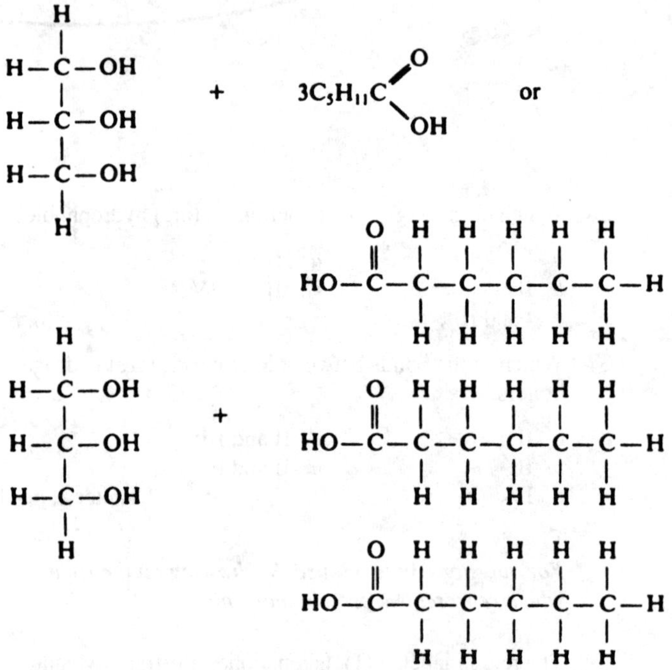

c

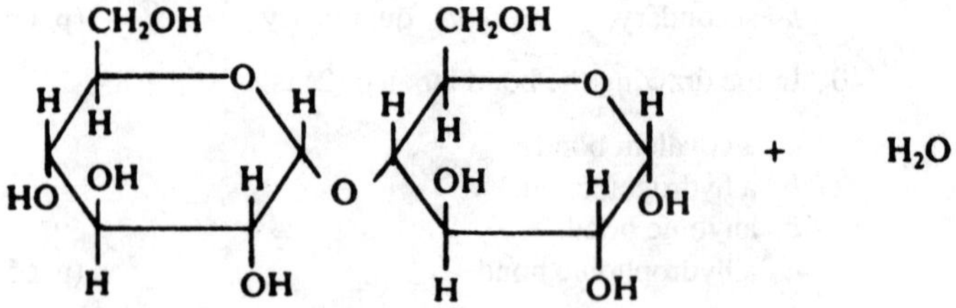

THE MOLECULES OF LIFE • 25

12	a	15	c	18	b	21	1
13	d	16	a	19	2, 3	22	6
14	c	17	d	20	1, 4, 5, 6	23	b

24 b. The completed structural formula would be

25	b
26	a
27	a
28	d
29	c
30	b
31	c
32	e
33	a

34	c
35	b
36	b
37	c
38	a
39	b
40	a
41	b
42	c

Chapter 4

THE FLOW OF ENERGY IN BIOCHEMICAL REACTIONS

A GENERAL GUIDE TO THE READING

This chapter focuses on some important principles governing chemical reactions and on the properties of the enzymes that catalyze the thousands of chemical reactions that occur in organisms. As you read Chapter 4, you will want to concentrate on the following topics:

1 Free energy. The concept of free energy (p. 79) must be understood because it lays the groundwork for material presented in Chapters 5, 7, and 8. The terms exergonic and endergonic should be learned.

2 Enzymes. Careful study of this section (pp. 82–86) is essential since enzymes and enzymatic activity will be referred to repeatedly throughout the text. Figures 4.10, 4.11, and 4.12 are particularly helpful in understanding enzymatic activity.

KEY CONCEPTS

1 The Laws of Thermodynamics govern chemical reations. The course of a chemical reaction depends on whether the free energy of the reactants is greater or less than the free energy of the products. If the reaction results in products with less free energy than the reactants possessed, the reaction is exergonic and will proceed spontaneously. If, however, the products have more free energy than the reactants, the reaction will require a net input of energy and is said to be endergonic. (p. 79)

2 Enzymes, like all catalysts, lower the activation energy needed for all reactions. An enzyme affects only the rate of the reaction; it speeds up the reaction but does not alter the direction of the reaction, its final equilibrium, or the reaction energy involved. (pp. 82–83)

3 Most enzymes are highly specific, and each can interact only with those ractants or substrates that fit spatially and chemically into the active site of the enzyme. Anything that alters the shape of the enzyme will alter its activity. (pp. 84–86)

CROSSLINKING CONCEPTS

1 The Laws of Thermodynamics underlie the physical and chemical reactions that occur in all living things.

2 The crucial importance of enzymes for the chemistry of life will be apparent throughout the text, particularly in the chapters concerned with photosynthesis, respiration, genetics, and the physiology of plants and animals.

3 The principle of feedback inhibition by which many enzyme systems are regulated is an important homeostatic mechanism. Analogous systems are involved in the control of gene expression (Chapter 12) and of many physiological processes (e.g., chemical control in animals, Chapter 23).

OBJECTIVES

1. Briefly describe the First and Second Laws of Thermodynamics and explain how they govern chemical reactions. (p. 78)

2. Define the term free energy. Explain how the change in free energy accompanying a chemical reaction is related to whether the reaction is exergonic or endoergonic. (p. 79)

3. Describe how coupled reactions enable endergonic reactions to occur in living systems, and how they are related to enzymatic pathways. (pp. 79–80)

4. Explain what is meant by activation energy and why the activation-energy barrier provides stability for high-energy molecules. Compare Figures 4.3, 4.4, and 4.7, and explain the role an enzyme plays in speeding up a chemical reaction; indicate whether the enzyme affects the free-energy change asociated with the reaction. (p. 80)

5. Explain why the three-dimensional strucure of an enzyme is the key to its activity. In doing so, include answers to the following questions: What is the active site of an enzyme? How are the enzyme and substrate molecules held together to form the enzyme substrate complex? What do we mean by "induced fit"? Why may changes in temperature or pH greatly reduce enzyme activity? What effect might an inhibitor moleucle have on enzyme activity? (pp. 83–86)

KEY TERMS

potential energy (p. 77)
kinetic energy (p. 77)
free energy (p. 79)
First Law of Thermodynamics (p. 78)
Second Law of Thermodynamics (p. 78)
reactants (p. 79)
products (p. 79)
exergonic (p. 79)
endergonic (p. 79)
activation energy (p. 80)
catalyst (p. 82)
enzyme (p. 82)
substrate (p. 82)
active site (p. 83)
induced-fit hypothesis (p. 84)
cofactor (p. 84)
coenzyme (p. 84)
enzyme inhibitor (p. 85)
feedback inhibition (p. 86)
enzymatic pathway (p. 86)

SUMMARY

The course of a chemical reaction depends on whether the *energy* required to break the covalent bonds of the *reactants* is greater or less than that released in the formation of covalent bonds in the products. Reactions that proceed spontaneously, releasing *free energy*, are said to be *exergonic*; reactions that require a net input of free energy to proceed are *endergonic*. In living systems an exergonic reaction is usually coupled with an endergonic reaction; energy released by the exergonic reaction is used to drive the endergonic reaction. Although exergonic reactions proceed spontaneously, initiating a reaction usually requires activation energy.

Chemical reactions can be speeded up by heat, by increasing the concentrations of the reactants, or by providing an appropriate *catalyst*. In living systems the catalysts are *enzymes*; as all other catalysts, they speed up reactions by lowering the activation energy barrier. Enzymes are large globular proteins that catalyze the thousands of chemical reactions in living organisms. Because the amino-acid sequence of each enzyme is different, every enzyme has a unique three-dimensional structure, which confers great specificity. Most enzymes can interact only with those reactants, or *substrates*, that fit spatially and chemically into the *active site*. When an enzyme reacts with its substrate, a short-lived *enzyme-substrate complex* is formed. This complex facilitates the reaction both by properly aligning the substrate's groups and by enhancing the reactivity of those groups. The substrate is converted into product and leaves the active site.

Since the formation of the enzyme-substrate complex requires the enzyme and its substrate to be complementary, anything that alters the shape of the enzyme will alter its activity and efficiency. Heat and change in pH break the weak bonds stabilizing an enzyme's conformation and alter its shape. If changes in the enzyme are large enough, the enzyme becomes *denatured* and can no longer function. Some *inhibitors* bind irreversibly to the active site of an enzyme and prevent the substrate from binding. Other inhibitors resemble the substrate molecules and compete for the active site. Still others bind to a second site on the enzyme and induce a change in the shape of the active site so the substrate can no longer bind.

Enzymes often work consecutively to catalyze a chain of reactions. In such a chain, the product of one enzyme-catalyzed reaction becomes the substrate for the next enzyme in the sequence, and so on.

CHAPTER 4

CONCEPT MAPS

Map 4.1

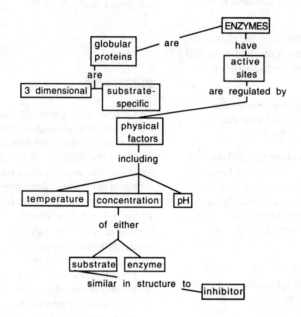

QUESTIONS

Testing recall

Decide whether the following statements are true or false, and correct the false statements.

1. The combustion of gasoline in a car is an example of an exergonic reaction. (p. 79)

2. A chemical reaction in which more energy is required to break the bonds of the reactants than is released in the formation of new bonds in the products is exergonic (p. 79)

3. All catalysts work by lowering the activation energy needed for a reaction to take place. (p. 82)

4. Chemical reactions that require a net input of free energy to proceed are endergonic reactions. (p. 79)

5. Most enzymes are proteins (p. 82)

Testing knowledge and understanding

Choose the one best answer.

6. A biochemist was working with an enzyme that she knew was a single polypeptide chain. When she lowered the pH of the enzyme slightly, she found that the enzyme's activity slowed. However, when she returned the pH to the enzyme's normal range, the enzyme resumed normal activity and speeded up a chemical reaction. Lowering the pH probably altered the enzyme's

   a primary structure.   c quaternary structure.
   b tertiary sturcture.   d peptide bonds.   (p. 84)

7. Which of the following characteristics does not apply to a structural protein such as silk?

   a peptide bonds
   b specific primary sturcture
   c active site
   d hydrogen bonds between separate polypeptide chains.
   e more than one kind of amino acid   (p. 83)

8 The reaction A + B → C + D is exergonic. All the following procedures are effective in accelerating the rate of the chemical reaction *except*

   a increasing the concentration of C and D.
   b increasing the concentration of A and B.
   c providing an appropriate catalyst.
   d heating A and B together. (p. 81)

9 Gl;ucose will not burn in the air unless it is strongly heated. Why is heat required?

   a Heat provides activation energy.
   b Heat acts as a catalyst.
   c Heat lowers the average energy content of the molecules.
   d Heat increases the net amount of energy released by the reaction.
   e Heat changes exergonic reactions to endergonic reactions. (p. 81)

Questions 10 to 14, use the diagram below, showing energy changes in the reaction X + W → Y + Z. Answers may be used once, more than once, or not at all.

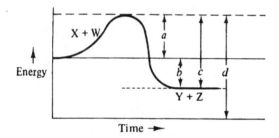

10 Which segment represents the energy of activation? (p. 80)

11 Which segment represents the amount of net free energy released by this reaction? (p. 80)

12 Which segment would be the same irrespective of whether the reaction was uncatalyzed or catalyzed by an enzyme? (p. 82)

13 The above reaction is

   a exergonic.
   b endergonic.
   c spontaneous.
   d *a* and *c*. (p. 79)

14 The above reaction

   a would have a negative free-energy change.
   b would have a positive free-energy change.
   c could have either a negative or a positive free-energy change depending on the reactants and the temperature. (p. 79)

15 Which one of the following statements is *true* of enzymes?

   a Enzymes lose some or all of their normal activity if their three-dimensional structure is disrupted.
   b Enzymes are composed of ribose, phosphate, and a nitrogen-containing base.
   c The activity of enzymes is independent of temperature and pH.
   d Enzymes provide the activtation energy necessary to initiate a reaction.
   e An enzyme acts only once and is then destroyed. (p. 84)

16 Enzymes

   a impart to substrates the kinetic energy they need to react.
   b may have at their active sites amino acids from widely different parts of a polypeptide chain.
   c help align substrates so that the latter are oriented to each other in a very precise way.
   d do two of the above.
   e do *a*, *b*, *c*. (p. 82)

Questions 17 to 18 refer to the following situation:

The enzyme succinic dehydrogenase normally catalyzes a reaction involving succinic acid. Another substance, malonic acid, sufficiently resembles succinic acid to form temporary complexes with the enzyme, although malonic acid itself cannot be catalyzed by succinic dehydrogenase.

17 In this example succinic acid is

   a the substrate.   d the product.
   b the active site.   e a coenzyme
   c an inhibitor. (p. 82)

18 Malonic acid is

   a the substrate.   d the product.
   b the active site.   e a coenzyme
   c an inhibitor. (p. 85)

ANSWERS

Testing recall

1. true
2. false—endergonic
3. true
4. true
5. true

Testing knowledge and understanding

6	b	10	a	13	d	16	d
7	c	11	b	14	a	17	a
8	a	12	b	15	a	18	c
9	a						

Chapter 5

AT THE BOUNDARY OF THE CELL

A GENERAL GUIDE TO THE READING

This chapter gives a brief history of our knowledge of cells. It then goes on to examine the outer boundaries of cells, notably the structure and function of the cell membrane. As you read Chapter 5 in your text, you will want to concentrate on the following topics.

1. The ratio of surface area to volume. This topic is introduced on page 93. Make sure you understand it, since it will come up repeatedly in later chapters.

2. Diffusion and osmosis. Though this material (pp. 94–98) is conceptually difficult at first, understanding it now is worth the effort. We will be using the terms osmosis, diffusion, osmotic concentration, osmotic potential, hypertonic, hypotonic, and isotonic frequently later.

3. The fluid-mosaic model. Pay close attention to the description of this model of the cell membrane on pages 98–100 and to Figure 5.10 (p. 99). Part of the great excitement in biology in recent years results from our new understanding of membrane structure and how it relates to membrane function.

4. Membrane channels and pumps. This section introduces some of the specific mechanisms by which molecules get into and out of cells. Figure 5.14 (p. 102) is an excellent summary diagram. It is particularly important to learn this material because you will encounter more detailed information on specific channels and pumps when you study photosynthesis (Chapter 7), respiration (Chapter 8), kidney function (Chapter 22), hormone control (Chapter 25), and nerve control (Chapter 28). Time spent learning the material now will save you time later on.

5. Facilitated diffusion and active transport. These processes, explained on pages 101 and 102, will be mentioned repeatedly throughout the text.

KEY CONCEPTS

1. The fundamental organizational unit of life is the cell. All living things are composed of cells; all cells arise from preexisting cells. (pp. 89–90)

2. A cell's interaction with its environment is crucial; materials necessary for life must be obtained from the environment, and waste products must be released into it. (pp. 94–98)

3. Membranes are barriers between cells and (as will be seen later in the text) around certain entities within cells. All substances moving into or out of a cell must pass through a membrane barrier, and the membrane of each cell can be quite specific about what is to pass through, at what rate, and in which direction. (pp. 98–102)

4. The plasma membrane consists of a bilayer of phospholipids oriented with their hydrophobic tails toward the interior of the membrane and their hydrophilic heads toward the aqueous surfaces. The proteins are distributed both on the surfaces and in the interior of the membrane. (pp. 98–100)

5 The lipid bilayer creates an effective barrier between the inside of the cell and the surrounding medium. Highly specialized channels and pumps control the passage of molecules into and out of the cell. (pp. 101–3)

6 The lipid bilayer of the cell membrane is relatively permeable to nonpolar and lipid-soluble substances and to small polar molecules such as water. It is quite impermeable to large molecules and ions. Such substances require the use of permeases to pass through the cell membrane. (pp. 101–3)

CROSSLINKING CONCEPTS

1 The Laws of Thermodynamics underlie diffusion and osmosis. We have seen that chemical reactions take place spontaneously only when reactants have more free energy than products. Likewise, diffusion and osmosis occur only when substances move from high free energy to low free energy.

2 Knowledge of the structure of large organic molecules and their building blocks helps to explain both the structure of the cell membrane and the movement of different particles across it. Lipid-soluble molecules cross the membrane easily, whereas charged or large molecules cross with difficulty. In particular, an understanding of the hydrophilic and hydrophobic parts of molecules helps to explain both the structure of the membrane and the movement of particles across the membrane.

3 The concepts of protein structure, R groups, and enzyme specificity from Chapters 3 and 4 explain the function of permeases, which control the function of the selectively permeable membrane.

4 In this chapter you learn that large particles enter the cell through endocytosis. In the next chapter you will learn more about how the endoplasmic reticulum, endosomes, and lysosomes work to process the resulting vesicles.

OBJECTIVES

1 State the two components of the cell theory and be familiar with the contributions made to our knowledge of cells and cell structure by Matthias Schleiden, Theodor Schwann, Rudolf Virchow, and Louis Pasteur. (p. 89)

2 Describe the process of osmosis and explain the relationship between osmosis and diffusion. In doing so, include answers to the following questions. How is the net movement of water through a membrane in osmosis related to the concentration of solute molecules? If two compounds had the same molecular weight but one ionized in solution and the other didn't, which one would have the higher osmotic concentration? Osmotic potential? (pp. 94–98)

3 Define the terms hypotonic, hypertonic, and isotonic. Explain what would happen to an animal cell placed in a hypertonic medium and to an animal cell placed in a hypotonic medium. Then explain what would happen to plant cells so placed. (pp. 96–98)

4 Using 5.10 (p. 99), describe the fluid-mosaic model of the cell membrane. In your description, point out the phospholipid molecules and indicate the hydrophobic and hydrophilic portions, point out the proteins that span the interior of the membrane and those that are confined to the surface, indicate the role of the cholesterol molecules, and explain why lateral movement of molecules within the membrane is possible. List substances to which the membrane is relatively permeable and those to which it is relatively impermeable. (pp. 98–100)

5 Describe the role that permeases play in moving material through membranes, mentioning membrane channels, gated channels, mobile carriers, and membrane pumps. Figure 5.14 (p. 102) may be helpful in doing this. (pp. 101–2)

6 Distinguish among simple diffusion, facilitated diffusion, and active transport. Indicate the role of these processes in the life of the cell. Figure 5.15 (p. 103) may be helpful in doing this. (pp. 101–3)

7 Using Figures 5.16 to 5.20 (pp. 103–7), describe the processes of endocytosis, phagocytosis, pinocytosis, receptor-mediated endocytosis, and exocytosis; explain the role of these processes in transporting substances into and out of the cell. (pp. 102–7)

8 Describe the formation and structure of the plant cell wall; include in your description the primary wall, secondary wall, middle lamella, pectin, lignin, and plasmodesmata. Indicate how the cell walls of bacteria and fungi differ from those of plants. (pp. 107–9)

9 Describe the structure of the glycocalyx in animal cells and name two of its functions. (pp. 110–11)

KEY TERMS

plasma membrane (p. 89)
cell theory (p. 89)
biogenesis (p. 90)
diffusion (p. 94)
osmosis (p. 95)
selectively permeable (p. 95)
osmotic concentration (p. 96)
osmotic pressure (p. 96)
hypertonic (p. 97)
hypotonic (p. 97)
isotonic (p. 97)
fluid-mosaic model (p. 98)
peripheral protein (p. 99)
integral protein (p. 99)

permease (p. 101)
membrane channel (p. 101)
facilitated diffusion (p. 101)
receptor (p. 101)
gated channel (p. 101)
mobile carrier (p. 101)
membrane pump (p. 101)
active transport (p. 102)
sodium-potassium pump (p. 102)
endocytosis (p. 102)
phagocytosis (p. 102)
pseudopodia (p. 102)
pinocytosis (p. 103)
receptor-mediated endocytosis (p. 105)
exocytosis (p. 107)
cellulose (p. 107)
primary wall (p. 108)
middle lamella (p. 108)
pectin (p. 108)
secondary wall (p. 108)
lignin (p. 108)
plasmodesmata (p. 108)
symplast (p. 108)
chitin (p. 109)
murein (p. 109)
turgor pressure (p. 109)
plasmolysis (p. 109)
glycocalyx (p. 110)
contact inhibition (p. 110)

SUMMARY

The fundamental organizational unit of life is the cell. According to the cell theory, all living things are composed of cells and all cells arise from preexisting cells.

Materials required by cells must enter by crossing the surface of the cell, and waste products must leave by the same route. As cell size increases, the volume increases much more rapidly than the surface area; this creates the problem of maintaining an adequate exchange surface to support the increased volume.

Functioning of a membrane The cell or plasma membrane is an active part of the cell; it regulates the movement of materials between the ordered interior of the cell and the outer environment. The general rule governing the movement of materials is that the net movement of particles of a particular substance is from regions of greater free energy (where there is an orderly, improbable arrangement) to regions of less free energy (where there is a disorderly, more probable arrangement) of that substance. This movement of particles is called *diffusion*. The plasma membrane is *selectively permeable*; it allows particles of some substances to pass through while restricting others. The movement of water through a selectively permeable membrane is called *osmosis*.

If two different solutions are separated by a selectively permeable membrane under constant conditions of temperature and pressure, the net movement of water will be from the solution with fewer dissolved particles per unit volume to the solution with more dissolved particles per unit volume, i.e., from the solution with the lower osmotic concentration to the solution with the higher osmotic concentration.

The plasma membrane Because the plasma membrane is selectively permeable, the processes of osmosis and diffusion are fundamental to cell life. Cell membranes are relatively permeable to water and to certain simple sugars, amino acids, and lipid-soluble substances; they are relatively impermeable to polysaccharides, proteins, and other very large particles. Their permeability to small particles varies, but in general uncharged particles cross more rapidly than charged ones.

The cell membrane cannot completely regulate the exchange of materials. The cell in a medium that is *hypertonic* (a medium with a higher osmotic concentration than the cell) tends to lose water and shrink. Conversely, a cell in a *hypotonic* medium (one with a lower osmotic concentration than the cell) tends to gain water and swell, and may even burst. A cell in an *isotonic* medium (one with an osmotic concentration that is in balance with it) neither gains nor loses appreciable water.

Cells are bounded by a membrane composed of lipids and proteins, with many small pores. According to the fluid-mosaic model, the cell membrane consists of a bilayer of phospholipids with their hydrophilic heads oriented toward the surfaces of the membrane, and their hydrophobic tails toward the interior. Because the individual lipid molecules are linked only by weak bonds, many of them have lateral mobility. The proteins are distributed both on the surfaces and in the interior of the membrane. The pores are thought to be bounded by protein; the distinctive properties of these proteins make the pores selective as to what can move through them.

Enzymelike protein carriers, *permeases*, control molecular traffic into and out of the cell. Some permeases act as membrane *channels*, providing openings through which specific substances can diffuse passively across the membrane, in a process called *facilitated diffusion*. Other permeases have a gated channel; a signal molecule combines with a receptor, which changes shape, and opens the channel. Mobile carrier molecules may transport molecules across a membrane. Still other permeases, known as pumps, carry on *active transport*; they use energy to move substances across the membrane against their concentration gradients.

Sometimes substances are taken into the cell by an active process called *endocytosis*, in which a substance is enclosed in a membrane-bound vesicle pinched off from the cell membrane. The reverse sequence, in which materials within vesicles are conveyed to the surface of the cell and discharged, is called *exocytosis*.

Cell walls and coats The plant cell wall is located outside the membrane and is composed mainly of *cellulose*. *The primary wall*, the first portion of the wall laid down, consists of a loose network of fibrils. Many plant cells add further layers, and so form a thicker, more compact *secondary wall*. Adjacent cells are bound together by the *middle lamella*.

CONCEPT MAPS

Map 5.1

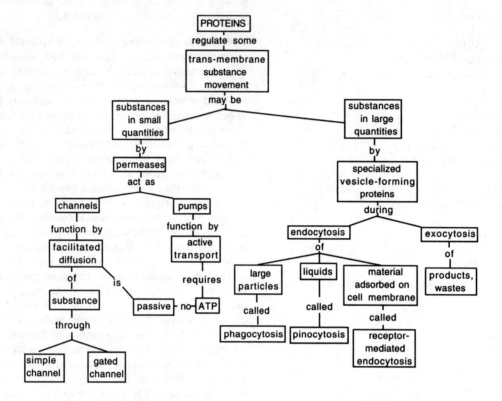

Map 5.2

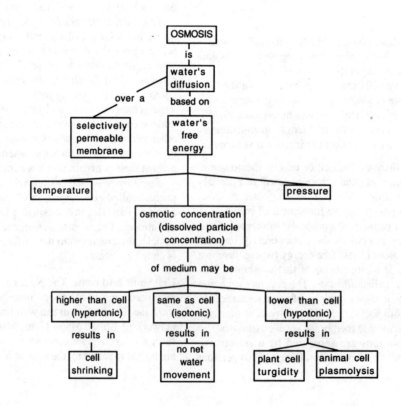

KEY DIAGRAM

This coloring exercise is designed to help you learn the basic features of membrane structure. You will need eight different colors for this picture. Note the number and arrangement of the different labels. Using the eight colors and the information in Figure 5.10 (p. 99), color the labels and the corresponding structures, using a different color for each structure.

Ask yourself:

1. Are peripheral proteins found only on the outer surfaces of membranes, or can they be found on the inner surface as well?
2. Do integral proteins, too, have hydrophilic and hydrophobic sections? If so, where are they located?
3. What factors alter the fluidity of membranes?

QUESTIONS

Testing recall

Match the names of the following scientists with their contributions.

a Pasteur
b Schleiden and Schwann
c Singer and Nicolson
d Virchow

1. formulated the cell theory (p. 89)
2. advanced the theory of biogenesis (p. 90)
3. disproved the theory of spontaneous generation (p. 90)
4. formulated the fluid-mosaic model (p. 98)

PHOSPHOLIPIDS:
HYDROPHILIC HEADS
HYDROPHOBIC TAILS

CHOLESTEROL

PROTEINS:
MEMBRANE CHANNEL
PERMEASE
INTEGRAL
PERIPHERAL

GLYCOCALYX

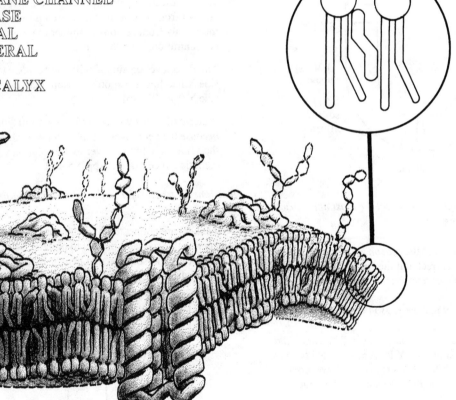

It is important not to confuse the following pairs of terms. In one sentence, distinguish between the words in each pair.

5   osmosis—diffusion (pp. 94–95)

6   hypertonic—hypotonic (pp. 96–97)

7   osmotic concentration—osmotic pressure (p. 96)

8   facilitated diffusion—active transport (pp. 101–2)

9   endocytosis—exocytosis (pp. 102–7)

10 pinocytosis—phagocytosis (p. 102–3)

11 cell wall–cell coat (pp. 107–10)

12 primary cell wall–secondary cell wall (pp. 107–8)

Testing knowledge and understanding

Questions 13 to 22 refer to the following situation.

The solutions in the two arms of the U-tube are separated at the bottom of the tube by a selectively permeable membrane. At the beginning of the experiment the volumes in both arms are the same, and the levels of the liquid are therefore at the same height. The membrane is permeable to water and to sodium and chloride ions but not to glucose. The apparatus is allowed to stand for three days.

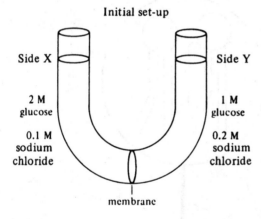

Initial set-up

Side X: 2 M glucose, 0.1 M sodium chloride
Side Y: 1 M glucose, 0.2 M sodium chloride

For each of the next 10 questions, select the most appropriate phrase using the following key.

  a Both the *statement* and the *reason* are correct.
  b The *statement* is correct, but the *reason* is incorrect.
  c The *statement* is incorrect, but the *reason* is a fact or principle.
  d Both the *statement* and the *reason* are incorrect.

13 The sodium chloride solution on side X will become more concentrated and that on side Y less concentrated *because* a substance tends to diffuse from regions of lower concentration to regions of higher concentration of that substance. (p. 94)

14 The concentrations of the glucose solutions on sides X and Y will remain unchanged *because* the membrane is impermeable to glucose and so glucose cannot diffuse from one side to the other. (pp. 95–96)

15 The concentration of sodium chloride on side X will eventually equal that on side Y *because* sodium and chloride ions will move by diffusion from one side to the other and gradually reach a uniform density; then the net movement of ions will stop. (pp. 95–96)

16 The concentration of glucose on side X will decrease and that on side Y will increase *because* water molecules will diffuse through the membrane from side Y to side X by osmosis, and thus lower the glucose concentration on side X. (pp. 95–96)

17 The fluid level will increase on side Y and decrease on side X *because* water molecules will move through the membrane from regions of higher to regions of lower concentration of water molecules. (pp. 95–96)

18 The fluid level on side X will rise *because* the water molecules on that side at the beginning of the experiment have more free energy than those on side Y. (pp. 95–96)

19 The net movement of water molecules will be from side X to side Y *because* water molecules will move from the solution with the lower osmotic pressure to the solution with the higher osmotic pressure when the two are separated by a selectively permeable membrane. (pp. 95–96)

20 Water molecules will move only from side Y to side X and not from side X to side Y *because* water molecules move only from regions of higher to regions of lower concentration. (pp. 95–96)

21 The fluid level on side X will rise *because* the solution in side X had lower osmotic pressure than the solution in side Y. (pp. 95–96)

22 Water molecules will tend to move from side Y to side X *because* the net movement of water molecules will be from the solution with the lower to the solution with the higher osmotic concentration. (pp. 95–96)

Questions 23 to 25 refer to the following situation. Choose the one best answer.

Two beakers are connected by a tube partitioned by a membrane permeable to water but not to protein. Beaker A contains a 2 percent protein solution and beaker B contains a 4 percent protein solution.

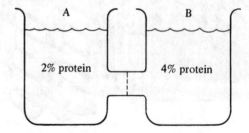

A: 2% protein; B: 4% protein

23 Assuming uniform temperature and pressure, which one of the following statements best describes what will happen in this system?

   a Water will move from A to B.
   b Water will move from B to A.
   c Water will move equally in both directions, so that there will be no net change in the system.
   d Water will move in both directions, but the net flow will be from A to B.
   e Water will move in both directions, but the net flow will be from B to A. (pp. 95–96)

24 Side A is _____ compared to side B.

   a hypertonic   c isotonic
   b hypotonic   (p. 97)

25 Suppose that instead of protein, we dissolve 1,000 molecules of NaCl in beaker A and 1,000 molecules of sucrose in beaker B. Suppose further that the membrane in the connecting tube is impermeable to both Na^+ and Cl^- ions and sucrose. What will happen in the system? (Select one of the answers listed under question 23.) (pp. 95–96)

Questions 26 to 30 refer to the following diagram of the plasma membrane.

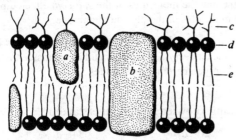

26 Of the five labeled parts of the diagram (*a, b, c, d,* and *e*), which one is now thought to play a critical role in cell recognition? (p. 110)

27 Which one of the labeled parts illustrates a globular protein that probably acts as a permease? (p. 101)

28 Which one of the labeled parts illustrates the hydrophilic portion of a phospholipid molecule? (p. 99)

29 When hydrolyzed, substance *b* would yield

   a amino acids.   d nucleotides.
   b fatty acids.   e phospholipids.
   c monosaccharides.   (pp. 65, 99)

30 Which one of the labeled parts shows the most hydrophobic region of the membrane? (p. 99)

31 Molecules move by diffusion. Diffusion is

   a faster in a liquid than in a gas.
   b faster in a solid than in a liquid.
   c from a region of higher concentration to a region of lower concentration.
   d from a region of lower concentration to a region of higher concentration. (p. 94)

32 Facilitated diffusion

   a involves a permease and requires energy.
   b involves a permease but does not require energy.
   c requires energy but does not involve a permease.
   d moves substances from a region of lower concentration to a region of higher concentration.
   e does not require energy or a permease. (p. 101)

33 Which of the following molecules would cross a cell membrane most easily? Assume there is no active transport or facilitated diffusion.

   a amino acid   d lipid-soluble substance
   b starch   e nucleotide
   c protein   (pp. 101, 102)

34 The concentration of potassium ions in a red blood cell is much higher than it is in the surrounding blood plasma, yet potassium ions continue to move into the cell. The process by which potassium ions move into the cell is called

   a osmosis.   d active transport.
   b simple diffusion.   e pinocytosis.
   c facilitated diffusion.   (pp. 102–3)

35 A dehydrated plant cell is placed in pure water. Which of the following statements best describes what will happen?

   a Water will enter the cell, but the cell will be prevented from bursting by the cell wall.
   b Water will enter the cell, and the cell will burst.
   c Water will be drawn out of the cell until the cell and the water are at equilibrium.
   d Water will be drawn out of the cell until the cell dies.
   e Water will not move either into or out of the cell. (pp. 107–9)

36 The only way in which a very large molecule such as a protein could cross a cell membrane is by

   a active transport.   d facilitated diffusion.
   b endocytosis.   e osmosis.
   c simple diffusion.   (p. 102)

37 According to the fluid-mosaic model of the cell membrane, the proteins are located

   a in a continuous layer over the outer surface of the membrane.
   b in a continuous layer over the inner surface of the membrane.
   c in a continuous layer over both the outer and inner surfaces of the membrane.
   d in the middle of the membrane, between the lipid layers.
   e in discontinuous arrangements, both on the surface and in the interior of the membrane. (pp. 98–100)

38 Among the cell structures and materials listed below, which one would be *innermost* if they were arranged in proper order?

   a glycocalyx   d secondary cell wall
   b plasma membrane   e pectin
   c primary cell wall   (p. 108)

39 You are studying the transport of a certain type of molecule into cells. You find that transport slows down when the cells are poisoned with a chemical that inhibits energy production. Under normal circumstances the molecule you are studying is probably transported into the cell by

    a simple diffusion.   *d* osmosis.
    b facilitated diffusion.   *e* exocytosis.
    c active transport.   (pp. 101–2)

For further thought

1 Distinguish between osmosis and diffusion and give a physical explanation for each. What is osmotic pressure? Discuss the implications of these processes for both an animal cell and a plant cell living in a hypotonic medium, a hypertonic medium, and an isotonic medium. (pp. 94–98)

2 Briefly describe the fluid-mosaic model of membrane structure. Explain why membranes are generally impermeable to charged molecules and to highly polar molecules such as sugars and amino acids. Discuss the importance of this impermeability to (a) osmosis and (b) the establishment of an electrochemical gradient across a membrane. (pp. 98–102)

3 A 5 percent glucose solution is frequently administered intravenously to persons who have undergone surgery. Would you expect the 5 percent glucose solution to be hypertonic, hypotonic, or approximately isotonic to the blood? Give reasons for your answer. (p. 97)

ANSWERS

Testing recall

1 *b*   3 *a*
2 *d*   4 *c*

5 *Osmosis* is the movement of water through a selectively permeable membrane; *diffusion* is the movement of any particles from high to low free energy and may or may not be through a membrane.

6 A *hypertonic* solution has a relatively higher concentration of solute and will gain water by osmosis; a *hypotonic* solution has a relatively lower concentration of solute and will lose water.

7 *Osmotic* pressure is a measure of the tendency of water to diffuse into a system; it varies directly with the concentration of osmotically active particles in the solution, the *osmotic concentration*.

8 *Facilitated diffusion* is passive movement, with the concentration gradient aided by channels; *active transport* involves pumps, moves substances across the membrane against the concentration gradient, and requires energy.

9 *Endocytosis* is a process by which a substance is drawn into a cell in a membrane-bounded vesicle. *Exocytosis* is essentially the reverse of endocytosis.

10 *Pinocytosis* and *phagocytosis* differ only in the size of the particles taken in. If the particles are large, the process is termed phagocytosis; if liquid or very small, pinocytosis.

11 The *cell wall* is entirely separate from the plasma membrane, but the molecules of the *cell coat* attach directly to the molecules of the plasma membrane.

12 The *primary cell wall* is found in all plant cells, whereas the thicker *secondary wall* is laid down inside the primary wall only in certain plant tissues.

Testing knowledge and understanding

13	*b*	20	*d*	27	*b*	34	*d*
14	*c*	21	*b*	28	*d*	35	*a*
15	*a*	22	*a*	29	*a*	36	*b*
16	*a*	23	*d*	30	*e*	37	*e*
17	*c*	24	*b*	31	*c*	38	*b*
18	*b*	25	*e*	32	*b*	39	*c*
19	*c*	26	*c*	33	*d*		

Chapter 6

INSIDE THE CELL

A GENERAL GUIDE TO THE READING

This chapter focuses on the internal structure of cells; it presents basic information on the various organelles. As you read Chapter 6, you will want to concentrate on the following topics.

1. Organelles. You will need to learn the basic structure and function of the various organelles, since we will be referring to them frequently throughout the book.

2. The differences between procaryotic and eucaryotic cells. These differences, summarized on pages 134 and 135 and in Table 6.1 (p. 135), will be referred to frequently in later chapters. The differences are quite simple to understand and should be mastered now.

KEY CONCEPTS

1. A living cell is an extraordinarily complex unit with an intricate internal structure; its activities are precisely integrated and controlled. (pp. 113–33)

2. Eucaryotic cells have a membrane-enclosed nucleus and internal membranous organelles, whereas procaryotic cells lack a membrane-enclosed nucleus and membranous organelles. (pp. 113–25)

CROSSLINKING CONCEPTS

1. Chapter 3 introduced the nucleic acid DNA. This chapter explains how DNA works in chromosomes and genes to control the production of proteins at ribosomes. More detail will come when you study how DNA is replicated and how it controls cellular function (Chapters 10 and 11).

2. To understand the function of lysosomes (digestion), it is important that you understand pH (Chapter 2), hydrolytic enzymes (Chapter 4), ion pumps and endocytosis (Chapter 5).

3. The structure of each of the subcellular organelles is particular to its function. What you learn about the structure of chloroplasts and mitochondria will guide your study of photosynthesis (Chapter 7) and energy production (Chapter 8).

4. Your understanding of the function of actin, myosin, and tubulin will enhance what you learned about pseudopods (Chapter 5) and underlie what you will learn when you study cell division (Chapter 9) and muscle function (Chapter 30).

5. The theme of evolution continues in this chapter, where the discussion focuses on the endosymbiotic hypothesis, which explains the origin of the eucaryotic cell.

OBJECTIVES

1. For each of the organelles listed below, describe its structure; give a major function; indicate whether it is surrounded by a single membrane, a double membrane, or no membrane; and state whether it is found in plant, animal, or bacterial cells. (pp. 113-30)

   nucleus
   chromosome
   nucleolus
   ribosome
   nuclear envelope
   endoplasmic reticulum (ER)
   Golgi apparatus
   lysosome
   peroxisome
   mitochondrion
   plastid (chromoplast and leucoplast)
   vacuole
   microfilament
   microtubule
   centrosome
   centriole
   basal body
   flagellum
   cilium

2. List four differences between procaryotic and eucaryotic cells. (pp. 134-35)

3. Describe the endosymbiotic hypothesis for the origin of the eucaryotic cell and give three lines of evidence for the validity of this hypothesis. (pp. 135-36)

KEY TERMS

cytoplasm (p. 113)
organelle (p. 113)
nucleus (p. 113)
procaryotic (p. 113)
eucaryotic (p. 113)
chromosome (p. 113)
nucleolus (p. 115)
nucleoplasm (p. 114)
nuclear envelope (p. 114)
chromatin (p. 114)
gene (p. 114)
nucleosome (p. 114)
endoplasmic reticulum (ER) (p. 116)
ribosome (p. 116)
rough ER (p. 116)
smooth ER (p. 116)
cytosol (p. 117)
Golgi apparatus (p. 120)
lysosome (p. 120)
microbody (p. 122)
peroxisome (p. 122)
catalase (p. 123)
vacuole (p. 123)
contractile vacuole (p. 123)
cell sap (p. 123)
turgid (p. 123)
anthocyanin (p. 124)
mitochondrion (p. 124)
cristae (p. 124)
plastid (p. 125)
chromoplast (p. 125)
leucoplast (p. 125)
chloroplast (p. 125)
chlorophyll (p. 125)
carotenoid (p. 125)
stroma (p. 125)
thylakoid (p. 125)
grana (p. 125)
amyloplast (p. 125)
cytoskeleton (p. 126)
microfilament (p. 126)
actin (p. 126)
myosin (p. 126)
intermediate filament (p. 127)
microtubule (p. 127)
tubulin (p. 127)
centrosome (p. 129)
centriole (p. 129)
basal body (p. 129)
flagellum (p. 130)
cilium (p. 130)
plasmid (p. 134)
endosymbiotic hypothesis (p. 136)

SUMMARY

Eucaryotic cells have a membrane-enclosed nucleus, whereas *procaryotic* cells (bacteria) lack a membrane-enclosed nucleus. The following discussion concerns only eucaryotic cells.

The *nucleus* contains the *chromosomes*, which contain the genes. It can therefore direct the cell's life processes. The nucleus also contains one or more *nucleoli*, where ribosomal RNA is synthesized and combined with proteins before moving into the cytoplasm to become part of the ribosomes. Separating the nucleus from the cytoplasm is a double *nuclear envelope* interrupted by pores. The nuclear envelope is continuous at places with the endoplasmic reticulum.

The *endoplasmic reticulum* (ER) forms a system of interconnected, membrane-enclosed spaces. Sometimes the membranes of the ER are "rough," with *ribosomes* on their outer surfaces; when no ribosomes are present, the ER is "smooth." Ribosomes are sites of protein synthesis. The ER functions both as a passageway for intracellular transport and as a manufacturing surface.

The *Golgi apparatus* consists of stacks of membrane-enclosed vesicles that function in the storage, modification, and packaging of secretory products.

Located within the cytoplasm are many other organelles. The *mitochondria* are the powerhouses of the cell; chemical reactions within the mitochondria provide energy for cellular activities. *Lysosomes* are membranous sacs that function as storage vesicles for powerful digestive enzymes; they may act as the cell's digestive system and hydrolyze materials taken in by endocytosis. *Peroxisomes* are also membranous sacs of enzymes; these are oxidative rather than digestive. Most plant cells have large membranous organelles called *plastids*. The two principal categories of plastids are the colored *chromoplasts* and the colorless *leucoplasts*. *Chloroplasts* are chromoplasts that contain the green pigment chlorophyll; they capture the energy of sunlight and utilize it in the manufacture of organic compounds. The leucoplasts' primary function is storage of starch, oils, or protein granules. Membrane-enclosed, fluid-filled spaces termed *vacuoles* are found in many cell types and perform a variety of functions. Most mature plant cells have a large central vacuole occupying much of the volume of the cell; it plays an important role in maintaining the turgidity of the cell and in storing important substances and wastes.

Microtubules and *microfilaments* function in intracellular movement and to support the cell. Microtubules also form the spindle fibers of dividing cells and are essential components of centrioles, cilia, and flagella. The *centrioles* are hollow cylindrical bodies located just outside the nucleus of most animal cells; they help organize the microtubular spindle for cell division. *Cilia* and *flagella* are hairlike projections from the cell's surface that move the cell or move materials across the cell's surface. Inside the stalk of cilia and flagella is a 9 + 2 arrangement of microtubules. At the base of the stalk is the *basal body*, whose structure is the same as that of the centriole.

Although procaryotic cells lack the membrane-surrounded nucleus and all the internal membranous organelles discussed above, they do have ribosomes and a single circular chromosome of DNA. Some cells have flagella but these lack microtubules.

Many biologists believe that at least two organelles found only in eucaryotes—mitochondria and chloroplasts—are the descendants of procaryotic organisms that took up residence in the precursors of eucaryotes. This view is called the *endosymbiotic hypothesis*.

CONCEPT MAPS

Map 6.1

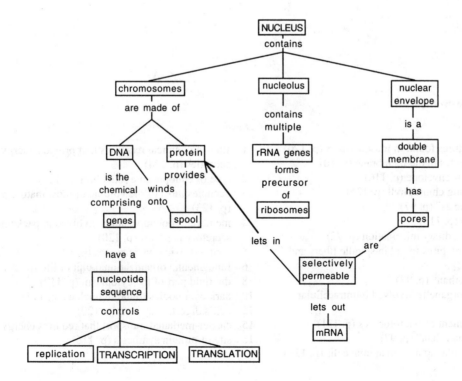

KEY DIAGRAM

In the diagram on the right, color each name and the appropriate structure(s) with the same color.

CELL MEMBRANE
NUCLEAR ENVELOPE
NUCLEUS
NUCLEOLUS
CYTOSOL
ENDOPLASMIC RETICULUM
RIBOSOMES
GOLGI APPARATUS
MITOCHONDRIA
VACUOLES
PINOCYTOTIC VESICLES
LYSOSOME
CENTRIOLES

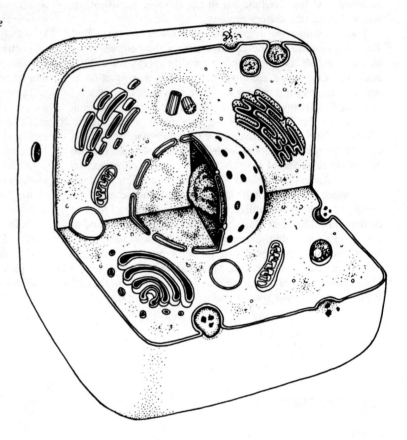

QUESTIONS

Testing recall

Do the following crossword puzzle.

Across

1. long threadlike structures found in muscle fibers (p. 126)
5. protein carrier molecules in the membrane (p. 101)
7. opening in the nuclear envelope (p. 116)
9. chromoplast containing chlorophyll (p. 125)
11. prefix meaning "same as" (p. 97)
13. control center of cell (p. 113)
14. chemical with the hereditary information (p. 72)
17. hollow cylindrical structures found in spindle fibers and centrioles (p. 127)
18. prefix meaning "less than" (p. 97)
19. abbreviation for the organelle involved in intracellular transport (p. 116)
20. have a 9 + 2 arrangement of microtubules (p. 131)
21. prefix meaning "greater than" (p. 97)
22. focus of the microtubular spindle in animal cells (p. 129)

Down

1. sites of chemical reactions that provide energy for cell's activities (p. 124)
2. type of endoplasmic reticulum (p. 116)
3. occupies most of the volume of most mature plant cells (p. 123)
4. membrane-bound vesicles involved in packaging secretory products (p. 120)
5. contains oxidative enzymes (p. 122)
6. fundamental organizational unit of life (p. 89)
8. the fluid part of the cytoplasm (p. 117)
10. dark oval bodies found in nucleus (p. 113)
12. cell's digestive system (p. 120)
15. carrier-mediated transport that requires energy (p. 102)
16. site of protein synthesis (p. 116)

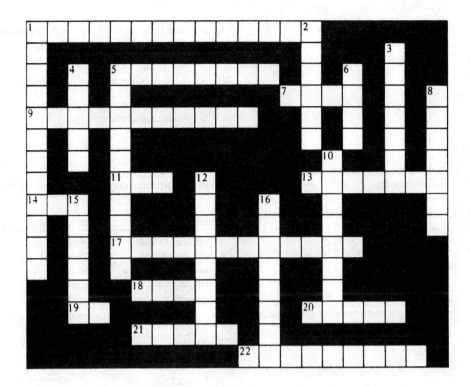

Listed below are 10 substances or organelles found in many cells. For each, write E if the item is found only in eucaryotic cells, P if it is found only in procaryotic cells, and B if it is found in both. (Table 6.1, p. 135)

1 nuclear membrane
2 ribosome
3 flagella with 9 + 2 structure
4 DNA in chromosome
5 mitochondrion
6 lysosome
7 plasma membrane
8 chlorophyll
9 endoplasmic reticulum
10 chloroplast

Testing knowledge and understanding

11 A particular cell steadily secretes a protein into the surrounding medium. The secretion is released from the cell by exocytosis from membranous vesicles derived from the Golgi apparatus. The cell also functions as a storage depot for glycogen. The cell can be identified as

a a bacterium.   *c* a cell from a terrestrial plant.
b a cyanobacterium.   *d* a cell from an animal.
(p. 135)

12 The antibiotic streptomycin is thought to combine with the ribosomes in bacteria and thus disrupt their normal functioning. In other words, this antibiotic destroys bacteria by

a preventing the synthesis of proteins.
b interfering with normal cell reproduction.
c slowing energy production within the cell.
d preventing transport within the endoplasmic reticulum.
e interfering with materials entering and leaving the cell.
(p. 116)

13 An organelle surrounded by a double membrane is the

a lysosome.   *d* centriole.
b Golgi apparatus.   *e* ribosome.
c chloroplast.   (p. 125)

14 Membranes are found as part of all the following subcellular structures *except* the

   a endoplasmic reticulum.
   b ribosomes.
   c Golgi apparatus.
   d mitochondria.
   e lysosomes. (pp. 116–24)

15 The protein tubulin

   a is a major component of the flagella of eucaryotic cells.
   b assembles into microfilaments.
   c is a major component of centrioles.
   d forms the major component of intermediate filaments.
   e is two of the above. (pp. 126–28)

16 All the following structures are found in both procaryotic and eucaryotic cells *except*

   a plasma membrane.
   b Golgi apparatus.
   c chromosomes.
   d cell wall.
   e ribosomes. (p. 135)

17 Which one of the following is *false* concerning lysosomes?

   a They act as the digestive system of the cell.
   b They function as storage vesicles for oxidative enzymes that catalyze certain condensation reactions.
   c Lipid-digesting lysosomes sometimes lack a particular enzyme; this results in Tay-Sachs disease.
   d They are produced by budding from the Golgi apparatus.
   e They possess a selectively permeable membrane that allows certain substances to pass through but is impermeable to the enzymes stored within it. (pp. 120–22)

18 The relatively homogeneous internal matrix of the chloroplast is called the

   a leucoplast.
   b stroma.
   c thylakoid.
   d grana.
   e vacuole. (p. 125)

19 A structure which is found in both plant and animal cells but which has its greatest development in plant cells is a

   a mitochondrion.   d Golgi apparatus.
   b lysosome.   e ribosome.
   c vacuole. (p. 123)

20 All the following are found in procaryotic cells *except*

   a mitochondrion.
   b plasma membrane.
   c DNA.
   d single circular chromosome.
   e ribosome. (p. 135)

21 In the following comparison of procaryotic and eucaryotic cells, which item is *incorrect*?

Characteristic	Procaryotes	Eucaryotes
a nuclear membrane	absent	present
b chromosomes	DNA	DNA and protein
c mitochondria	present	present
d ribosomes	small	large
e flagella	lack 9 + 2 structure	have 9 + 2 structure

(p. 135)

22 All the following organelles have chromosomes *except*

   a ribosomes.   c chloroplasts.
   b nucleus.   d mitochondria.
(pp. 113–25)

23 It has been proposed that mitochondria and chloroplasts (and flagella) are modern descendants of primitive forms of procaryotic cells that took up residence within primitive cells and evolved independently there. Which one of the following is *not* evidence of this view?

   a Like procaryotic cells, mitochondria and chloroplasts have DNA that is *not* wound on protein spools.
   b The ribosomes in procaryotic cells, mitochondria, and chloroplasts are very similar, and differ from the cytoplasmic ribosomes of eucaryotes.
   c Mitochondria and chloroplasts are known to have their own genes and ribosomes and to conduct protein synthesis.
   d Mitochondria and chloroplasts build their own membranes.
   e The chromosomes of mitochondria and chloroplasts have genes that code for all their enzymes; the organelles are completely independent of the host cell's genes. (p. 136)

ANSWERS

Testing recall

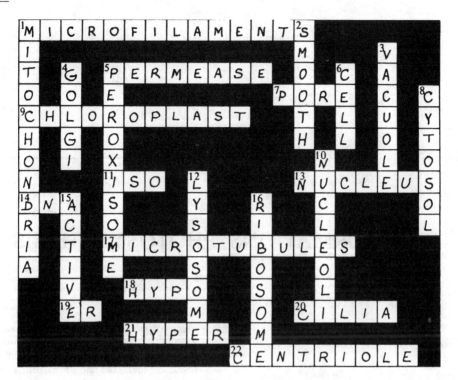

1 E	6 E	
2 B	7 B	
3 E	8 B	
4 B	9 E	
5 E	10 E	

Testing knowledge and understanding

11	*d*	14	*b*	17	*b*	20	*a*	22	*a*
12	*a*	15	*e*	18	*b*	21	*c*	23	*e*
13	*c*	16	*b*	19	*c*				

Chapter 7

ENERGY TRANSFORMATIONS: PHOTOSYNTHESIS

A GENERAL GUIDE TO THE READING

The focus of this chapter is photosynthesis, the process by which green plants capture the sun's radiant energy and use it to synthesize complex organic molecules. As you read Chapter 7 in your text, special attention to the following topics will help you concentrate on the primary aspects of the process.

1. Oxidation and reduction. These terms, defined on p. 142, will be used frequently in this chapter and the next; you will want to understand them completely.

2. Leaf anatomy is described on pages 143 to 144. Try to relate each part of the leaf to its function in enabling the plant to carry out photosynthesis.

3. Photosynthesis: the light reactions. Be sure you understand the organization of the photosystems (p. 145) and the effect of light on chlorophyll (pp. 145–46) before you go on to read the sections on cyclic and noncyclic photophosphorylation.

4. Nonyclic and cyclic photophosphorylation. You will want to read these sections slowly since crucial conceptual material is presented. It is important that you understand the relationships among the flow of electrons along the electron-transport carriers, the movement of H^+ ions, and the synthesis of ATP. Study Figure 7.17 (p. 151) carefully, because it shows how, in the noncyclic pathway, the electrons from water are moved along the two electron-transport chains and how this leads to the accumulation of H^+ ions inside the thylakoid and the establishment of an electrochemical gradient. It also shows how the H^+ ions flow through the ATP synthetase site to produce ATP. You will need to remember that the products of the light reactions provide the energy for the synthesis of carbohydrate in the dark reactions.

5. The compound adenosine triphosphate (ATP). The synthesis and function of ATP, the universal energy currency of living things, is first discussed on pages 149 to 150. You need not learn the structure of the molecule, but you will want to be familiar with the symbols ATP, ADP, AMP, Ⓟ, and P_i and with the relationships among the substances they represent.

6. Photosynthesis: the dark reactions. The Calvin cycle is discussed on pages 153 to 155. Figure 7.21 (p. 154) is particularly helpful in understanding this process.

KEY CONCEPTS

1. The ultimate energy source for most organisms is sunlight; green plants transform light energy into chemical energy, which can be used directly or passed to other organisms. (pp. 139–41)

2   When a chlorophyll molecule absorbs a photon of light, the photon's energy raises one of the chlorophyll's electrons to a higher energy level and the excited state is passed through the photosystem. The energy released in the transfer of excited electrons from one acceptor to the next is converted into a form that can be used by the cell. (pp. 144–46)

3   During noncyclic photophosphorylation, light energy is used to pull electrons and hydrogen away from water. The electrons are passed through two electron-transport chains, and eventually reach the final acceptor, NADP. (pp. 146–47)

4   The energy stored in complex organic molecules must be transformed into the energy of ATP—the universal energy currency of living organisms—in order to be used by the cell; this transformation must occur in every living cell. (pp. 149–51)

5   The energy released as the electrons flow along the electron-transport chains is used to move H$^+$ ions into the inside of the thylakoid, and thus creates an H$^+$ ion gradient across the thylakoid membrane. ATP is synthesized when the H$^+$ ions flow down the concentration gradient, through ion channels in the membrane. (pp. 148–49)

6   During cyclic photophosphorylation light energy is used to move electrons from chlorophyll, through a series of electron-acceptor molecules, and back to the chlorophyll molecule. The free energy released in the electron transfer is used to pump H$^+$ ions across the thylakoid membrane. The energy of the H$^+$ ion gradient is used to synthesize ATP. (p. 152)

7   A cyclic series of reactions, the Calvin-Benson cycle, reduces CO_2 to carbohydrate. NADPH and ATP formed in the light reactions drive the dark reactions. (pp. 153–55)

8   The hydrogen and electrons necessary for the reduction of CO_2 come from water. The water is split, oxygen is formed as a by-product and is released as a gas. (pp. 153–55)

CROSSLINKING CONCEPTS

1   The sun's energy flows through ecosystems (Chapter 34) from autotrophs (using photosynthesis) to heterotrophs.

2   Study of the evolutionary development of biochemical pathways has shown that anaerobic respiration (glycolysis and fermentation, discussed in Chapter 8) evolved first, then photosynthetic pathways. (Cyclic photophosphorylation was the first photosynthetic pathway; later noncyclic photophosphorylation and carbon fixation evolved.) Finally, when enough oxygen accumulated in the atmosphere, the aerobic respiratory pathways evolved.

3   A theme throughout biology is the complementarity of structure and function. A good example of this is the structure of the leaf and the functions of each part.

4   Water's central role in living systems (Chapter 2) can be seen in photosynthesis; water serves as an electron donor in noncyclic photophosphorylation as well as a medium for biochemical reactions.

5   An understanding of osmotic concentration and osmosis (Chapter 4) underscores the significance of storing glucose (produced through photosynthesis) as starch.

OBJECTIVES

1   Write a summary equation for the reactions of photosynthesis, with glucose as the end product. (p. 142)

2   Explain why almost all organisms depend directly or indirectly on photosynthesis to satisfy their energy needs. Mention the terms autotroph and heterotroph in your answer. (pp. 139–41)

3   Define oxidation and reduction in terms of gain and loss of electrons and gain and loss of hydrogen. Indicate whether reducing a substance stores or releases energy in that substance and do the same for oxidizing a substance. (p. 142)

4   On a drawing or photograph of a leaf cross section (see Fig. 7.6, p. 143), point out the cuticle, the palisade mesophyll, a guard cell, a vein, the epidermis, the spongy mesophyll, a stoma, and a bundle sheath. (pp. 143–44)

5   Sketch a chloroplast; include in your diagram the thylakoids, grana, and stroma. Indicate where the various photosynthetic reactions take place, where the H$^+$ ion concentration is the highest, and where it is the lowest. (pp. 144–54)

6   Describe the general organization of a photosystem. Explain why most of the chlorophyll molecules in a photosynthetic unit are referred to as "antenna" molecules and what is meant by a reaction center. (pp. 144–46)

7   Using Figure 7.8 (p. 145), describe the absorption spectrum of chlorophyll and explain why it is advantageous for the chloroplast to contain more than one form of chlorophyll as well as accessory pigments.

8 Using a diagram that shows the structure of an ATP molecule (Fig. 7.15, p. 150), identify the adenine and the ribose portions and the parts constituting adenosine monophosphate, adenosine diphosphate, and adenosine triphosphate, respectively. Explain how ATP is formed from ADP and inorganic phosphate, and state whether the reaction involved is exergonic or endergonic. Describe the role ATP plays in the transfer of energy. (pp. 149–50)

9 Using Figure 7.17 (p. 151), trace the process (noncyclic photophosphorylation) that results in the flow of electrons from water to NADP. In doing so, start with the lights being absorbed by Photosystem I and trace the flow of electrons activated by that event and specify what the final electron acceptor molecule in that system is called. Explain how the electron holes in Photosystem I are filled by means of another light event, how the resulting electron holes in Photosystem II are filled, and what happens to the released oxygen when, during this process, water is split. Point out the portion of the pathway that is indirectly linked to ATP synthesis (e.g., where the H^+ ions are pumped across the membrane to create an H^+ ion gradient). Point out the enzyme complex that can use the energy of the gradient of H^+ ions to phosphorylate ADP. List the products of noncyclic photophosphorylation. (p. 150)

10 Using Figure 7.18 (p. 153), trace the flow of electrons during cyclic photophosphorylation; point out the portion of the pathway that is indirectly linked to ATP synthesis (e.g., where H^+ ions are transported across the membrane to create an H^+ ion gradient), and explain what cyclic photophosphorylation accomplishes. List the products of cyclic photophosphorylation. (p. 152)

11 Explain how the ATP and NADP produced in the light reactions are used to reduce CO_2 in the Calvin cycle to form PGAL, and describe the fate of this PGAL. Figure 7.21 (p. 154) and 7.22 (p. 155) may be helpful. (pp. 153–55)

KEY TERMS

chemosynthesis (p. 139)
aerobic respiration (p. 139)
photosynthesis (p. 139)
autotroph (p. 140)
heterotroph (p. 140)
light reactions (p. 142)
dark reactions (p. 142)
reduction (p. 142)
oxidation (p. 142)
redox reaction (p. 142)
parts of the leaf (pp. 143–44):
  petiole
  blade
  epidermis
  cuticle
  palisade mesophyll
  spongy mesophyll
  stomata
  guard cell
  phloem
  bundle sheath
thylakoid (p. 144)
granum (p. 144)
stroma (p. 144)
chlorophyll (p. 145)
Photosystem I (p. 145)
Photosystem II (p. 145)
reaction center (p. 145)
photon (p. 145)
acceptor molecule (p. 146)
electron-transport proteins (p. 146)
ATP synthetase (p. 146)
NADP, NADPH (p. 146)
electochemical gradient (p. 149)
ATP, ADP, AMP (pp. 149–150)
photophosphorylation (p. 150)
chemiosmotic hypothesis (p. 150)
noncyclic photophosphorylation (p. 150)
cyclic photophosphorylation (p. 152)
Calvin-Benson cycle (p. 154)
RuBP (p. 154)
PGA (p. 154)
Rubisco (p. 155)
PGAL (p. 155)

SUMMARY

Within the living cell, a constant supply of energy is required to drive the various chemical reactions that maintain life. The first living organisms obtained energy by metabolizing abiotically synthesized organic molecules. Later, *chemosynthetic* organisms evolved that obtained their energy from inorganic materials. Today, the ultimate energy source for most living things is sunlight, transformed by green plants into chemical energy in the process called photosynthesis. The green plants utilize the energy of light to remove the electrons and hydrogen from water and use them to reduce carbon dioxide to energy-rich organic molecules such as carbohydrates. As oxygen formed as a by-product of photosynthesis accumulated in the atmosphere, some organisms evolved the ability to use oxygen in *aerobic respiration*, which became the main mechanism of extracting energy from food. Photosynthesis and aerobic respiration are linked through an exchange of end products: photosynthesis utilizes the carbon dioxide and water produced by aerobic respiration, and aerobic respiration utilizes the food and oxygen produced by photosynthesis.

The initial storage of energy in compounds such as carbohydrates and the release of energy in respiration involve oxidation and reduction *(redox)* reactions. *Reduction* is the addition of one or more electrons to an atom or molecule, while *oxidation* is the removal of electrons. Reduction stores energy in the reduced compound; oxidation liberates energy from the oxidized substance. During photosynthesis, CO_2 is reduced to form energy-rich carbohydrate.

ATP is a universal energy currency used by cells to do all their work. It is composed of adenosine bonded to three phosphate groups in sequence:

$$\text{Adenosine}-\text{P}-\text{P}-\text{P}$$

If the terminal phosphate group is removed by hydrolysis, energy is released and the compounds ADP and inorganic phosphate are left. New ATP can be synthesized from ADP and inorganic phosphate in an energy-demanding process called *phosphorylation*.

The summary equation for the photosynthetic process is

$$6\ CO_2 + 12\ H_2O + \text{light} \xrightarrow{\text{chlorophyll}} 6\ O_2 + C_6H_{12}O_6 + 6\ H_2O$$

The reactions of photosynthesis can be divided into two parts: the light reactions, in which light energy is trapped and stored in specialized energy-transfer molecules, and the dark reactions, in which the stored energy is used to reduce carbon dioxide into carbohydrates like glucose.

The leaves of higher green plants are the principal organs of photosynthesis. The outer surfaces of a leaf are made up of epidermal tissue, covered with a waxy cuticle impermeable to water. The region between the upper and lower epidermis, the mesophyll region, is filled with parenchyma cells. The mesophyll cells have many chloroplasts and, being loosely packed, leave air spaces that communicate with the outside by tiny holes in the epidermis called stomata. Veins, which contain the xylem and phloem cells for transport, branch profusely within the mesophyll. The veins are usually surrounded by tightly packed parenchyma cells making up a bundle sheath.

Different wavelengths of light, especially red and violet light, are trapped by various pigment molecules organized in two kinds of photosystems within the chloroplasts. When a photon of light is absorbed by a chlorophyll molecule, the photon's energy raises an electron to a higher energy level and the excited state is passed from pigment molecule to pigment molecule, and eventually reaches a specialized reaction-center molecule, where it is trapped. The energized electron is then passed through a series of acceptors, and releases energy at each step. This energy is used for generating an energy-storage compound.

Noncyclic photophosphorylation Noncyclic photophosphorylation begins when light energy excites electrons in the pigments of the P700 antenna system and the excited state reaches the reaction-center molecule P700. In noncyclic photophosphorylation the energized electrons flow through another series of electron-transport molecules, and eventually reach the final acceptor, $NADP^+$, which retains two electrons and one of their associated protons. The antenna molecules, the P700 reaction center, and the electron-transport chain constitute Photosystem I.

Photosystem I is now short of electrons; it has electron "holes." Another light event occurs in a different photosystem; light energy is trapped by pigment molecules in the P680 antenna system, and the excited state eventually reaches the reaction-center molecule P680. Next the excited electrons are passed to an electron-acceptor Q, which in turn passes them, via a series of transport molecules, step by step down an energy gradient to the electron holes in the P700 system. As the electrons move down the transport chain, some of the energy released along the way is used to transport H^+ ions across the thylakoid membrane to create an H^+ ion gradient across the membrane. The energy of the gradient can be used to synthesize ATP. The antenna molecules and the P680 reaction center, plus its special set of electron-transport molecules, constitute Photosystem II.

Photosystem II is now short of electrons; the deficit is filled by electrons pulled from water. The splitting of water also produces free protons and molecular oxygen.

$$2\ H_2O \rightarrow 4e^- + 4\ H^+ + O_2$$

Because the electrons are not passed in a circular chain in this process (some leave the system via NADPH and others enter from water), this series of reactions is termed *noncyclic photophosphorylation*; NADPH, ATP, and O_2 are the direct end products.

The antenna pigments, reaction centers, and electron-transport chain molecules are precisely arranged within the membranes of flattened sacs called *thylakoids*. Disc-shaped thylakoids often lie close together in stacks called *grana*. The thylakoid membrane serves as a barrier between the interior of the thylakoid and the interior of the chloroplast, which is known as the stroma. Like the inner mitochondrial membrane, the thylakoid membrane makes possible an electrochemical gradient, which supplies energy for the synthesis of ATP. During the light reactions of photosynthesis, water is split and the resulting H^+ ions remain in the interior of the thylakoid, and thereby increase the H^+ ion concentration. In addition, as electrons flow through Photosystems I and II during noncyclic photophosphorylation, some of the energy is used to move H+ ions from the stroma to the thylakoid interior, and thereby add to the H+ ion concentration. When the electrons reach NADP+ they reduce it, and thus remove H+ ions from the stroma. The net effect is that H+ ions accumulate in the interior of the thylakoids, while the outer compartment, the stroma, becomes negatively charged. The membrane of the thylakoid contains numerous ATP synthetase complexes that can utilize the energy of the gradient to phosphorylate ADP to ATP.

Cyclic photophosphorylation In cyclic photophosphorylation, light energy is trapped by pigment molecules in the P700 antenna system and the excited state eventually reaches the reaction-center molecule P700. Energized electrons from P700 move from one acceptor molecule to the next; they release free energy at each step and eventually return to the P700 molecule. Some of the energy released as the electron is eased down the energy gradient is used by the cell to transport H+ ions across the thylakoid membrane and create an H+ ion gradient. The energy of this gradient is used to phosphorylate ADP into ATP. Because the electrons are returned to the chlorophyll molecules from which they originated, this process is termed *cyclic photophosphorylation*. ATP is the only product of this process.

The dark reactions The ATP and NADPH produced in the light reactions are used to reduce carbon dioxide to carbohydrate in a series of reactions called the Calvin-Benson cycle. First, a five-carbon sugar, ribulose bisphosphate (RuBP), is combined with CO_2 to form an unstable six-carbon molecule that splits into 2 three-carbon molecules. These are then phosphorylated by ATP and reduced by NADPH to form PGAL, a three-carbon sugar. PGAL may be used to synthesize more RuBP, as a source of energy, or for the synthesis of glucose, fatty acids, and amino acids. The reactions of the Calvin-Benson cycle occur in the stroma of the chloroplast.

CONCEPT MAPS

Map 7.1

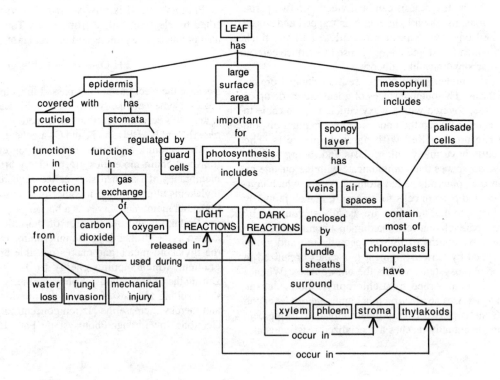

ENERGY TRANSFORMATIONS: PHOTOSYNTHESIS • 51

Map 7.2

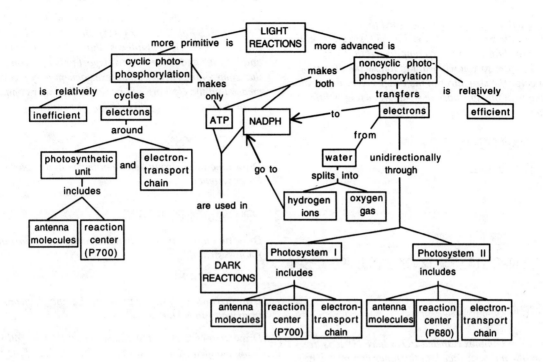

Map 7.3

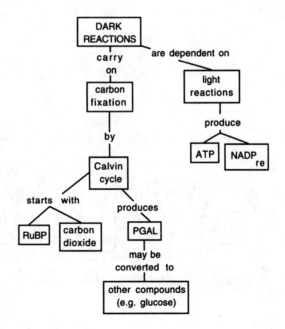

KEY DIAGRAM

The following coloring exercise is designed to help you understand and learn the processes involved in ATP synthesis in the chloroplast. The diagram is more complicated than any you have colored so far; so you will want to follow the directions closely and color the structures in the order given. First color the name and then the appropriate structure, using a different color for each structure. (Fig. 7.17, p. 151)

HYDROGEN IONS (H^+)

ATP SYNTHETASE

H_2O

O_2

PHOTONS OF LIGHT

REACTION-CENTER MOLECULES (P680 AND P700)

H^+ PUMP COMPLEX (PQ)

Color the enzyme complex (labeled PQ) where H^+ ions are pumped from the outer surface into the inner compartment of the thylakoid. Note that enzyme PQ is a shuttle molecule that accepts electrons and H^+ ions on the outer surface and transports them to the inner surface, where the electrons are passed to cytochrome f and the H^+ ions are released into the interior of the thylakoid.

ELECTRON-TRANSPORT MOLECULES

Color the remaining electron-transport molecules (labeled Z, Q, Cyt f, PC, FeS, Fd, and FAD). Note that we have colored in two types of carrier complexes: those that transport electrons and H^+ ions and those that transport only electrons. The fact that some of the carrier complexes accept only electrons means that the H^+ ions will be left in the inner compartment. What is the function of this H^+ ion gradient?

PHOTOSYSTEM I

Draw an outline around each molecule that is part of Photosystem I.

PHOTOSYSTEM II

Draw an outline around each molecule that is part of Photosystem II.

Draw in NADPH

Draw arrows to show the pathway that the electrons follow from water to NADPH

Draw arrows to show the pathway that H^+ ions follow when being pumped from the stroma to the interior of the thylakoid.

Ask yourself

1. In what part of the chloroplast is the pH the lowest?
2. Note that the H^+ ions in the interior of the thylakoid come from two sources: name the two sources.

Stroma (pH8)

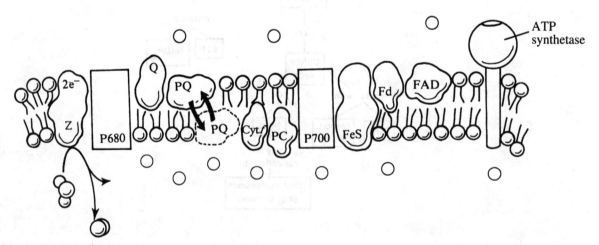

Thylakoid Interior (pH4)

QUESTIONS

Testing recall

Select the correct term or terms to complete each supplement.

1. Reactant A in the reaction A + H → AH is (oxidized, reduced). (p. 142)

2. The reaction ADP + P_i → ATP (stores, releases) energy. (p. 149)

3. The wavelengths of light effective in driving photosynthesis are those in the (red, yellow, blue-violet, green) part of the visible spectrum. (p. 145)

4. The only product of cyclic photophosphorylation is (O_2, NADPH, ATP, PGAL, glucose). (p. 152)

5. The *direct* products of noncyclic photophosphorylation are (O_2, PGAL, ADP, ATP, NADPH). (p. 150)

6. The products of the light reactions necessary to drive the dark reactions are (ATP, NADPH, O_2, CO_2). (p. 155)

7. Products of the Calvin-Benson cycle, or dark reactions, include (O_2, PGAL, ADP, $NADP^+$). (pp. 153–55)

8. The light reactions of photosynthesis take place in the (stroma, thylakoid membrane, thylakoid interior) of the chloroplast whereas the dark reactions take place in the (stroma, thylakoid membrane, thylakoid interior). (p. 154)

9. The molecular oxygen evolved during photosynthesis comes from (cyclic photophosphorylation, carbon dioxide, water). (p. 150)

10. The pH in the interior of the thylakoid is (higher, lower) than that in the stroma. (p. 149)

For questions 11 to 19:

Of the basic processes of photosynthesis

  a cyclic photophosphorylation
  b noncyclic photophosphorylation
  c both light reactions
  d Calvin-Benson cycle
  e both light and dark reactions

which process (or group of processes) involves

11. utilization of CO_2? (p. 153)
12. oxidation-reduction reactions? (pp. 145–55)
13. light-energized electrons? (p. 145)
14. production of NADPH? (p. 146)
15. synthesis of PGAL? (p. 155)
16. production of O_2? (p. 150)
17. occurrence in stroma of chloroplasts? (p. 154)
18. splitting of water? (p. 149)
19. production of ATP? (pp. 149–53)
20. chlorophyll as both the initial electron donor and the ultimate electron acceptor? (p. 152)

Match each item below with the associated part of the leaf shown in cross section in the drawing below. Answers may be used once, more than once, or not at all.

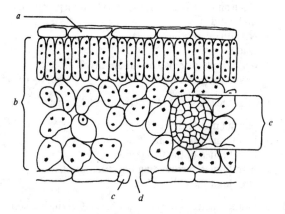

21. cells containing large numbers of chloroplasts that are active in photosynthesis (p. 144)
22. system that delivers water and minerals to cells of the leaf (p. 144)
23. place where CO_2 enters the leaf (p. 144)
24. structure containing xylem and phloem (p. 144)
25. layer that protects internal tissues of the leaf (p. 143)
26. mesophyll tissues (p. 143)
27. cell that regulates the size of the stomatal opening (p. 144)

Testing knowledge and understanding

28. Suppose you are studying a plant that is an unusual blue color. If you extracted its photopigments, which of the three photopigments whose absorption spectra are shown below would you predict would be present?

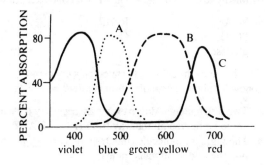

  a pigment A   d pigments A and C
  b pigment B   e pigments A and B
  c pigments B and C
                                                    (p. 145)

29 In the *reaction centers* during photophosphorylation,

   a water is oxidized.
   b NADP is reduced.
   c light energy is converted into chemical energy.
   d ATP is synthesized from ADP and P_i.
   e an electrochemical gradient is established. (p. 145)

30 During cyclic photophosphorylation,

   a electron flow causes H^+ ions to be transported into the thylakoid.
   b NADPH is produced.
   c water is split.
   d both Photosystems I and II are involved.
   e oxygen is generated. (p. 152)

31 The light reactions of photosynthesis take place within the

   a plasma membrane of the cell.
   b membranes of the mitochondria.
   c membranes of the thylakoids.
   d membranes surrounding the chloroplast.
   e stroma of the chloroplast. (p. 154)

32 The dark reactions of photosynthesis (Calvin-Benson cycle) take place within the

   a membranes surrounding the chloroplast.
   b thylakoids of the chloroplast.
   c cytoplasm outside the chloroplast.
   d stroma of the chloroplast.
   e vacuole. (p. 154)

33 In photosynthesis the reduction of CO_2 to carbohydrate requires that both energy currency and a strong reducing substance (hydrogen donor) be in ample supply. The process of noncyclic photophosphorylation provides these requisites; it uses light energy to synthesize

   a ADP and ATP.   d ADP and $NADP^+$
   b ATP and P700.   e P700 and P680.
   c ATP and NADPH. (p. 150)

34 If photosynthesizing green algae are provided with CO_2 synthesized with heavy oxygen (^{18}O), later analysis will show that all but one of the following compounds produced by the algae contain the ^{18}O label. That one exception is

   a PGA.   d RuBP.
   b PGAL.   e O_2.
   c glucose. (p. 154)

35 Which one of the following statements concerning noncyclic photophosphorylation is *false*?

   a Two different light-driven events are necessary if electrons are to be moved all the way from H_2O to NADP.
   b The pigment molecules that trap light energy are built into the thylakoid membranes.
   c There are at least two different places in the overall noncyclic pathway where energized electrons are passed energetically downhill via a series of electron-carrier substances.

   d One of the products of noncyclic photophosphorylation that help make possible the dark reactions of the Calvin-Benson cycle is NADPH.
   e Some of the energy released during electron transport is used to hydrolyze ATP to ADP and inorganic phosphate. (pp. 146–50)

36 The first step in photosynthesis is the

   a formation of ATP.
   b energizing of an electron of chlorophyll by a photon of light.
   c splitting of water into H and O components.
   d addition of CO_2 to a five-carbon sugar.
   e combining two molecules of PGAL to form a molecule of glucose. (p. 145)

37 The light reactions of photosynthesis

   a provide CO_2 for the dark reactions.
   b produce carbohydrate.
   c provide the energy required for the dark reactions.
   d use O_2 in the production of ATP.
   e include two of the above. (p. 153)

38 What is the source of the electrons that reduce $NADP^+$ during photosynthesis?

   a O_2 d ATP
   b water   e PGAL
   c light (pp. 145–46)

39 In biology a limiting factor is a condition or substance that, by its absence or short supply, limits the rate at which a biological process can proceed. Which one of the following would be *least* likely to be a limiting factor for photosynthesis?

   a oxygen   d light
   b carbon dioxide   e chlorophyll
   c water (p. 141)

40 Which of the following occurs in noncyclic but *not* in cyclic photophosphorylation?

   a flow of electrons
   b synthesis of ATP through an H^+ ion gradient
   c synthesis of NADPH
   d absorption of light by chlorophyll (p. 152)

41 The oxygen in our atmosphere is a product of

   a the splitting of CO_2 during photosynthesis.
   b cyclic photophosphorylation.
   c noncyclic photophosphorylation.
   d both cyclic and noncyclic photophosphorylation.
   e the Calvin-Benson cycle. (p. 140)

42 In an oak tree, most photophosphorylation takes place in

   a the parenchyma cells of the roots.
   b the xylem cells of the stem.
   c the epidermal cells of the leaves.
   d the mesophyll cells of the leaves. (pp. 143–44)

43 The electron-transport molecules of photophosphorylation are

   a built into the thylakoid membrane.
   b built into the outer membrane of the chloroplast.
   c located in the interior of the thylakoid.
   d located in the inner membrane of the mitochondrion.
   e located in the stroma. (p. 146)

44 In the dark reactions of photosynthesis,

   a PGAL is synthesized.
   b oxygen is produced.
   c water is split.
   d ATP is synthesized.
   e electrons are returned to the chlorophyll molecule. (p. 155)

45 Which one of the following compounds is involved in *both* noncyclic photophosphorylation and the Calvin-Benson cycle?

   a O_2
   b NADPH
   c CO_2
   d chlorophyll
   e PGAL (p. 153)

46 Which of the following represents a likely sequence in the evolution of biochemical pathways?

   a noncyclic photophosphorylation → cyclic photophosphorylation → aerobic respiration → Calvin-Benson cycle
   b cyclic photophosphorylation → noncyclic photophosphorylation → Calvin-Benson cycle → aerobic respiration
   c noncyclic photophosphorylation → cyclic photophosphorylation → Calvin-Benson cycle → aerobic respiration
   d aerobic respiration → cyclic photophosphorylation → noncyclic photophosphorylation → Calvin-Benson cycle (pp. 139–40)

For further thought

1 Describe the principal events in both cyclic and noncyclic photophosphorylation, and contrast these two processes. (pp. 146–50)

2 Describe the Calvin-Benson cycle, by which carbohydrates are synthesized from CO_2 in the dark reactions. Indicate the starting materials and the end products, and describe the relationship between photophosphorylation and carbon fixation. (pp. 153–55)

3 Discuss structural features of a typical leaf that are important in making it an efficient organ for carrying out photosynthesis. (pp. 143–44)

ANSWERS

Testing recall

1 reduced
2 stores
3 mostly blue-violet; some in red
4 ATP
5 O_2, NADPH
6 ATP, NADPH
7 PGAL, ADP, $NADP^+$
8 thylakoid membrane, stroma
9 water
10 lower

11	d	16	b	20	a	24	e
12	e	17	d	21	b	25	a
13	c	18	b	22	e	26	b
14	b	19	c	23	d	27	c
15	d						

Testing knowledge and understanding

28	c	33	c	38	b	43	a
29	c	34	e	39	a	44	a
30	a	35	e	40	c	45	b
31	c	36	b	41	c	46	b
32	d	37	c	42	d		

Chapter 8

ENERGY TRANSFORMATIONS: RESPIRATION

A GENERAL GUIDE TO THE READING

This chapter discusses how the energy stored in complex organic compounds is released and converted into a form that the cell uses to do work. You will find that the content of the chapter is readily understandable, despite the large amount of organic chemistry, if you focus your attention on the following aspects of the process.

1. Stages in the breakdown of glucose. Notice that the complete breakdown of glucose involves four stages, discussed in successive sections of the text: glycolysis (Stage I), oxidation of pyruvic acid to acetyl-CoA (Stage II), the Krebs citric acid cycle (Stage III), and oxidative phosphorylation (Stage IV). (pp. 163–69)

2. Glycolysis (Stage I). This stage, described on pages 163 to 166, occurs in all living organisms and is therefore a fundamental characteristic of life. Focus your attention on Figure 8.2 (p. 163) the summary on page 164, and summary Figure 8.7 (p. 169).

3. Fermentation. This process is covered in the first paragraph in the section on fermentation (pp. 165–66). Reread this paragraph until it is clear to you. In particular, note that fermentation allows a cell to continue making ATP when oxygen is absent; without fermentation all glycolysis would cease under anaerobic conditions.

4. Oxidation of pyruvic acid to acetyl-CoA (Stage II). This stage is a single reaction (p. 167).

5. The Krebs citric acid cycle (Stage III). Study Figure 8.6 (p. 168) to understand the basic idea of the cycle.

6. Oxidative phosphorylation (Stage IV). This stage (pp. 168–69) may be most difficult for you to grasp. Study Figure 8.8 (p. 170) and the accompanying caption.

7. The anatomy of respiration. Note where each of the four stages occurs and their relation to one another. Figure 8.9 (p. 171) is particularly helpful in visualizing how the stages fit together. Also note how a chemiosmotic gradient is built up across the inner membrane of the mitochondrion and how this gradient is related to ATP production. Figure 8.12 (p. 174) will help you understand the similarities and differences between H^+ ion movement and ATP synthesis in the chloroplast and the mitochondrion.

8. Summary of respiration energetics (pp. 169–72). You should learn the ATP yield from the various stages. Figure 8.10 (p. 172) summarizes the essential information for you. Concentrate on ATP production—on where and how much ATP is produced at each stage.

9. Respiration of fats and proteins (pp. 173–74). Recognize that glycolysis and the Krebs cycle do not occur only in carbohydrate metabolism; actually, fats and proteins, using various metabolic pathways, also feed into the Krebs cycle, as Figure 8.13 (p. 175) makes clear. In addition, many of the pathways are reversible. For instance, glucose can be converted into fat, and amino acids can be converted into glucose or fat.

KEY CONCEPTS

1 Within a living cell, a constant supply of energy is required to drive the various chemical reactions that maintain life. (p. 161)

2 The energy stored in complex organic molecules must be transformed into the energy in ATP—the universal energy currency of living organisms—in order to be used by the cell; this transformation must occur in every living cell. (Chapter 7, pp. 149–53)

3 The energy stored in complex organic molecules is not liberated through a single large reaction; rather, the universal catabolic process by which the molecules are broken down occurs as a series of small reactions, each catalyzed by its own specific enzyme. (p. 161)

4 All living cells break down sugars by the process of glycolysis. (pp. 163–65)

5 Fermentation enables a cell to continue the reactions of glycolysis in the absence of oxygen, by providing reactions in which NAD^+ is regenerated from NADH. (pp. 165–66)

6 Considerably more energy can be extracted from glucose if oxygen is present. Consequently a plentiful supply of oxygen is essential for most organisms if their energy demands are to be met. (pp. 166–72)

7 Cells living under aerobic conditions obtain most of their energy from aerobic respiration, a process in which pyruvic acid is oxidized to acetyl coenzyme A (acetyl-CoA), which is oxidized in the Krebs citric acid cycle. (pp. 167–69)

8 The NADH and FADH synthesized during Stages I, II, and III pass their electrons to oxygen indirectly by way of a series of electron carrier molecules. The energy thus liberated is used to pump H^+ ions from the inner compartment of the mitochondrion into the outer compartment, and so create a chemiosmotic gradient across the inner membrane. ATP is produced when H^+ ions move back across the membrane. (pp. 168–69)

9 More than half the energy liberated by the metabolism of high-energy compounds is released as heat (thermal energy). Ectothermic organisms promptly lose most of this heat to the environment, whereas endotherms have evolved mechanisms for retaining this heat and can thus maintain a high body temperature and metabolic rate. (pp. 175–77)

CROSSLINKING CONCEPTS

1 Thermodynamics (Chapter 4) governs the energy transfers covered in Chapters 7 and 8. Although individual steps may vary, respiration is basically an exergonic process in which carbohydrates are oxidized to release energy. Some of this energy is used to synthesize ATP, which is used to carry out cellular processes. Photosynthesis is basically an endergonic process in which light energy reduces carbon dioxide to carbohydrate energy.

2 Knowledge of enzyme structure and function from Chapter 4 is vital to an understanding of the step-by-step reactions of cellular respiration.

3 The concept of evolution can be applied to more than just species; biochemical pathways can be said to have evolved as well. The first pathways were those of anaerobic respiration; then came photosynthetic pathways, and finally, aerobic respiration.

4 Different forms of energy are analogous to different forms of currency. That is, they are available over different periods of time. Fats are long-term certificates of deposit; glucose is a day-in, day-out savings account; ATP is cash.

5 The fluid-mosaic model, as well as the concepts of diffusion and active transport, underlies the structure and function of mitochondrial membranes, the electron-transport chain, and chemiosmosis.

6 Note similarities and differences between the electron-transport chains of photosynthesis (Chapter 7) and aerobic respiration. Both chains depend on proteins embedded in cellular membranes to carry on a series of oxidation-reduction reactions. In both the chloroplast and the mitochondrion, an inner membrane separates the organelle into two environments, which differ considerably in pH levels. Pumps transport H^+ ions across the inner membrane of both organelles to create an electrochemical gradient, which is used to synthesize ATP through chemiosmosis. In respiration, H^+ ions are pumped from the matrix into the outer compartment of the mitochondrion. In photosynthesis, H^+ ions are pumped from the stroma into the thylakoid interior.

OBJECTIVES

1 Use Figure 8.2 (p. 163) to point out and name the starting reactant and the end products of glycolysis. Point out

(a) the reaction in which phosphates from ATP are transferred to glucose, (b) oxidation-reduction reaction, and (c) the two reactions in which molecules of ATP are synthesized. Explain what is meant by substrate-level phosphorylation. Summarize the ATP production of glycolysis, specifying the number of ATP molecules used in the preparatory reactions, the number of ATP molecules synthesized, and the net gain. (pp. 163–65)

2. Explain why NAD^+ must be regenerated from NADH in order for glycolysis to continue. Next describe how NAD^+ is regenerated in the absence of oxygen. Specify what the end product of fermentation would be in your body cells in the absence of oxygen and what it would be in most plant cells. (pp. 165–66)

3. Summarize in an equation the conversion of pyruvic acid into acetyl-CoA. Notice that two of the six carbon atoms in the original glucose molecule have been released as CO_2. Is this a redox reaction? (p. 168)

4. Using Figure 8.6 (p. 168), point out the reactions in which CO_2 is produced, specify how many CO_2 molecules are produced in each "turn" of the Krebs cycle, and indicate how many turns of the cycle are necessary to oxidize the four carbons remaining from the original molecule of glucose. Point out (a) redox reactions and (b) the reaction in which ATP is produced, and indicate whether substrate-level phosphorylation is involved here. (p. 168).

5. List the products of the oxidation of two acetyl-CoA molecules in the Krebs cycle, as shown in Figure 8.10 (p. 172).

6. Using Figures 8.8 (p. 170) and 8.9 (p. 171), trace the pathway the electrons follow as they move down the respiratory electron-transport chain and indicate the sites where hydrogen ions are pumped into the outer compartment. Indicate what is meant by a chemiosmotic gradient and explain why this gradient is important. State the number of ATP molecules formed per molecule of glucose resulting from electron transport and chemiosmotic synthesis of ATP. (pp. 169–72)

7. Sketch a mitochondrion and label the outer membrane, the inner membrane, the inner compartment, and the outer compartment. Indicate where (a) the Krebs cycle takes place, (b) the electron-transport enzymes are located, (c) the concentration of H^+ would be highest, and (d) the concentration of H^+ would be lowest. (pp. 168, 171)

8. Using a diagram such as Figure 8.10 (p. 172), summarize the ATP yield from the complete breakdown of glucose to carbon dioxide and water. Indicate the net number of ATP molecules found in glycolysis, in the Krebs cycle, and by chemiosmotic synthesis. Compare the number of ATP molecules produced from the metabolism of one glucose molecule under anaerobic conditions with the number produced under aerobic conditions. Which process is more efficient? (pp. 172–73)

9. Using Figure 8.13 (p. 175), explain how fats and proteins can be metabolized to yield energy in the form of ATP and how proteins and carbohydrates can be converted into fats for storage. (pp. 173–74)

10. Using Figure 8.14 (p. 176), describe the relationship between temperature and metabolic rate. Explain the consequences of this relationship for ectotherms, endotherms, homeotherms, and heterotherms.

KEY TERMS

aerobic respiration (p. 161)
metabolism (p. 161)
anabolism (p. 161)
catabolism (p. 161)
anaerobic metabolism (p. 163)
aerobic metabolism (p. 163)
glycolysis (p. 163)
substrate-level phosphorylation (p. 164)
PGAL (p. 164)
NAD (p. 164)
pyruvic acid (p. 164)
fermentation (p. 165)
lactic acid (p. 166)
ethyl alcohol (p. 166)
acetyl-CoA (p. 168)
Krebs citric acid cycle (p. 168)
FAD (p. 168)
respiratory electron-transport chain (p. 169)
oxidative phosphorylation (p. 169)
chemiosmosis (p. 169)
ATP synthetase (p. 169)
ectothermic (p. 175)
endothermic (p. 176)
homeotherm (p. 177)
heterotherm (p. 177)

SUMMARY

Metabolism, a process embracing the myriad enzyme-mediated reactions of a living cell, can be divided into two phases: *anabolism*, the assembling of organic molecules, and *catabolism*, the breaking down of organic molecules to yield energy.

Before the energy stored in lipids, proteins, and carbohydrates can be used by the cell to do work, the molecules must be broken down in a series of chemical reactions and the energy produced used to synthesize ATP. The complete degradation of an energy-rich compound such as glucose to carbon dioxide and water involves many enzymatically controlled reactions.

Anaerobic respiration The first series of reactions in the degradation of glucose is termed *glycolysis*; it is the breakdown of glucose to two molecules of pyruvic acid, with the production of two molecules of NADH and a net gain of two ATP molecules. The glycolytic pathway, which is common to all living cells, is *anaerobic*; i.e., it does not require molecular oxygen. It takes place in the cytoplasm of the cell.

The fate of the pyruvic acid depends on the oxygen supply. In the absence of sufficient O_2, the pyruvic acid may be reduced by NADH to form CO_2 and ethyl alcohol in most plants and many unicellular organisms, or lactic acid in animal cells and some unicellular organisms. The NAD^+ molecules formed in this reaction are then available to be reused in glycolysis. The process whereby the glycolytic pathway leads to the production of alcohol or lactic acid from pyruvic acid is called *fermentation*; it enables the cell to continue synthesizing ATP by the breakdown of nutrients under anaerobic conditions.

Aerobic respiration Under aerobic conditions—when molecular oxygen is available—pyruvic acid can be further oxidized, with the accompanying synthesis of ATP. The process begins when pyruvic acid moves from the cytoplasm into the inner compartment of the mitochondrion. Two pyruvic acid molecules are then oxidized to form two molecules each of acetyl-CoA, CO_2, and NADH. The two-carbon acetyl-CoA is fed into a complex circular series of reactions, the *Krebs citric acid cycle*. Here the acetyl-CoA combines with a four-carbon compound to form the six-carbon molecule citric acid. In subsequent reactions two carbons are lost as CO_2; this leaves a four-carbon molecule that can combine with another acetyl-CoA and start the cycle over again. In the course of the cycle, a molecule of ATP is synthesized, and eight hydrogens are removed and picked up by carrier compounds to form three molecules of NADH and one of $FADH_2$. Since one molecule of glucose gives rise to two molecules of acetyl-CoA, two turns of the cycle occur for each molecule of glucose oxidized.

The final stage of respiration involves the passage of the electrons from the carrier molecules NADH and $FADH_2$ down a *respiratory chain* of electron-transport molecules to oxygen, with which the electrons and H^+ ions from the medium combine to form water. The electron-transport chain molecules are found in the inner membrane of the mitochondrion. The transfer of electrons along the electron-transport chain results in the pumping of H^+ ions from the inner compartment of the mitochondrion to the outer compartment. As a result, the H^+ concentration increases in the outer compartment and an electrostatic and osmotic concentration gradient is built up across the inner membrane. Special enzyme complexes, called ATP synthetases, act as H^+-ion channels in the inner mitochondrial membrane. As the H^+ ions move down the electrochemical gradient through the complex, energy is released and can be used to synthesize ATP.

The total number of new ATP molecules produced by the complete metabolic breakdown of glucose is usually 36: two from glycolysis, two from the Krebs cycle, and 32 from chemiosmotic synthesis in the mitochondrion. This yield of ATP represents about 39 percent of the energy of glucose; the rest of the energy is released, mostly as heat.

Cells can also extract energy in the form of ATP from fats and proteins, which are hydrolyzed to form products that can be fed into the carbohydrate metabolic pathways.

Ectothermic ("externally heated") animals promptly lose most of their heat to the environment. The body temperature and metabolic rate of such organisms fluctuate with the temperature of the environment. *Endothermic* ("internally heated") animals retain the heat generated by metabolism and hence maintain a constant high body temperature. Their metabolic rate and activity can accordingly be maintained at a uniformly high level.

Animals that maintain a more or less constant body temperature are referred to as *homeotherms* ("same temperature"). Organisms whose body temperature varies on a regular basis are known as *heterotherms* ("different temperature").

CONCEPT MAPS

Map 8.1

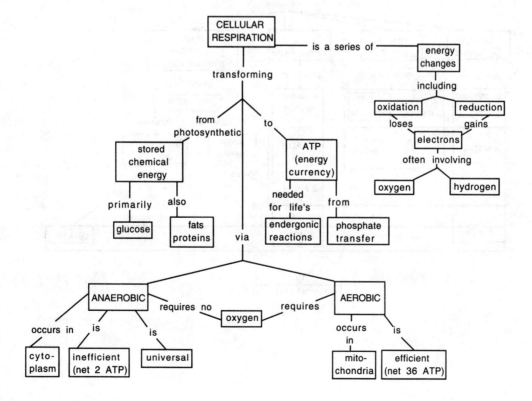

Map 8.2

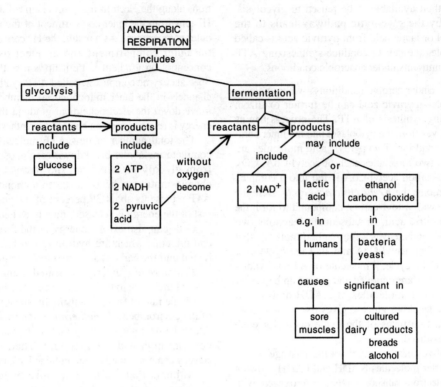

Map 8.3

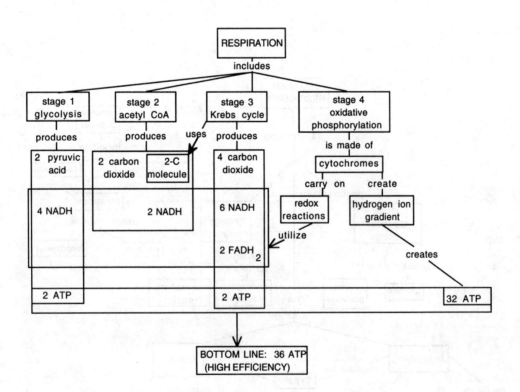

KEY DIAGRAM

The following coloring exercise is designed to help you understand and learn the processes involved in ATP synthesis in the mitochondrion. First color the title and then the appropriate structure, using a different color for each structure. (Fig. 8.9, p. 171)

OUTER COMPARTMENT OF MITOCHONDRION

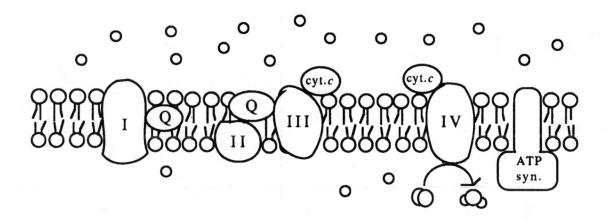

INNER COMPARTMENT OF MITOCHONDRION

HYDROGEN IONS (H^+)

ATP SYNTHESIS COMPLEX

H_2O

O_2

H^+ PUMP COMPLEXES

Color the three respiratory enzyme complexes where H^+ ions are pumped from the inner compartment to the outer compartment (respiratory enzymes labeled I, Q, II, and IV). Note that Q is a shuttle molecule that accepts electrons and hydrogen ions on the inner surface and transports them to the outer surface, where the electrons are passed to cytochrome b_2 and the H^+ is released into the outer compartment.

ELECTRON-TRANSPORT MOLECULES

Color the remaining electron-transport molecules (labeled III and cyt. c). Note that we have colored in two types of carrier complexes, those that transport electrons and H^+ ions, and those that transport only electrons. The fact that some of the carrier complexes accept only electrons means that the H^+ ions will be left in the outer compartment. Do you see why this is vital for the functioning of the pump?

Draw in NADH and the pathway that the electrons follow from NADH to O_2.

Draw in the pathways that hydrogen ions follow when being pumped from the inner compartment to the outer compartment.

Ask yourself

1. In what part of the mitochondrion is the pH the lowest?
2. What would happen if the inner mitochondrial membrane were freely permeable to hydrogen ions? Could ATP be synthesized?
3. Compare ATP generation in the mitochondrion with that in the chloroplast, showing how H^+-ion movement is coupled to electron transport and how ATP is produced.

QUESTIONS

Testing recall

1 Complete the following chart; it will be a useful study aid.

	Starting Compound(s)	*Products*
Stage I (pp. 163–65)		
Fermentation (pp. 165–66)		
Stage II (pp. 167–68)		
Stage III (p. 168)		
Stage IV (pp. 168–73) _____ NADH from Stage I _____ NADH from Stage II _____ NADH from Stage III _____ FADH$_2$ from Stage III → _____ ATP		

Decide whether the following statements are true or false, and correct the false statements.

2 The end product of glycolysis is lactic acid. (p. 164)

3 Alcoholic fermentation produces CO$_2$ and ethyl alcohol in most plant cells and in yeasts. (p. 166)

4 The reactions of the Krebs cycle take place within the mitochondrion. (p. 168)

5 Most of the ATP yield from the complete oxidation of glucose results from the electron-transport chain. (p. 169)

6 Aerobic respiration is a less efficient process for generating ATP than glycolysis. (p. 173)

Questions 7 to 10 refer to the following diagram of the mitochondrion.

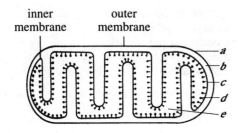

7 Where is ATP synthesized? (p. 171)

8 Where is the concentration of H^+ ions highest? (p. 170)

9 Where do the reactions of the Krebs cycle take place? (p. 171)

10 Where are the molecules of the respiratory electron-transport chain located? (p. 169)

Testing knowledge and understanding

Choose the one best answer.

11 All the following take place in *both* photosynthesis and respiration *except*

   a electron flow along an electron-transport chain.
   b splitting of water molecules.
   c synthesis of ATP via a chemiosomotic gradient.
   d transfer of electrons to acceptor molecules.
   e establishment of an H^+ ion gradient across a membrane. (p. 173)

12 Which of the following statements *best* describes the relationship between photosynthesis and respiration?

   a Respiration is the exact reversal of the biochemical pathways of photosynthesis.
   b Photosynthesis stores energy in complex organic molecules, and respiration releases it.
   c Photosynthesis takes place only in the light, and respiration takes place only in the dark.
   d Photosynthesis occurs only in plants, and respiration only in animals.
   e ATP molecules are produced in photosynthesis and used up in respiration. (pp. 169–73)

13 The glycolytic pathway from glucose to pyruvic acid involves a lengthy series of different chemical reactions. Each individual reaction requires

   a a molecule of ATP.
   b a molecule of NAD.
   c a molecule of ADP.
   d a molecule of a specific enzyme.
   e a molecule of NADP. (p. 164)

Questions 14 to 17 refer to the following diagram.

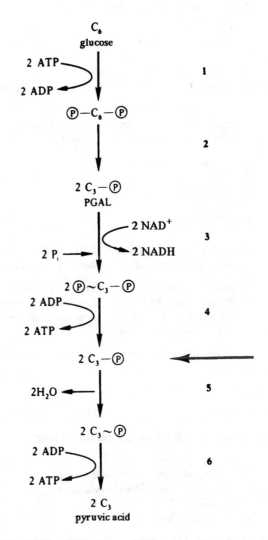

14 The diagram shows a series of reactions involved in glycolysis. How many net molecules of ATP can be produced from the products of the breakdown of one glucose molecule by the time it is degraded to $2C_3 \rightarrow$ ⓟ (note arrow)? *Assume that no oxygen is present.*

   a 2   d 8
   b 4   e 0
   c 6   (p. 164)

15 The reactions shown in the diagram take place in the

   a chloroplast.   d nucleus.
   b cytoplasm.   e ribosomes.
   c mitochondria.   (p. 163)

16 What is the *net gain* in ATP molecules produced during the reactions of glycolysis under anaerobic conditions?

   a 2   d 8
   b 4   e 10
   c 6   (p. 164)

17 Which of the numbered steps in the diagram show an oxidation-reduction reaction?

   a 1, 2
   b 1, 2, 3
   c 3, 5
   d 1, 2, 3, 4, 5, 6
   e 3 only   (pp. 163–65)

18 When a muscle cell is metabolizing glucose in the complete absence of molecular oxygen, which one of the following substances is *not* produced?

   a PGAL
   b ATP
   c pyruvic acid
   d lactic acid
   e acetyl-CoA   (p. 165)

For questions 19 to 21 use the following three answers. An answer may be used once, more than once, or not at all.

   a Process would be inhibited under anaerobic conditions.
   b Process would be promoted under anaerobic conditions.
   c Process would occur at the same rate under aerobic and anaerobic conditions.

19 formation of ethyl alcohol in certain microorganisms (p. 166)

20 Krebs citric acid cycle (p. 168)

21 chemiosmotic synthesis of ATP (pp. 169–73)

22 An important function of fermentation is to

   a regenerate NAD^+
   b produce alcohol as a nutrient source.
   c prevent oxidative phosphorylation.
   d produce NADH.
   e synthesize glucose.   (p. 165)

23 Which one of the following statements concerning glycolysis is *false*?

   a It proceeds in a step-by-step series of chemical reactions, each catalyzed by an enzyme.
   b Phosphorylation occurs during the process.
   c Oxygen is not required for the process to occur.
   d The end products are carbon dioxide and water.
   e ATP is formed.   (pp. 163–65)

24 How much molecular oxygen is required in the fermentation of one molecule of glucose?

   a none
   b 1 molecule
   c 24 molecules
   d 36 molecules
   e 38 molecules   (p. 165)

25 Which metabolic pathway is a common pathway for both anaerobic and aerobic metabolism?

   a the electron-transport chain
   b the citric acid cycle
   c the oxidation of pyruvic acid
   d glycolysis
   e none of the above   (p. 165)

Questions 26 to 29 refer to the following diagram of the Krebs citric acid cycle.

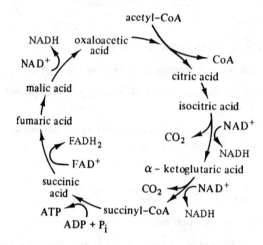

26 If two acetyl-CoA molecules are fed into the cycle, how many ATP molecules are synthesized directly in the cycle?

   a 1   d 12
   b 2   e more than 20
   c 4   (p. 168)

27 If two acetyl-CoA molecules are fed into the cycle, how many NADH molecules are produced?

   a 0   d 12
   b 3   e 32
   c 6   (p. 168)

28 If citric acid has six carbon atoms, how many carbon atoms does succinic acid have?

   a 1   d 6
   b 4   e 12
   c 5   (p. 168)

29 Besides supplying the cell with ATP and NADH, the Krebs citric acid cycle

   a breaks down glucose to CO_2 and water.
   b produces ATP from the NADH.
   c provides intermediates from which various other compounds can be made.
   d utilizes oxygen in its stepwise reactions.
   e converts lactic acid to pyruvic acid.   (p. 168)

30 The final electron acceptor in respiratory electron transport is

   a O_2.
   b H_2O.
   c CO_2.
   d NAD.
   e FAD. (p. 170)

31 Cyanide blocks the respiratory electron-transport chain. As a result

   a the Krebs cycle speeds up.
   b electrons and hydrogens cannot flow from NADH to oxygen.
   c three ATPs are produced for every pair of electrons.
   d production of water increases.
   e glycolysis is inhibited. (p. 173)

32 According to the chemiosmotic hypothesis,

   a H^+ ions diffuse out of the mitochondrion; this movement releases energy that can be used to make ATP.
   b H^+ ions are actively transported from the cytoplasm into the inner compartment of the mitochondrion.
   c the hydrogen reservoir in the mitochondrion accumulates in the inner compartment of the mitochondrion.
   d the electron-transport molecules synthesize ATP.
   e ATP is produced in the mitochondrion when the hydrogen ions stored in the outer compartment flow back into the inner compartment. (p. 170)

33 Which one of the following summary reactions (each a part of the process of metabolic breakdown of glucose) ultimately yields the most new ATP molecules under aerobic conditions? Assume that all electrons complete their trip down the respiratory electron chain.

   a 2 pyruvic acid → 2 acetyl-CoA
   b glucose → 2 pyruvic acid
   c 2 PGAL → 2 pyruvic acid
   d 2 acetyl-CoA → 4 carbon dioxide + water
   e 2 acetyl-CoA → 2 citric acid (pp. 169–73)

34 For a living animal, which of the following compounds has the greatest amount of energy per molecule?

   a ATP
   b ADP
   c H_2O
   d NADH
   e pyruvic acid (p. 172)

35 When a molecule of glucose is completely broken down in a cell to water and carbon dioxide, some ATP molecules are synthesized by substrate-level phosphorylation and some by chemiosmotic phosphorylation via the electron-transport system. What percentage of the total number of ATP molecules formed comes from the latter process?

   a 94 percent
   b 89 percent
   c 83 percent
   d 78 percent
   e 6 percent (p. 172)

36 When glucose is broken down to carbon dioxide and water during aerobic respiration, more than 60 percent of its energy is released as

   a oxygen.
   b carbon dioxide.
   c heat.
   d ATP.
   e NAD. (p. 172)

37 Which one of the following is *not* associated with ectothermic animals.

   a body temperature close to environmental temperature
   b low metabolic rate
   c insulating hair or feathers
   d metabolic rate that varies with environmental temperatures
   e sluggish behavior in cold temperature (p. 175)

For further thought

1 Compare the energy levels of glucose, pyruvic acid, NADH, ATP, CO_2, and H_2O.

2 Give an account of the formation of ATP, referring to H_2O, CO_2, O_2, six-carbon carbohydrate, three-carbon carbohydrate, two-carbon molecule, four-carbon molecule, five-carbon molecule, and six-carbon molecule.

3 Why must glycolysis take place in all cells?

4 Summarize the complete respiratory breakdown of one molecule of glucose, explaining (*a*) the main stages in the process, (*b*) the role of oxygen in the process, (*c*) the significance of the electron-transport chain, and (*d*) the fate of the energy contained in the glucose.

5 What is the relative efficiency of glycolysis and cellular respiration?

6 Explain how the energy stored in NADH is converted into energy stored in ATP.

ANSWERS

Testing recall

1.

	Starting Compound(s)	*Products*
Stage I	Glucose 2 ATP 2 NAD	2 pyruvic acid 2 NADH 4 ATP (2 net ATP)
Fermentation	2 pyruvic acid 2 NADH	2 ethyl alcohol + 2 CO_2 + 2 NAD^+ or 2 lactic acid + 2 NAD^+
Stage II	2 pyruvic acid 2 CoA 2 NADH	2 acetyl-CoA 2 CO_2 2 NADH
Stage III	2 acetyl-CoA 2 four-carbon compounds (oxaloacetic acid)	2 CoA 4 CO_2 6 NADH 2 $FADH_2$ 2 ATP 2 four-carbon compound (oxaloacetic acid)

Stage IV and V

$\left.\begin{array}{l}\underline{\quad 2 \quad}\text{ NADH from Stage I}\\ \underline{\quad 2 \quad}\text{ NADH from Stage II}\\ \underline{\quad 6 \quad}\text{ NADH from Stage III}\\ \underline{\quad 2 \quad}\text{ FADH}_2\text{ from Stage III}\end{array}\right\} \rightarrow \underline{\quad 32 \quad}\text{ ATP}$

2. false—pyruvic acid or glycolysis to fermentation
3. true
4. true
5. true
6. false—more efficient
7. d
8. b
9. e
10. c

Testing knowledge and understanding

11	b	18	e	25	d	32	e
12	b	19	b	26	b	33	d
13	d	20	a	27	c	34	e
14	e	21	a	28	b	35	b
15	b	22	a	29	c	36	c
16	a	23	d	30	a	37	c
17	e	24	a	31	b		

Chapter 9

HOW CELLS REPRODUCE

A GENERAL GUIDE TO THE READING

This chapter begins the second part of the text, "The Perpetuation of Life." This unit focuses on genetic information and how it is transmitted from one generation to another. Whereas Chapter 9 is primarily concerned with reproduction at the cellular level, Chapter 10 examines the molecular structure of DNA and how it replicates. Chapter 11 discusses how the genetic code carried by DNA is used as a template for protein synthesis, and Chapter 12 focuses on different mechanisms for controlling gene replication, transcription, and translation. Chapter 13 explains how inherited characteristics can be manipulated using the new and powerful techniques of genetic engineering that stem from the principles of molecular genetics described in the previous three chapters. Chapter 14 looks at the transmission of inherited characteristics over generations, traditional "Mendelian genetics."

This chapter discusses how cells themselves reproduce, passing genetic information from parent to offspring. As you read Chapter 9 in your text, you will want to give special attention to the following aspects of the topic.

1 The flow of genetic information. Study carefully Figure 9.1 (p. 180), since it provides a summary of the material presented in Part II of the text.

2 The process of mitosis. Figure 9.10 (p. 185) graphically presents the stages of mitosis. Know these stages thoroughly. One mnemonic device for learning their order is the sentence "I put my arm there"; the first letter of each word is the first letter of a stage. (Interphase, Prophase, Metaphase, Anaphase, and Telophase). Another device is simply the acronym IPMAT.

3 The structure of a late-prophase chromosome in mitosis (pp. 184–86). Notice that, as shown in Figure 9.8 (p. 183), the chromosome in late prophase is double stranded; each strand is a chromatid. The DNA for the duplicate strand is produced during the S stage of interphase, when the chromosomal material is replicated. At the end of mitosis and cytokinesis, each chromosome in the newly formed cells is once again single stranded.

4 The structure of the late-prophase chromosome in meiosis. Notice that the twin chromatids are not so obviously separated as they are in mitosis. The chromatids are held together by protein axes (see Fig. 9.20, p. 192).

5 The process of meiosis. You may have difficulty with this topic, so you will want to study pages 189 to 196 carefully. Figure 9.19 (p. 191) is central for understanding this process.

6 Differences between mitosis and meiosis. Compare metaphase of mitosis (Fig. 9.10.5, p. 185) with metaphase of meiosis (Fig. 9.19.4, p. 191), with special attention to the way the chromosomes line up totally independently during mitosis, but in synapsed pairs during meiosis. This difference is crucial. Other differences are summarized in Figure 9.25 (p. 196).

7 Synapsis and crossing over. You will need to read pages 190 to 193 carefully and study Figures 9.20 and 9.21 (p. 192) in order to thoroughly understand this process. Notice that crossing over is not a rare event; it is a frequent and organized mechanism for producing genetic variation.

8 The adaptive significance of sexual reproduction. Make sure you understand the section on this subject (pp. 199–201), since it lays important groundwork for Chapters 14 and 31. Note that sexual reproduction can be viewed both as a source of variation and a source of stability.

KEY CONCEPTS

1 When a cell divides, it must replicate its DNA to make a complete copy of the genetic information in its nucleus and then, as it divides, in mitosis, give one complete set to each daughter cell (mitosis). (p. 179)

2 Mitosis produces new cells with exactly the same chromosomal endowment as the parent. It offers stability rather than variability to life's processes. (pp. 179–180)

3 Meiosis reduces the number of chromosomes by half so that when the egg and sperm unite in fertilization, the normal diploid number is restored. (pp. 189–190)

4 Sexual reproduction increases variation in the population by making possible genetic recombination. (pp. 199–200)

CROSSLINKING CONCEPTS

1 The concept of the flow of chemical information was introduced in Chapter 1. In Chapter 9, you add to your understanding of the flow of information when you focus on the transmission of genetic information to daughter cells during cell division. Chapter 10 will emphasize what goes on at a chemical level when chromosomes double to form two sister chromatids.

2 Centrioles, microtubules, and actin and myosin microfilaments, first introduced in Chapter 6, produce the movement of the chromosomes and the cleavage of the cell membrane in animal-cell cytokinesis.

3 Faults in the control of the cell cycle appear to be closely linked to the control failures that generate cancer cells (Chapter 12).

6 Evolution (Part IV) provides the key to understanding why eucaryotic cells do not reproduce by binary fission, but rather go through the complicated processes of mitosis and meiosis, which provide for genetic variation.

7 The concept of crossing over introduced in this chapter will facilitate your study of chromosomal linkage in Chapter 14.

OBJECTIVES

1 Using Figure 9.1 (p. 180), describe the transfer of information in a cell. (pp. 179–81)

2 Using Figure 9.2 (p. 180), explain how cell division takes place in procaryotic cells. Explain how the chromosomes of procaryotic cells differ from those of eucaryotic cells. (pp. 179–81)

3 Describe the various stages in the cell cycle. Specify the stage in which genetic replication occurs and the stage in which cell division occurs, and indicate which of the stages in the cell cycle compose interphase. In addition, explain what is meant by G_1 arrest and by G_2 arrest, and name a type of cell that shows each type of arrest. (pp. 182–84)

4 Explain the difference between a chromatid and a chromosome. For a cell in prophase that has six chromosomes, specify the number of chromatids and the number of centromeres. (pp. 183–84)

5 List the stages of mitosis and, using a diagram such as Figure 9.10 (p. 185), describe the principal events that occur during each stage. Explain how you would recognize a cell in prophase, metaphase, anaphase, and telophase. Distinguish plant-cell mitosis from animal-cell mitosis. (pp. 182–83)

6 Define cytokinesis, and show how it differs in algal cells, higher-plant cells, and animal cells. Using Figure 9.14 (p. 187) or the micrograph in Figure 9.16 (p. 189), point out the cell plate. Specify the organelles thought to be involved in the production of the cell plate. (pp. 187–89)

7 List the stages of meiosis and, using Figure 9.19 (p. 191), describe the principal events that occur during each stage. Define the terms haploid and diploid. Explain what happens during synapsis, showing what is meant by a synaptonemal complex and by a chiasma. Specify how and when crossing over occurs, and indicate the significance of this process. (pp. 190–93)

8 Compare and contrast meiosis with mitosis. Explain the importance of each process in the life of the organism. Compare the behavior of chromosomes in meiosis I and in mitosis (Fig. 9.25, p. 196, may be helpful). Compare the behavior of chromosomes in meiosis II and in mitosis. Compare the end products of meiosis and mitosis. (pp. 194–96)

9 Using Figure 9.28 (p. 198), compare and contrast spermatogenesis with oogenesis. (p. 197)

10 Using Figure 9.31 (p. 199), describe the two representative life cycles. In doing so, explain how the products of meiosis differ in plants and in animals and how the life cycle of a higher plant differs from that of an animal. (pp. 196–99)

11 Discuss reasons why natural selection has favored the more complicated process of sexual reproduction over the simpler one of asexual reproduction in a wide variety of organisms. Describe the circumstances under which asexual reproduction is advantageous. (pp. 199–201)

KEY TERMS

binary fission (p. 180)
nucleosomes (p. 180)

homologous pairs (p. 180)
diploid (p. 182)
interphase (p. 182)
centrosome (p. 182)
centrioles (p. 182)
cell cycle (G_1, S, G_2, and M stages) (p. 183)
centromere (p. 183)
chromatid (p. 183)
cyclins (p. 184)
mitosis (p. 184)
cytokinesis (p. 184)
prophase (p. 184)
polar microtubules (p. 186)
astral microtubules (p. 186)
spindle (p. 186)
metaphase (p. 186)
anaphase (p. 186)
telophase (p. 186)
cleavage furrow (p. 187)
cell plate (p. 188)
gametes (p. 189)
zygote (p. 189)
meiosis (p. 189)
reduction division (p. 190)
haploid (p. 190)
tetrad (p. 190)
synapsis (p. 190)
crossing over (p. 190)
hybrid (p. 192)
chiasmata (p. 192)
interkinesis (p. 194)
primordial germ cells (p. 196)
oogonia (p. 196)
spermatogonia (p. 196)
primary spermatocytes (p. 196)
primary oocytes (p. 196)
spermatogenesis (p. 197)
oogenesis (p. 197)
secondary oocyte (p. 198)
polar body (p. 198)
ovum (p. 198)
spore (p. 198)
clone (p. 199)

SUMMARY

The genes of an organism's DNA contain all the information necessary to determine the organism's characteristics and direct its biochemical activities. The entire set of genetic information must be duplicated and transferred when each new cell is formed.

Procaryotic cells divide by *binary fission*; the single circular chromosome replicates and a ring of new plasma membrane and wall material grows inward and separates the replicates.

Eucaryotic chromosomes differ from those of procaryotes in that they are linear, the DNA is wound on *nucleosome cores*, and the chromosomes occur in homologous pairs.

Mitotic cell division Cell division in eucaryotic cells involves two processes: the division of the nucleus *(mitosis)* and the division of the cytoplasm *(cytokinesis)*. Nuclear division entails, first, precise duplication of the genetic material and, second, distribution of a complete set of the material to each daughter cell.

Divisible cells pass through a series of stages known as the *cell cycle*. During the *interphase* stage, composed of the G_1, S, and G_2 stages, the cell is nondividing. The nucleus of the cell is visible, and one or more nucleoli are prominent. The chromosomes are long and thin and cannot be seen as distinct structures.

Just after a cell has completed the division process, but before replication of the genetic material begins, there is a gap in time, called the G_1 stage. Next comes the S stage, during which new DNA is synthesized. Another time gap, the G_2 stage, separates the end of replication from the onset of mitosis proper. Some animal cells, such as skeletal muscle cells and nerve cells, become arrested in the G_1 stage and a few in G_2; these cells normally will not divide. After the cell has passed through G_1, S, and G_2, it enters the M stage (mitosis). Mitosis is customarily divided into four stages:

1 *Prophase*. As the two pairs of *centrioles* move to the opposite sides of the nucleus, the chromosomes condense into visible threads. Each chromosome consists of *chromatids* held together by their centromeres. The polar and astral microtubules appear near each pair of centrioles and begin to form the spindle apparatus. The nuclear membrane and nucleoli disappear. Plant cells, unlike animal cells, lack centrioles and astral microtubules.

2 *Metaphase*. The chromosomes, attached at their centromeres to spindle microtubules, line up along the middle of the spindle. Metaphase ends when the centromeres of each pair of twin chromatids split.

3 *Anaphase*. The two sets of single-stranded chromosomes are pulled toward their respective poles by spindle microtubules. Cytokinesis often begins here.

4 *Telophase*. The chromosomes reach the poles, the nuclear membrane and nucleoli are reformed, cytokinesis is completed, and the chromosomes become longer, thinner, and less distinct.

Cytokinesis frequently accompanies mitosis. In animal cells, cytokinesis begins with the formation of a *cleavage furrow* running around the cell; the furrow deepens until it divides the cell in two. In plants, a *cell plate* forms in the center of the cytoplasm and enlarges until it cuts the cell in two. In many fungi and algae, new plasma membrane and wall grow inward down the midline. Mitosis without cytokinesis is common in some fungi and invertebrate animals.

Mitosis produces two new cells with exactly the same chromosomal endowment as the parent cell.

Meiotic cell division *Meiosis* is a special process of cell division that reduces the number of chromosomes by half so that when the egg and sperm unite in fertilization the normal number is restored. During the *reduction division* of meiosis, the chromosomal pairs are partitioned so that each gamete contains one of each type of chromosome. (It is *haploid*.) When the two gametes unite in fertilization, the resulting zygote is *diploid*; it has received one chromosomal type from each parent.

Meiosis involves two successive divisional sequences, which produce four new haploid cells. The first division is the reduction division; the second separates the chromatids. Both divisional sequences can be divided into four stages. In meiosis I these are:

1. *First prophase.* The events are similar to the mitotic prophase with one crucial difference: in meiosis the homologous chromosomes move together and lie side by side in a process called *synapsis*. The twin chromatids are held tightly together by a pair of protein axes; the protein axes of the two homologous chromosomes then join by protein cross-bridges. *Crossing over* then takes place; chromatids are clipped and parts of separate chromatids are spliced together. When the synaptic complex breaks up, the *chiasmata* hold the spliced chromatids together.

2. *First metaphase.* The two chromosomes of each homologous pair attach to the same spindle microtubule along the equator of the cell. The centromeres do not split.

3. *First anaphase.* The two double-stranded chromosomes of each synaptic pair move to opposite poles.

4. *First telophase.* Two new nuclei are formed, each with half the chromosomes present in the parental nucleus. The nuclei are not identical. The chromosomes become indistinct.

A short period called *interkinesis* follows the first telophase. No replication of genetic material occurs during this stage. The second division sequence of meiosis is essentially mitotic; each double-chromatid chromosome attaches to a *separate* microtubule, the centromeres split, and the new single-stranded chromosomes move to the poles. Four new haploid cells containing single-chromatid chromosomes are produced.

The first meiotic division then produces two haploid cells containing double-chromatid chromosomes. Each of these cells divides again in the second meiotic division; this produces a total of four new haploid cells containing single-chromatid chromosomes.

Most higher animals are diploid. During reproduction, meiosis produces haploid gametes, which unite to produce the diploid zygote. *Gametes* are haploid cells specialized for sexual reproduction. In male animals, *spermatogenesis* produces four functional sperm cells from each diploid cell. In females, *oogenesis* produces only one mature ovum; the *polar bodies*, also formed in oogensis, are nonfunctional.

Meiosis in plants produces haploid cells called *spores* (Stage 1), which often divide mitotically to form haploid multicellular plants (Stage 2). Eventually these haploid plants produce gametes (Stage 3) by mitosis. Two of the gametes unite to form the diploid zygote (Stage 4), which develops into a diploid multicellular plant (Stage 5), and the cycle is complete. Most multicellular plants have all five stages in their life cycles.

Asexual reproduction produces a *clone* of organisms genetically identical to the parent organism, whereas sexual reproduction increases variation within the population by creating organisms with new combinations of characteristics. Each diploid cell undergoing meiosis can produce 2^n different chromosomal combinations, where n is the haploid number. Crossing over greatly increases the number of possible variants.

CONCEPT MAPS

Map 9.1

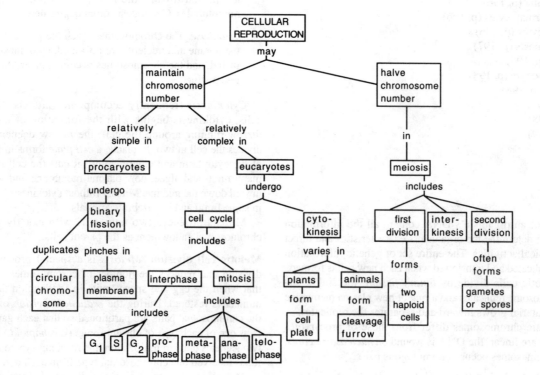

QUESTIONS

Testing recall

Match each event described below with the stage in mitosis in which it occurs. Answers may be used once, more than once, or not at all.

- a anaphase
- b interphase
- c metaphase
- d prophase
- e telophase

1 Astral microtubules appear (p. 186)

2 Chromosomes migrate to poles. (p. 186)

3 Centromeres divide. (p. 186)

4 Centrioles move to poles. (p. 182)

5 Chromosomes reach poles. (pp. 186–87)

6 G_1, S, and G_2 stages occur. (pp. 183–84)

7 Nuclear membrane and nucleoli disappear. (pp. 184–85)

8 Chromosomal replication occurs. (pp. 184–85)

Match each event described below with the type of cell division in which it occurs. Some of the events may occur in more than one type of cell division.

- a mitosis
- b first division of meiosis
- c second division of meiosis

9 Twin-chromatid chromosomes move to poles. (p. 186)

10 Chromosomes shorten and thicken, and are double-stranded. (p. 184)

11 Centromeres divide. (pp. 186. 194)

12 Single-chromatid chromosomes move to poles. (pp. 186, 194)

13 Nuclear membrane and nucleolus disappear. (pp. 184, 190)

14 Haploid cells are produced. (p. 194)

15 Crossing over occurs. (p. 190)

16 Cells genetically identical to the parent cell are produced. (p. 187)

17 Synapsis occurs. (p. 190)

Decide whether the following statements are true or false, and correct the false statements.

18 If a cell in prophase of mitosis has 20 centromeres, it has 20 chromosomes. (pp. 183–84)

19 Centromeres do not divide during meiosis I. (pp. 190–94)

20 Duplication of the genetic material (that is, the synthesis of new genetic material) occurs during prophase. (pp. 184, 190)

21 Centrioles are essential for cell division in both higher plant and animal cells. (p. 182)

22 A cell in prophase I of meiosis has half as many chromosomes as a cell in prophase II. (pp. 190–95)

23 Synapsis occurs during both mitosis and meiosis. (p. 190)

24 Crossing over is a rare event in human chromosomes. (pp. 190–93)

25 The vesicles of the cell plate formed during cytokinesis of a plant cell give rise to the middle lamella and the beginnings of the primary cell wall. (pp. 187–88)

Testing knowledge and understanding

Choose the one best answer.

26 During the S stage of the cell cycle, the cell

- a undergoes cytokinesis.
- b undergoes meiosis.
- c replicates its DNA.
- d undergoes mitosis.
- e enters interphase.   (p. 183)

27 If there are 12 single-chromatid chromosomes in a cell in G_1 of the cell cycle, what is the diploid number of chromosomes for the organism?

- a 6
- b 12
- c 24
- d 36
- e 48   (p. 183)

28 Which one of the following occurs during mitosis?

- a Homologous chromosomes synapse.
- b Twin-chromatid chromosomes move to the poles.
- c Crossing over takes place between homologous chromosomes.
- d Chromosome number is reduced from diploid to haploid.
- e Cells genetically identical to the parental cell are produced.   (p. 187)

29 Cytokinesis

- a always accompanies mitosis.
- b occurs in animal cells by a pinching-in of the plasma membrane.
- c occurs in higher-plant cells by an inward growth of new wall and membrane.
- d does two of the above.   (p. 187)

72 • CHAPTER 9

Questions 30 to 36 refer to the diagram below of a diploid cell in prophase.

30 How many *chromatids* are present in this cell?

   a 2
   b 4
   c 8   (p. 183)

31 How many *chromosomes* are present in this cell?

   a 1
   b 2
   c 4   (p. 183)

32 Which diagram shows how the chromosomes would be arranged during metaphase of mitosis? (p. 186)

 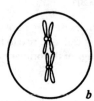 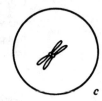  

33 In G_1 after mitosis, which of the following would one daughter cell contain?

   a two double-chromatid chromosomes   c one double-chromatid chromosome
   b two single-chromatid chromosomes   d one single-chromatid chromosome   (p. 184)

34 Which diagram shows how the chromosomes would be arranged during the first metaphase of meiosis? (p. 193)

35 Which diagram shows how the chromosomes would be arranged at the start of the second division of meiosis following interkinesis? (p. 194)

 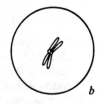  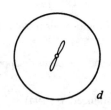

36 Which diagram shows what the chromosomes in one daughter cell would look like at the end of meiosis? (p. 194)

  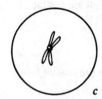 

37 If there are 12 chromosomes in a cell that has just completed meiosis, what is the diploid number of chromosomes for that organism?

   a 6
   b 12
   c 24
   d 36
   e 48
   (p. 183)

38 Which one of the following occurs in meiosis but *not* in mitosis?

   a Double-chromatid chromosomes move to the poles.
   b Chromosomes shorten and thicken and are double-stranded.
   c Single-chromatid chromosomes move to the poles.
   d Nuclear membrane and nucleolus disappear.
   e Centromeres divide.   (p. 194)

39 Which one of the following occurs in the first divisional sequence of meiosis?

   a Diploid daughter cells are produced.
   b Single-chromatid chromosomes move to the poles.
   c Centromeres divide.
   d Homologous chromosomes synapse.
   e Two of the above are correct.   (pp. 189–94)

40 If a cell in the process of meiosis is haploid and its chromosomes are becoming individually visible, which stage is it in?

   a first meiotic prophase
   b first meiotic anaphase
   c second meiotic prophase
   d second meiotic telophase
   e interkinesis   (pp. 190–94)

41 At what stage of cell division does synapsis occur?

   a anaphase of mitosis and meiosis
   b prophase of mitosis and meiosis
   c metaphase of mitosis and meiosis
   d prophase of the first meiotic division
   e metaphase of the second meiotic division   (p. 190)

42 Which one of the following statements is *false* concerning meiosis?

   a DNA is replicated between each cell division.
   b Each chromosome is double-stranded during prophase.
   c Each chromosome pairs with a homologous chromosome during meiosis I.
   d Cell division follows chromosome migration.
   e Each chromosome may exchange a part of a chromosome with the equivalent part of a homologous chromosome.   (pp. 189–94)

43 D, E, F, and G are the four daughter cells resulting when cell A undergoes meiosis, as shown in the following diagram. If no crossing over has occurred, which cells listed below, if any, are genetically identical?

   a cells B and C
   b cells A and D
   c cells D, E, F, and G
   d cells D and E; cells F and G
   e no two cells   (pp. 190–94)

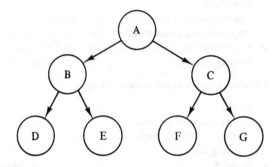

44 The drawing below shows a cell whose diploid chromosome number is four. This cell is in

   a metaphase.
   b anaphase of mitosis.
   c first anaphase of meiosis.
   d second anaphase of meiosis.
   e telophase of mitosis.   (p. 193)

45 The cell below is from a(n)

   a plant.
   b animal.
   c plant or animal.
   d bacterium.   (p. 187)

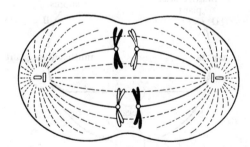

46 From each primary oocyte that undergoes meiosis, the number of functional egg cells produced is

   a one.
   b two.
   c three.
   d four.
   e eight.   (p. 198)

74 • CHAPTER 9

47 If the following events in the formation of the synaptic complex and crossing over were arranged in order, which event would be third?

   a Homologous chromosomes are brought into perfect alignment and protein cross-bridges form.
   b Recombination nodules appear.
   c Axial proteins gather the DNA of the chromatids into paired loops.
   d Chromatids from one chromosome exchange fragments with chromatids of the other chromosome.
   e Chromosomes shorten and thicken; replication has already occurred. (pp. 190–92)

48 Genetic recombination occurs during

   a mitosis.
   b first division of meiosis.
   c second division of meiosis.
   d crossing over.
   e b and d. (pp. 190–93)

49 Suppose the diploid chromosome number of a particular organism is 10. How many different chromosomal combinations could be produced by meiosis in this organism (i.e., how many different kinds of gametes could be formed)? Exclude combinations resulting from crossing over.

   a 5   d 32
   b 16   e 64
   c 25   (pp. 199–200)

Questions 50 to 52 refer to the following diagram of a plant life cycle.

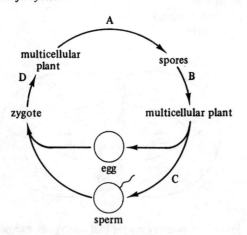

50 Which type of cell division occurs during part D of the life cycle?
   a mitosis   b meiosis   (pp. 198–99)

51 Which type of cell division occurs during part C of the life cycle?
   a mitosis   b meiosis   (pp. 198–99)

52 Which type of cell division occurs during part A of the life cycle?
   a mitosis   b meiosis   (pp. 198–99)

For further thought

1 In a single sentence, how would you explain the basic significance of mitosis?

2 Which of the phases of mitosis do you think would require the greatest expenditure of energy? Explain your answer.

3 Suggest a hypothesis to explain the adaptive value of having genetic material packaged in chromosomes.

4 In a single sentence, state the fundamental difference between mitosis and meiosis in terms of chromosomal behavior.

5 Without looking at your notes or textbook, take a sheet of paper and make diagrams of mitosis and meiosis in an organism with three pairs of chromosomes. Make a list of the most important differences between mitosis and meiosis.

ANSWERS

Testing recall

1	d	6	b	11	a, c	16	a
2	a	7	d	12	a, c	17	b
3	c	8	b	13	a, b, c	18	true
4	d	9	b	14	b, c	19	true
5	e	10	a, b, c	15	b		

20 false—interphase
21 false—neither plant nor animal
22 false—twice as many
23 false—only during meiosis
24 false—frequent
25 true

Testing knowledge and understanding

26	c	33	b	40	c	47	a
27	b	34	a	41	d	48	e
28	e	35	b	42	a	49	d
29	b	36	d	43	d	50	a
30	b	37	c	44	c	51	a
31	b	38	a	45	b	52	b
32	b	39	d	46	a		

Chapter 10

THE STRUCTURE AND REPLICATION OF DNA

A GENERAL GUIDE TO THE READING

This chapter discusses the chemical nature of DNA and describes how it replicates. It is important to learn the material in this chapter since the following chapters build on this crucial information. As you read Chapter 10 in your text, you will want to concentrate on the following topics.

1. Bacteriophage reproduction. The relationship between bacteria and certain viral parasites is presented early in the chapter, as part of a historical account of the original experiments that showed DNA to be the genetic substance. Figures 10.3 and 10.4 (p. 205) will help you learn and remember how bacteriophages replicate; this will aid your study of recombinant DNA technology (Chapter 13) and viruses (Chapter 37).

2. The molecular structure of DNA. This topic is crucial; you must understand this section (pp. 206–10) before you continue reading.

3. The replication of DNA. Make sure you understand this process thoroughly described on pages 210 to 212.

4. DNA repair and mutation. It is important to know that there are special mechanisms to locate and correct errors that occur due to mutations or during replication. (pp. 212–13)

KEY CONCEPTS

1. DNA is the genetic material that makes up the genes; it contains all the information needed for the cell's growth, operation, and division into two similar cells. (pp. 203–206)

2. DNA is a polymer of nucleotides arranged in a double helix. Sugars and phosphates form the sides of this ladderlike structure. The rungs are made of nitrogenous bases, which carry the hereditary message. (pp. 206–10)

3. During replication, an exact copy of the DNA is made. (pp. 210–11)

4. Replication in eucaryotic cells is initiated simultaneously at many independent sites on the DNA and proceeds in both directions away from the initiation sites. (p. 211)

5. Special mechanisms have evolved in both procaryotic and eucaryotic cells to locate and correct mutations and replication errors. (pp. 212–13)

CROSSLINKING CONCEPTS

1. The research that established DNA as the hereditary material is an excellent illustration of the scientific method (Chapter 1).

2. A knowledge of enzyme function (Chapter 4) is necessary for understanding DNA replication.

4. A knowledge of DNA structure is essential for understanding replication, transcription, and translation.

5. The role of mutations in evolution will be discussed further in Chapter 31.

OBJECTIVES

1. Summarize the transfer of genetic information along two lines: from DNA to protein synthesis, and from DNA to new daughter cells. (p. 203)

2. Cite two pieces of experimental evidence that support the conclusion that DNA is the genetic material. (pp. 203–206)

3. Given a diagram of a nucleotide, point out the phosphate group, the sugar group and the nitrogenous base. Indicate whether the base is a purine (double-ring structure) or pyrimidine (single-ring structure). (pp. 206–207)

4. Name the four nitrogenous bases found in DNA, and indicate which are pyrimidines and which are purines. Explain what the "D" in DNA stands for. (pp. 206–208)

5. Using Figure 10.10 (p. 208), point out an individual nucleotide and the three components of which it is made, the purine and pyrimidine bases, and the number of hydrogen bonds in a GC pair and an AT pair. (pp. 208–209)

6. Given a sequence of bases on one DNA strand, give the sequence on the complementary strand. (p. 209)

7. Show how the Watson-Crick model accounts for precise replication of genetic material, and name the enzymes that catalyze DNA replication. (pp. 210–12)

8. Describe similarities and differences between procaryotic and eucaryotic DNA structure and replication. (pp. 210–12)

9. Describe how repair enzymes act to correct mutations and errors that occur during replication. (Fig. 10.16, p. 212, may be helpful.) (pp. 212–13)

KEY TERMS

replication (p. 203)
transformation (p. 204)
bacteriophage (p. 204)
virions (p. 204)
adenine (p. 206)
guanine (p. 206)
purine (p. 206)
cytosine (p. 206)
thymine (p. 206)
pyrimidine (p. 206)
double helix (p. 210)
DNA polymerase (p. 211)
replication origins (p. 211)
mutation (p. 212)

SUMMARY

Experiments on bacterial transformations and radioactively labeled bacteriophage demonstrated that DNA, not proteins, constitutes the genetic material. DNA is composed of nucleotides that are made up of a five-carbon sugar, deoxyribose, attached by covalent bonding to a phosphate group and a nitrogenous base.

There are four different nucleotides in DNA; they differ in their nitrogenous bases, which may be the double-ring purines, *adenine* and *guanine*, or the single-ring pyrimidines, *cytosine* and *thymine*.

The molecular structure of DNA Using information gained from chemical analysis and X-ray diffraction studies, James Watson and Francis Crick formulated a model of the DNA molecule. According to this model, the nucleotides are joined together by covalent bonds between the sugar of one nucleotide and the phosphate group of the next nucleotide in the sequence; the nitrogenous bases are side groups of the chains. DNA molecules are double-chained structures, with the two chains held together by hydrogen bonds between adenine and thymine and between guanine and cytosine from opposite chains. The sequence of bases in one segment determines the complementary sequence in the other. The ladderlike double-chained molecule is coiled into a *double helix*, which is stabilized by hydrogen bonds.

The replication of DNA The Watson-Crick model of DNA explains how genetic replication can occur. The process produces two complete double-chained molecules, each identical in base sequence to the original double-chained molecule. First, the hydrogen bonds linking the strands of the two chains of DNA and stabilizing the helical shape are broken and the chains separate. Each chain then acts as a template for the synthesis of a new partner; complementary nucleotides are paired with nucleotides of each existing chain. The new nucleotides are covalently linked to form a chain. Each step is catalyzed by specific enzymes.

Replication of DNA in both procaryotes and eucaryotes begins at particular sequences in the DNA called replication origins and proceeds in both directions from these sites. Unlike replication in procaryotes, eucaryotic replication also involves the assembly of chromosomal proteins (histones).

Eucaryotic chromosomal replication occurs in the S phase of the cell cycle.

DNA repair Special mechanisms have evolved in both procaryotes and eucaryotes to locate and correct mutations and replication errors. Enzymes find and bind to faulty or damaged sequences and clip out the faulty sequences; the intact complementary strand guides repair. The repair process is not perfect, however; some errors are missed, others cannot be detected, and still others may be created by the repair system. Nevertheless, the rate at which mutations accumulate is kept relatively low by the repair system.

THE STRUCTURE AND REPLICATION OF DNA

CONCEPT MAP

Map 10.1

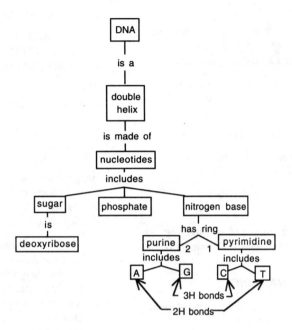

KEY DIAGRAM

In this diagram of a segment of a DNA molecule, color in the following, using a different color for each structure. (Fig. 10.10, p. 208)

PHOSPHATE GROUPS

DEOXYRIBOSE GROUPS

HYDROGEN BONDS

Remembering the number of hydrogen bonds between an AT pair and a GC pair, and which bases are purines and which are pyrimidines, color the following:

ADENINE

GUANINE

CYTOSINE

THYMINE

QUESTIONS

Testing recall

Below are listed some of the important events that led to our present knowledge of the nature of the gene and its functions. Match each of these with the investigators associated with it.

 a Chargaff
 b Griffith
 c Hershey and Chase
 d Watson and Crick
 e Wilkins and Franklin

1 proposal of the double-helix model of DNA (p. 209)
2 transformation of bacteria by material extracted from heat-killed virulent cells (p. 204)
3 infection of bacteria using radioactively labeled bacteriophage (p. 206)
4 discovery that in any DNA the amount of A equals the amount of T and the amount of G equals the amount of C (p. 206)
5 X-ray diffraction studies of DNA (p. 207)

Match each item below with the appropriate term. A term may be used more than once or not at all.

 a sugar-phosphate groups
 b purine(s)
 c pyrimidine(s)
 d covalent bonds
 e hydrogen bonds

6 backbone of the DNA molecule (p. 209)
7 forces between the two polynucleotide chains (p. 209)
8 single-ring nitrogenous bases (p. 206)
9 double-ring nitrogenous bases (p. 206)
10 adenine and guanine (p. 206)
11 cytosine and thymine (p. 206)

Testing knowledge and understanding

Choose the one best answer.

12 A nucleic acid is composed of a chain of nucleotides. Nucleotides themselves are made of three components. Which of these components could be removed from a nucleotide that is part of a nucleic acid chain without breaking the chain?

 a sugar
 b phosphate
 c nitrogenous base (pp. 208–10)

13 The structure of DNA as proposed by Watson and Crick *depended* on all the following observations *except*

 a that DNA is capable of replicating itself precisely.
 b that DNA base sequences vary from organism to organism.
 c that DNA contains nitrogenous bases, sugars, and phosphates.
 d X-ray periodicities of 3.4, 2, and 0.34 nm.
 e Chargaff's finding that DNA contained equal amounts of A and T nucleotides and G and C nucleotides. (pp. 206–10)

14 If a segment of nucleic acid is CATCATTAC, the complementary DNA strand is

 a CATTACTAC.
 b CAUCAUUAC.
 c GUAGUAAUG.
 d GTAGTAATG.
 e CUACUACAT. (p. 209)

15 The DNA of a certain organism has guanine as 30 percent of its bases. What percentage of its bases would be adenine?

 a 10
 b 20
 c 30
 d 40
 e 50 (p. 206)

16 Suppose that you provide an actively dividing culture of *E. coli* bacteria with radioactive thymine. What would you expect to happen if a cell replicated its DNA and divided once in the presence of the radioactive base?

 a One of the daughter cells, but not the other, would have radioactive DNA.
 b Neither daughter cell would have radioactive DNA.
 c All four bases of the DNA would be radioactive.
 d Radioactive thymine would pair with nonradioactive guanine.
 e DNA in both daughter cells would be radioactive. (p. 210)

17 Which of the following statements about DNA structure and replication in procaryotes is correct?

 a There is a single circular chromosome composed of DNA.
 b Chromosomes are wound around nucleosome cores.
 c Replication begins at multiple initiation sites.
 d Chromosomes line up in the center of the cell during replication. (p. 211)

For further thought

1 Explain why the DNA repair enzymes can detect and remove a mutation in which cytosine has been converted into uracil, but are not able to detect a mutation in which cytosine has been converted into thymine.

2 Describe the biological research (questions, methods, results, and conclusions) that contributed to our understanding of DNA structure.

ANSWERS

Testing recall

1	d	5	e	9	b
2	b	6	a	10	b
3	c	7	e	11	c
4	a	8	c		

Testing knowledge and understanding

12	c	14	d	16	e
13	b	15	b	17	a

Chapter 11

TRANSCRIPTION AND TRANSLATION

A GENERAL GUIDE TO THE READING

The last chapter described how the DNA in chromosomes was replicated; in this chapter we see how the information in DNA is used to produce proteins. As you read the chapter, you will want to concentrate on the following topics.

1. Transcription. You will want to understand how transcription works and be able to predict the nucleotide sequence of the mRNA coded for a section of DNA. You should note the differences between transcription in procaryotes and eucaryotes and learn what introns and exons are. Figure 11.10 (p. 220) is an excellent summary of messenger RNA processing in eucaryotes.

2. Translation. You should learn the process of translation, and how to use the genetic code to predict the amino acid sequence in the protein to be produced (Table 11.1, p. 222). Figure 11.16 (p. 225) is a helpful summary diagram.

3. Mutation. This concept, introduced in Chapter 10 (see p. 212), is enlarged upon here (pp. 227–29).

KEY CONCEPTS

1. The information encoded in DNA is used to produce both the proteins that form cellular structure and the enzymes that direct cellular metabolism; these determine the inherited characteristics of the organism. (p. 216)

2. During transcription, the DNA serves as a template for the formation of RNA. (p. 217)

3. Contrary to procaryotic and organelle RNA, eucaryotic RNA has specific regions within the transcript called introns that must be removed while the RNA is still in the nucleus before becoming a functional mRNA molecule. Only after these introns are removed can the mRNA go through the nuclear envelope to the cytoplasm. (pp. 220–21)

4. The sequence of the bases in DNA determines the sequence of nucleotide bases in transcribed mRNA, which dictates the order of the amino acids to be linked in protein synthesis. (pp. 216–18)

5. The genetic code consists of triplet coding units; a combination of three nucleotides in the mRNA specifies one amino acid in the polypeptide. (p. 221)

6. Translation of the information contained in DNA into functional proteins is indirect, it involves three types of RNA: mRNA, tRNA, and rRNA. (p. 218)

7. Mutations are changes in the sequence of bases in DNA. (pp. 227–29)

8. Mitochondria and chloroplasts have their own DNA and are capable of self-replication. (pp. 229–30)

CROSSLINKING CONCEPTS

1. The flow of information from DNA to protein, originally introduced in Chapter 1 and expanded in Chapter 10, become more detailed in Chapter 11.

2. The endosymbiotic hypothesis, introduced in Chapter 6, is reinforced by the similarities between DNA structure, transcription, and translation in procaryotes and cell organelles.

3. The possible evolutionary significance of transposons and of eucaryotic gene organization will be explored further in Chapter 31.

5. The connection between a substance's ability to cause mutations and its ability to cause cancer will be enlarged upon in later chapters.

OBJECTIVES

1. State three ways in which RNA is different from DNA, name the three types of RNA and indicate where each is synthesized, and describe the function of each type of RNA. (pp. 216–18)

2. Given a sequence of bases on one DNA strand, write the sequence of mRNA that would be transcribed from this DNA. State the role of the promoter, RNA polymerase, the "start" signal, and the "stop" signal in the process. (pp. 218–20)

3. Explain how transcription in eucaryotes differs from that in procaryotes. Figure 11.10 (p. 220) may be helpful. (pp. 220–21)

4. Describe in some detail the process of translation; be sure to show how the sequence of bases in DNA determines the sequence in which amino acids are linked in protein synthesis. (pp. 221–27)

5. Using the genetic code in Table 11.1 (p. 222), give the amino acid sequence for any given mRNA.

6. Sketch a typical tRNA molecule, indicating the location of the anticodon and the position at which an amino acid becomes attached. (p. 224)

7. Describe in detail how the genetic code is read by the ribosome. In doing so, explain the terms codon, anticodon, initiation codon, and termination codon. (p. 224)

8. Define mutation, give three examples of different types of mutations, and indicate how mutations can be induced. Then, explain the relationship between mutagenicity and carcinogenicity. (pp. 227–28)

9. Describe the characteristics of the DNA and protein-synthesis machinery of mitochondria and chloroplasts that are consistent with the endosymbiotic hypothesis for the origin of these organelles. (pp. 229–30)

KEY TERMS

one gene–one polypeptide hypothesis (p. 216)
transcription (p. 218)
translation (p. 218)
messenger RNA (mRNA) (p. 218)
ribosomal RNA (rRNA) (p. 218)
transfer RNA (tRNA) (p. 218)
RNA polymerase (p. 218)
promoter (p. 218)
transcription factors (p. 219)
enhancer elements (p. 219)
primary RNA transcript (p. 220)
exon (p. 220)
intron (p. 220)
poly-A tail (p. 220)
codon (p. 221)
polyribosome (p. 223)
Shine-Delgarno sequence (p. 224)
initiation codon (p. 224)
termination codon (p. 224)
anticodon (p. 224)
addition mutation (p. 227)
deletion mutation (p. 227)
frameshift mutation (p. 227)
base substitution (p. 227)
transposition (p. 227)
transposons (p. 227)
mutagenic agent (p. 228)

SUMMARY

According to the *one gene–one enzyme hypothesis,* each gene directs the synthesis of an enzyme that controls a chemical reaction in the cell; the chemical reactions, in turn, determine the phenotypic characteristics. Because some proteins are composed of two or more different polypeptide chains, each determined by its own gene, the hypothesis has been restated as the *one gene–one polypeptide hypothesis.* The sequence of bases in DNA determines the sequence in which amino acids must be linked in protein synthesis via a two-step pathway. First, the information in the genes is transcribed into ribonucleic acid (RNA); second, the transcripts are translated into protein.

RNA differs from DNA in three ways: (1) The sugar in RNA is ribose whereas that in DNA is deoxyribose. (2) RNA has uracil where DNA has thymine. (3) RNA is usually single stranded where DNA is double stranded. There are three types of RNA: *messenger RNA (mRNA), ribosomal RNA (rRNA),* and *transfer RNA (tRNA).* All three are synthesized from DNA.

Transcription During transcription, RNA polymerase binds to the promotor region in the DNA and partially uncoils the two nucleotide chains. One of the chains acts as the template for the synthesis of single-stranded messenger RNA. The mRNA then leaves the chromosome and moves into the cytoplasm where it becomes associated with the ribosomes.

In eucaryotic cells, the mRNA must be processed before it leaves the nucleus, the noncoding introns must be cut out, and a "cap" of modified guanine and a tail of adenines is added to the transcript. Only the exons, the coding sequences, survive the editing process.

Translation During translation, the sequence of bases in mRNA is used to order the amino acids to form a polypeptide. The coding unit or *codon* in nucleic acids is three nucleotides long. All but three of the 64 possible triplet codons code for one of the 20 amino acids; these three exceptions are termination codons, which mark the end of the message. Most amino acids are represented by more than one codon. The genetic code is essentially universal.

The mRNA acts as a template for synthesis of polypeptide chains. As the ribosomes move along the mRNA, they read the codons. Amino acids to be incorporated into the polypeptide chains are picked up by molecules of transfer RNA specific for each of the 20 amino acids. The tRNAs bring their amino acids to the ribosome as it moves along the mRNA. Each tRNA attaches to the mRNA at the point where a triplet of mRNA bases (a codon) is complementary to the exposed triplet (anticodon) on the tRNA. This ordering of the tRNAs along the mRNA automatically orders the amino acids, which are then linked by peptide bonds. This process is called *translation*; the nucleic acid message has been translated into an amino acid sequence of a protein.

Synthesis of the polypeptide chain proceeds one amino acid at a time in an orderly sequence as the ribosomes move along the mRNA. As each tRNA donates its amino acid to the growing polypeptide chain, it uncouples from the mRNA and moves away to pick up another load. When a ribosome reaches a termination codon, it releases the completed polypeptide chain.

Mutations Mutations are alterations in the DNA that change its information content. Several types of mutations are possible. The *addition* or *deletion* of nucleotides in DNA often results in the production of inactive enzymes because of frame shifts in the translation process. In *base substitution* (point mutation), one nucleotide is exchanged for another. Base-substitution mutations are not as serious as additions or deletions because only one codon (and hence one amino acid) is involved, and often the change in the codon does not even result in a change in the amino acid. *Transposition* mutations occur when DNA is moved from one position on a chromosome to another, or to another chromosome.

High-energy radiation and a variety of chemicals can cause genetic mutations. Ionizing radiations sometimes induce only point mutations, but frequently produce more damaging large deletions of genetic material. Some mutagenic chemicals convert one base into another. There is a strong relationship between the mutagenicity of a chemical and its cancer-inducing activity (carcinogenicity).

Organelle Heredity Cytoplasmic organelles such as chloroplasts and mitochondria have their own DNA and replicate themselves independently of the nucleus. The DNA in these organelles is usually circular and is not wound on nucleosomes. Transcription and translation in these organelles are similar to the corresponding processes in procaryotes.

CONCEPT MAPS

Map 11.1

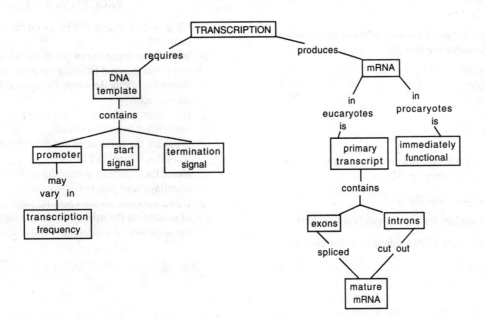

QUESTIONS

Testing recall

For questions 1 to 11, indicate whether the structural or functional features are true for:

a DNA only
b RNA only
c both DNA and RNA
d neither DNA nor RNA

1 is usually single stranded (p. 217)
2 contains pyrimidines (p. 216)
3 contains deoxyribose (p. 216)
4 is coiled in a double helix (p. 217)
5 contains thymine (p. 216)
6 contains cytosine (p. 216)
7 contains uracil (p. 217)
8 brings amino acids to the ribosome (p. 218)
9 is present in the ribosome (p. 218)
10 is involved in transcription (p. 218)
11 is involved in translation (p. 218)
12 Fill in the blanks.

   DNA (strand 1)   TGT   GCA   CGT
   DNA (strand 2)   ACA   CGT   GCA
   mRNA (from strand 2) UGU   GCA   CGU
   tRNA anticodons   ACA   CGU   GCA   (p. 217)

For questions 13 to 17, indicate whether the structural or functional features are true for:

a mRNA only
b rRNA only
c tRNA only
d all RNA
e no RNA

13 has an anticodon at one end and a binding site for an amino acid at the other (p. 224)
14 forms part of the ribosome (p. 222)
15 serves as a template for protein synthesis (p. 220)
16 is synthesized from a DNA template in the nucleus (p. 219)
17 carries the code for a particular protein to the ribosome (p. 221)

Testing knowledge and understanding

18 Which of the following steps of transcription occurs in *both* procaryotic and eucaryotic cells?

   a A length of mRNA is "tagged" to produce the primary transcript.
   b The mRNA crosses the nuclear envelope to reach the ribosomes.
   c Exons are snipped out from the primary transcript.
   d Exons are bonded together to make the mRNA that will be transcribed.
   e RNA polymerase bonds to the promoter on the DNA.
   (p. 220)

19 Which of the following statements best describes the relationship between the primary mRNA transcript and functional mRNA?

   a The primary transcript goes to the nucleus, where it is snipped and pasted to become functional mRNA.
   b The primary transcript must have both the "tagged" ends and introns removed in order to become functional mRNA.
   c The primary transcript must be activated by ATP to become functional mRNA.
   d The introns are removed from the primary transcript and the remaining pieces spliced together to form functional mRNA.
   e Introns are snipped out of the primary transcript and joined together to form functional mRNA.
   (p. 220)

20 One strand of a DNA molecule has the sequence of bases

   TACCTTCAGCGT

   a What is the sequence of bases on the complementary strand of DNA? (p. 219)
   b What is the sequence of bases on the strand of mRNA that is synthesized from the original strand? (p. 218)
   c Name the organelle where the synthesis of mRNA takes place. (p. 216)
   d How many different codons are there in the mRNA transcribed from this strand? (p. 221)
   e What are the anticodons for the mRNA transcribed from this segment of DNA? (p. 224)
   f Name the organelle where the codon and anticodon couplings take place. (p. 223)
   g Using the table on the opposite page and the sequence of codons on the mRNA strand (*b*, above), determine the sequence of amino acids coded for by this strand. (pp. 222–23)

TRANSCRIPTION AND TRANSLATION • 83

The genetic code (messenger RNA)

First base in the codon	Second base in the codon				Third base in the codon
	U	C	A	G	
U	Phenylalanine	Serine	Tyrosine	Cysteine	U
	Phenylalanine	Serine	Tyrosine	Cysteine	C
	Leucine	Serine	Termination	Termination	A
	Leucine	Serine	Termination	Tryptophan	G
C	Leucine	Proline	Histidine	Arginine	U
	Leucine	Proline	Histidine	Arginine	C
	Leucine	Proline	Glutamine	Arginine	A
	Leucine	Proline	Glutamine	Arginine	G
A	Isoleucine	Threonine	Asparagine	Serine	U
	Isoleucine	Threonine	Asparagine	Serine	C
	Isoleucine	Threonine	Lysine	Arginine	A
	Methionine	Threonine	Lysine	Arginine	G
G	Valine	Alanine	Aspartic acid	Glycine	U
	Valine	Alanine	Aspartic acid	Glycine	C
	Valine	Alanine	Glutamic acid	Glycine	A
	Valine	Alanine	Glutamic acid	Glycine	G

Choose the one best answer.

21 If mRNA is transcribed from a chain of DNA with the base sequence CATTAG, the mRNA will have the sequence

   a GUAAUC.   *d* TUCCUA.
   b GTAATC.   *e* GTUUTC.
   c TGCCGA.   (p. 218)

22 DNA, but not RNA, contains

   a adenine.   *d* cytosine.
   b guanine.   *e* uracil.
   c thymine.   (p. 217)

23 Transfer RNA functions in

   a carrying RNA from the ribosomes to mRNA.
   b attaching RNA to the ribosomes.
   c joining proteins to form the ribosomes.
   d carrying mRNA from the nucleus to the cytoplasm.
   e carrying amino acids to the correct site on the mRNA.
   (p. 224)

Questions 24 to 26 refer to the following segment of the base sequence of a gene: ACGTGCCCGGAT

24 How many amino acids would the polypeptide encoded by this gene segment have?

   a 1   *d* 12
   b 3   *e* 36
   c 4   (p. 221)

25 The *second* codon on the mRNA derived from the segment will be

   a TAC.   *d* CAC.
   b ACG.   *e* UGC.
   c ATG.   (p. 221)

26 The anticodon of the tRNA for the *first* codon will be

   a TCG.
   b ACG.
   c TCG.
   d UGC.
   e GCA.   (p. 224)

Questions 27 to 30 refer to the short mRNA sequence:

AUGCCCUACUAC

Use this sequence and the table in question 20 to answer the questions.

27 The sequence of the template strand of DNA of the gene that codes for this message will be

   a AUGCCCAACUAC.
   b TACGGGATGATG.
   c UTGCCCUUCTAC.
   d UACGGGUUGAUG.
   e ATGCCCTACTAC.   (p. 218)

28 The anticodon in the tRNA that attaches to the first codon will be

   a UAC.
   b TAC.
   c AUG.
   d GUA.
   e ATG.   (p. 224)

29 The protein coded for by this message will have _____ amino acids.

   a 1   *d* 5
   b 3   *e* 12
   c 4   (p. 222)

84 • CHAPTER 11

30 The second amino acid in the polypeptide chain will be (use the table from question 20)

  a lysine.
  b proline.
  c glycine.
  d aspartic acid.
  e phenylalanine.
  (p. 222)

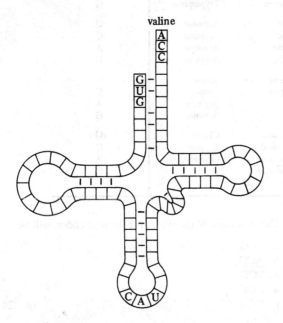

31 The diagram above represents tRNA, which recognizes and binds a particular amino acid (in this instance, valine). Which one of the following triplets of bases on the *DNA* strand, or template, codes for this amino acid?

  a TGG
  b GUG
  c GUA
  d ACC
  e CAT
  (p. 224)

32 According to current ideas about the DNA genetic code, which one of the following statements is *false*?

  a The codon is three nucleotides long.
  b Every possible triplet codes for some amino acid.
  c The code is redundant (i.e., it contains "synonyms").
  d The code is read in a regular sequence, beginning at one end.
  e The code is nonoverlapping.
  (pp. 221–22)

33 A polyribosome is

  a a ribosome engaged in the synthesis of a polypeptide.
  b a number of ribosomes simultaneously translating one mRNA molecule.
  c a number of ribosomes linked together by rRNA
  d a number of ribosomes linked together by DNA
  e a number of ribosomes linked together by tRNA
  (p. 223)

34 Which one of the following statements is *false*?

  a tRNA binds to an amino acid and activates it with energy from ATP.
  b In base substitution mutations only a single nucleotide of a gene is altered.
  c Molecules of mRNA are synthesized on the ribosomes from nucleotides brought by tRNA.
  d Some amino acids are specified by several "synonymous" codons.
  (pp. 224–25)

35 Though a gene codes ultimately for all aspects of a protein's structure, it codes directly only for

  a primary structure.
  b secondary structure.
  c tertiary structure.
  d quaternary structure.
  (p. 226; Chapter 3, p. 64)

36 As proteins are being synthesized, tRNA molecules are constantly being released from the site of amino acid incorporation. What happens to these tRNA molecules?

  a They return to the nucleus and bind to DNA again.
  b They are used to code for synthesis of a protein.
  c They immediately bind to another mRNA.
  d They pick up another amino acid of the same type that they had before.
  e They pick up an amino acid of another type, specifically the amino acid coded for by the codon next to the one to which they originally bound.
  (p. 226)

37 The direction of transfer of genetic information in most living organisms is

  a protein → DNA → mRNA.
  b DNA → mRNA → protein.
  c DNA → tRNA → protein.
  d protein → tRNA → DNA.
  e RNA → DNA → mRNA → protein.
  (p. 218)

38 A certain procaryotic gene codes for a polypeptide that is 126 amino acids long. The gene is probably how many nucleotides long?

  a 42
  b 126
  c 252
  d 381
  e 504
  (p. 221)

39 Suppose a gene has the DNA nucleotide sequence

  1 2 3 4 5 6 7 8 9 10 11 12 13 14 15 16 17
  C T G G C A T G C T T C G G A A A

(No real gene could be this short, but for our purposes this will suffice.) Which one of the following mutations would probably produce the greatest change in the activity of the protein for which this gene codes?

  a substitution of A for G in position 3
  b deletion of the C at position 5
  c deletion of the A at position 16
  d addition of a G between positions 14 and 15
  (p. 227)

40 An *intron* is

   a a foreign RNA sequence inserted in the normal message for a protein.
   b an RNA sequence that is edited from a transcript before translation.
   c a DNA sequence that is used to link a plasmid with a foreign DNA.
   d a DNA sequence that codes for the protein product of the gene.
   e a DNA sequence that is not transcribed. (p. 220)

41 Which of the following statements about DNA structure and protein synthesis in procaryotes and cell organelles is correct?

   a Organelle ribosomes closely resemble those of procaryotes.
   b Chromosomes are wound around nucleosome cores.
   c The chromosome of an organelle is approximately the same size as that of a procaryotic cell.
   d Cell organelle DNA encodes all the proteins the organelle needs. (p. 220)

For further thought

1 A nonsense mutation is one in which the substitution of one nucleotide for another results in a triplet that does not code for any amino acid, but instead codes for a termination codon. A missense mutation is one in which such a substitution results in a triplet that codes for a different amino acid. What effect would each of these mutations have on protein synthesis? Which of them would be expected to have the more severe effect on enzyme activity?

2 Active bovine insulin is a small protein composed of 51 amino acids. How many nucleotide units (exclusive of those which code for introns) does it take to code for this protein? Using the diagram of beef insulin on page 66 of your text and the genetic code (use the table from question 20), determine the nucleotide sequence in the DNA that codes for the shorter of the two chains (again excluding introns).

3 What would be the effect on protein synthesis of changing a single nucleotide in the anticodon of a tRNA molecule?

4 Chlorofluorocarbons (CFCs) released into the atmosphere by refrigerator and air-conditioner leakage damage the ozone layer that screens out much of the UV radiation. What are the biological implications of a widening stratospheric ozone hole and increased radiation?

ANSWERS

Testing recall

1	b	4	a	7	b	10	c
2	c	5	a	8	b	11	b
3	a	6	c	9	b		

12 DNA   T G T   G C A   C G T
    DNA   A C A   C G T   G C A
    mRNA U G U   G C A   C G U
    tRNA A C A   C G U   G C A

| 13 | c | 15 | a | 17 | c | 19 | d |
| 14 | b | 16 | a | 18 | e | | |

Testing knowledge and understanding

20 a ATGGAAGTCGCA
    b AUGGAAGUCGCA
    c nucleus
    d 4
    e UAC, CUU, CAG, CGU
    f ribosome
    g methionine, glutamic acid, valine, alanine

21	a	27	b	32	b	37	b
22	c	28	a	33	b	38	d
23	e	29	c	34	c	39	b
24	c	30	b	35	a	40	b
25	b	31	e	36	d	41	a
26	b						

Chapter 12

THE CONTROL OF GENE EXPRESSION

A GENERAL GUIDE TO THE READING

This chapter describes the control mechanisms that determine for each cell when and how it will act on its inherited genetic instructions. Read the chapter slowly and carefully and allot plenty of time to study it. You will want to concentrate on the following topics.

1. Control of gene expression in bacteria. Focus your attention on the Jacob-Monod model and learn the different parts of the model and how it works; Figure 12.3 (p. 235) will be particularly helpful in understanding gene expression. Be sure to distinguish negative control (illustrated in Figures 12.3, p. 235, and 12.4, p. 237) from positive control (see Figure 12.5, p. 238). Note that negative controls are typical of procaryotic cells.

2. Control of gene expression in eucaryotic cells. Note that controls in eucaryotic cells are usually positive controls and are far more complex than those in procaryotic cells. You should be able to distinguish between pretranscription controls and posttranscription controls and understand the function of transcription factors, enhancer regions, and methylation. Note visual evidence of control of gene expression: euchromatin, heterochromatin, lampbrush chromosomes, and chromosomal puffs. Figure 12.7 (p. 237) is a useful summary of the material on the organization of eucaryotic chromosomes.

3. Cancer. The section on cancer is interesting to most students. The material on the genetic basis and environmental causes of cancer is up-to-date and fascinating. Note that it is the ability of cancer cells to spread throughout the body (metastasize) that makes them life-threatening.

KEY CONCEPTS

1. Although every cell in the body of a multicellular organism has identical genetic information, individual cells have different structural and functional characteristics. Various control mechanisms determine when and how each cell will act on its inherited genetic instructions. (pp. 233–34)

2. Each gene codes for only one kind of messenger RNA; regulators determine whether and when each gene will be transcribed. (pp. 233–34)

3. Most chromosomal DNA of eucaryotes is never transcribed, and only about 1 percent of the eucaryotic DNA that codes for mRNA will be translated. In procaryotic cells approximately 90 percent of the DNA is translated. (p. 233)

4. Gene activation in eucaryotes involves the unwinding of particular sections of the nucleosomes, followed by activation of a specific gene or groups of genes in the unwound regions. (p. 238)

5. Transcription factors, whose binding to activator regions is controlled by enhancer regions, help RNA polymerase bind to the promoter and are important regulators of gene transcription in eucaryotic cells. (pp. 236–38)

6. In procaryotic cells, most control of gene transcription is by inhibition of specific genes, in contrast to eucaryotic cells where most control is by activation of specific genes. (pp. 242–43)

7. There are numerous points in eucaryotes other than transcription where control can be exerted on the flow of

information. In some cases a gene's expression can be controlled at every step from before transcription until after translation. (pp. 243–44)

8 The single most distinctive feature of cancer cells is their unrestrained proliferation. The conversion of a normal cell into a cancer cell involves several changes: loss of fixed-number-of-divisions control, loss of contact inhibition, loss of anchorage dependence, and cell-surface changes. (pp. 244–48)

9 Oncogenes, genes that cause cancer, are produced when certain normal genes, called proto-oncogenes, which stimulate cell proliferation, undergo a mutation that converts them into cancer-causing genes. (pp. 251–52)

CROSSLINKING CONCEPTS

1 You need to understand thoroughly the physical properties of both DNA and enzymes (Chapter 3), including the structure of nucleic acids and proteins and the effects of hydrogen bonding and polarity, in order to follow the processes by which gene transcription is controlled.

2 The negative control of gene transcription by corepressor substances is an example of a feedback control mechanism that contributes to homeostasis: the maintenance of constant internal conditions in a cell or organism. This concept was introduced in Chapter 1 (see "Central Concepts in the Study of Life," p. 26) and will recur in many later chapters.

3 The differences between procaryotic and eucaryotic transcription described in Chapter 11 (such as the presence of introns in primary mRNA and the need for mRNA to pass through the nuclear envelope into the cytoplasm) explain why there are so many more control possibilities in eucaryotic gene expression than in procaryotic.

4 The need for multiple rRNAs in developing eggs, introduced in this chapter, will be important when you study animal development (Chapter 27).

OBJECTIVES

1 Describe the Jacob-Monod model of gene induction. In doing so, explain the roles of inducers, operators, operons, promoters, repressor proteins, regulator genes, and structural genes. Using Figure 12.3 (p. 235), show how expression of the *lac* operon of *E. coli* is regulated. (pp. 234–36)

2 Using Figure 12.4 (p.237), describe the function of a repressible operon and explain how it differs from an inducible operon. (pp. 234–36)

3 Differentiate between positive control and negative control of gene transcription, and give two examples of each. Then, using Figure 12.5 (p. 238), describe how transcription factors activate genes. (pp. 236–38)

4 Explain why the models for the control of gene transcription in bacteria are not directly applicable to eucaryotic cells. (p. 240)

5 Review Figures 6.3 (p. 114) and 9.4B and C (p. 181) to refresh your memory about the structure of nucleosomes and describe the relationships between histone proteins and DNA. Explain the role of histone proteins in the control of eucaryotic gene activity. (p. 238)

6 Differentiate among highly repetitive DNA, moderately repetitive DNA, and single-copy DNA, and give the approximate proportions of each type in the eucaryotic genome. Explain why there are so many copies of some genes, whereas one copy is sufficient for other genes. (pp. 239–40)

7 Give the percentage of the eucaryotic genome that is both transcribed and translated and contrast that with the percentage of the procaryotic genome that is transcribed and translated. (p. 239)

8 Explain what lampbrush chromosomes are and how they are formed; describe the structure of the giant chromosomes of *Drosophila*, and summarize the evidence indicating that the chromosomal puffs seen in them are sites of intense transcriptional activity. Differentiate between euchromatin and heterochromatin. (pp. 240–41)

9 Describe how transcription factors exert gene control. Be sure to include the function of activator and enhancer regions. (pp. 242–43)

10 Explain what is meant by posttranscriptional control and give three examples of points where cellular control can be applied in the path of information flow from the genes to their phenotypic expression. (pp. 243–44)

11 Give three principal differences between cancer cells and normal cells, and show how these differences are related to the tendency of cancer cells to metastasize. (pp. 245–46)

12 Describe the evidence supporting the hypothesis that cancer originates as a result of several independent genetic events occurring in the same cell. (pp. 248–51)

13 Describe the relationship between oncogenes and proto-oncogenes and discuss the ways in which these genes might function to convert a normal cell into a cancer cell. (pp. 251–52)

14 Describe how exposure to radiation and carcinogenic chemicals might lead to cancer. Review the Ames test (pp. 228–29) and relate the mutagenicity of a chemical to its carcinogenicity. (pp. 252–54)

KEY TERMS

inducer (p. 234)
structural gene (p. 234)
regulator gene (p. 234)
repressor protein (p. 234)
operator (p. 234)
promoter (p. 234)
operon (p. 234)
corepressor (p. 236)
negative control (p. 236)
positive control (p. 236)
transcription factor (TF) (p. 236)
histone (p. 238)
nonhistone protein (p. 238)
chromatin (p. 238)
methylation (p. 238)
highly repetitive DNA (p. 239)
moderately repetitive DNA (p. 239)
polytene (p. 239)
nucleolus (p. 239)
pseudogene (p. 239)
euchromatin (p. 240)
heterochromatin (p. 240)
lampbrush chromosome (p. 241)
chromosomal puff (p. 241)
enhancer region (p. 242)
tumor (p. 244)
benign (p. 244)
malignant (p. 244)
carcinoma (p. 244)
sarcoma (p. 244)
leukemia (p. 244)
lymphoma (p. 244)
metastasis (p. 244)
translocation (p. 245)
oncogenes (p. 251)
proto-oncogenes (p. 251)
tumor-supressor genes (p. 252)

SUMMARY

Although every cell in the body of a multicellular organism has identical genetic information, individual cells have different structural and functional characteristics. Only a small percentage of the genetic material in cells is active at any one time. Various control mechanisms determine when and how each cell will act on its genetic instructions; these control mechanisms usually involve chemicals that bind directly or indirectly to DNA or mRNA.

Control of gene expression in bacteria In bacteria the Jacob-Monod model of gene induction proposes that three parts of the chromosome are involved in controlling transcription of the *structural genes:* the *regulator* gene, the *operator* region, and the *promoter* region. The regulator gene codes for the *repressor* protein. When the repressor binds to the operator, it blocks the promoter's binding sites for RNA polymerase and thus prevents transcription of the structural genes. In this system, the genes specifying particular enzymes are inactive until turned on by an inducer substance (usually a substrate).

In a *negative control* system such as the *lac* operon, a repressor protein binds to the operator and turns off transcription. When the repressor protein is inactive, the operator is turned on and transcription and translation automatically occur. In a *positive control* system, proteins called *transcription factors* bind to the promoter and activate transcription.

Control of gene expression in eucaryotes Control mechanisms in eucaryotes differ from those in bacteria. Eucaryotes have many more genes than procaryotes, and their DNA has introns that must be removed from the mRNA transcript. Also, the DNA of eucaryotic chromosomes is tightly wrapped around histone-protein cores to form *nucleosomes,* the basic packing unit of the chromosome. These nucleosomes must be unwound before transcription can begin.

About 10 percent of the eucaryotic genome contains base sequences that are repeated thousands of times. The function of much of this *highly repetitive DNA* is not well understood, but one class consists of long, tandemly repeated units that code for one of the ribosomal RNAs. Because a single copy of this sequence would be inadequate to meet the needs of the cell for ribosomal RNA, most eucaryotes have about 25,000 copies of it. Another 20 percent of the genome consists of *moderately repetitive DNA,* each base sequence is repeated hundreds of times. The genes for the other three ribosomal RNAs are not only tandemly repeated hundreds of times, but the portion of the chromosome in which they are situated can be replicated independently of the rest of the chromosome to form a polytene region, which makes up the nucleolus. The remaining 70 percent of the genome consists of single-copy sequences. Most chromosomal eucaryotic DNA is never transcribed and only about 1 percent of the DNA codes for mRNA that is translated. In contrast, 90 percent of the DNA of procaryotic cells is translated.

Control of gene transcription in eucaryotic cells Gene activation in eucaryotes involves the unwinding of particular sections of the nucleosomes, followed by activation of a specific gene or groups of genes in the unwound regions. Gene activity can be visually observed.

When stained, eucaryotic chromosomes can be differentiated into *euchromatic* and *heterochromatic* regions. The euchromatic regions contain active genes; heterochromatic regions are inactive. *Lampbrush chromosomes* and *chromosomal puffs* are further visible evidences of gene activity. In both, the DNA of active regions of the chromosomes loops out laterally from the main chromosome and exposes the maximum surface for transcription. Lampbrush chromosomes show feathery projections where genes are being repeatedly transcribed. Chromosomal puffs on polytene chromosomes consist of hundreds of parallel copies of the chromosomal DNA. The pattern of chromosomal puffs varies with the developmental stage and with changes in the extranuclear environment, a feature which demonstrates that changes in the cytoplasm can alter gene activity.

The degree of methylation of cytosines appears to play a role in determining whether genes are active or inactive: active mammalian genes have fewer methylated cytosines than inactive genes.

Eucaryotes exhibit primarily positive control: transcription does not occur without aid from a transcription factor that helps the RNA polymerase bind to the promoter region. Control proteins that bind to distant *enhancer regions* also influence transcription by helping to load the transcription factors onto the promoter, helping the polymerase to bind, or, in some cases, inhibiting polymerase binding.

Posttranscriptional control There are numerous points other than transcription where control can be exerted on the flow of information. In some cases a gene's expression can be controlled at every step from before transcription until after translation. Primary transcripts may be edited to yield slightly different proteins. Almost half the mRNA molecules never get to the cytoplasm, and the mRNA that does may be prevented from being translated by inhibitor substances that block either ribosomal binding or complete translation. Control may also be exerted by regulation of the rate at which mRNA is broken down. Finally, the activity of the enzymes created by translation is frequently regulated y activators or inhibitors.

Cancer: A failure of normal cellular controls The single most distinctive feature of cancer cells is their unrestrained proliferation, which often results in invasion of surrounding tissue and the *metastasis,* or spread, of cells from the original site of growth to other sites. To metastasize, cells must break away from tumor mass, burrow through the wall of a blood vessel or lymphatic vessel, travel to a distant site, move through the vessel wall, establish themselves in a new area, stimulate blood vessel formation, and proliferate to form a new tumor.

Cancer cells have a high rate of mutation and frequently show specific chromosomal rearragements such as *translocations,* deletions, and minute chromosomes. The cells are more spherical and more mobile than normal cells, and do not show normal contact inhibition of cell division or fixed-number-of-division control.

Cancer cells have an abnormal cell surface which prevents them from recognizing other cells of their own tissue type as normal cells do. The body's immune system probably destroys most new cancer cells as fast as they arise.

Several independent genetic events are required to induce a cell to become cancerous. Thus, the development of cancer takes many years, dating from the time the first mutational event took place.

Oncogenes are genes that cause cancer. Sometimes the oncogene is brought into the cell by a virus. Other oncogenes are certain normal cellular genes *(proto-oncogenes)* that have undergone a genetic change to become oncogenes. More than one active oncogene is necessary to transform a normal cell into a malignant cell.

CONCEPT MAPS

Map 12.1

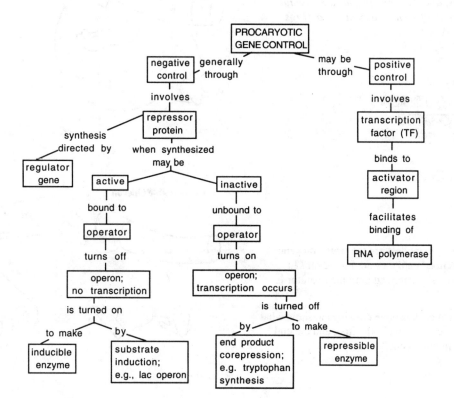

90 • CHAPTER 12

Map 12.2

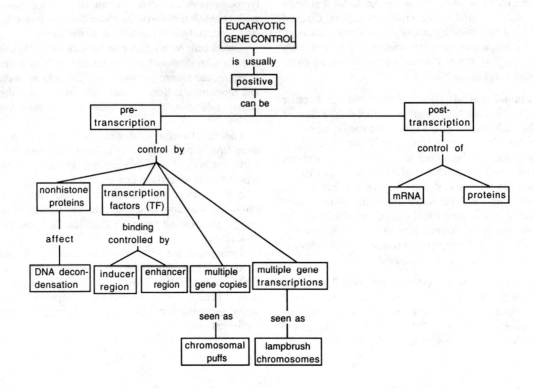

KEY DIAGRAM
The following coloring exercise is designed to help you learn the similarities and differences in the structure of an inducible operon, such as the lac operon, and a repressible operon. First color in the names, and then the appropriate structure, noting the similarities and differences as you do so. (Figs. 12.3A and 12.4A, p. 235)

REGULATOR GENE
OPERATOR GENE
PROMOTER SEQUENCE
STRUCTURAL GENES
RNA POLYMERASE
mRNA
REPRESSOR PROTEIN
RIBOSOMES

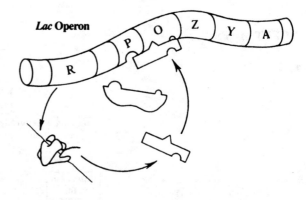

Lac Operon

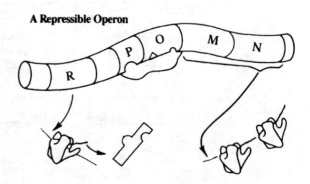

A Repressible Operon

Ask yourself

1. In which one of the two models are the structural genes automatically transcribed until they are turned off? In which are they normally repressed until the operator is turned on?

2. Can you redraw each of the above drawings to show what happens when lactose is present in the first, and the end product is present in the second?

QUESTIONS

Testing recall

For each of the processes listed in questions 1 to 6, write yes if it is associated with increased gene transcription and no if it is not.

1. binding of the inducer to the repressor (p. 234)
2. binding of the RNA polymerase to the promoter (p. 234)
3. binding of a transcription factor to an activator region (p. 236)
4. puffing of the chromosomes (p. 241)
5. formation of lampbrush chromosomes (pp. 240–41)
6. binding of the repressor to the operator (p. 236)

For each of the items listed in questions 7 to 14, write C if it is more characteristic of cancer cells and N if it is more characteristic of normal cells.

7. extra chromosomes (p. 245)
8. anchorage dependence (pp. 245–46)
9. unrestrained proliferation (p. 244)
10. fewer glycoproteins and glycolipids on cell surface (p. 246)
11. recognition of other cells of their own tissue type (p. 246)
12. spherical shape (p. 245)
13. contact inhibition of growth and cell division (p. 245)
14. high mobility (p. 246)

Testing knowledge and understanding

Choose the one best answer.

The next four questions refer to the diagram of the lac *operon from* E. coli, *shown below. Each letter may be used more than once.*

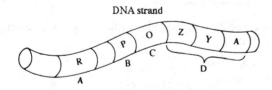

15. Which letter indicates the binding site for the repressor protein? (p. 235)
16. Which letter indicates the binding site for RNA polymerase? (p. 235)
17. Which letter indicates the structural genes? (p. 235)
18. Which letter indicates the gene that codes for the repressor protein? (p. 235)

19. In the lactose operon system in *E. coli*, the repressor is

    a a product of a structural gene locus.
    b bound to the promoter sequence.
    c lactose.
    d a protein.
    e a short length of DNA. (pp. 234–36)

20. Which of the following statements concerning the regulator gene (R) associated with the *lac* operon is correct?

    a mRNA is transcribed from the R gene whether lactose is present or not.
    b mRNA is transcribed from the R gene only when lactose is present.
    c mRNA is transcribed from the R gene only when lactose is not present.
    d Lactose inhibits the translation of R gene mRNA.
    e Lactose binds to the promoter of the lac operon. (pp. 234–36)

21. According to the Jacob-Monod (*lac* operon) model of gene regulation, inducer substances in bacterial cells probably combine with

    a operator regions and activate the associated operons.
    b structural genes and stimulate them to synthesize messenger RNA.
    c repressor proteins and inactivate them.
    d promoter regions and activate RNA polymerase.
    e nucleoli and trigger the production of more ribosomes. (p. 234)

22. The promoter region of a bacterial operon

    a codes for repressor proteins.
    b codes for inducer substances.
    c codes for corepressor substances.
    d is a binding site for inducers.
    e is a binding site for RNA polymerase. (p. 234)

23. The regulator gene of a bacterial operon

    a codes for inducer substances.
    b codes for repressor proteins.
    c acts as an on-off switch for the structural genes.
    d is a binding site for RNA polymerase.
    e is a binding site for inducers. (pp. 234–36)

24. The sugar lactose induces synthesis of the enzyme lactase. What happens when an *E. coli* (bacterial) cell runs out of lactose?

    a Repressor protein binds to the operator.
    b Repressor protein binds to the promoter.
    c RNA polymerase attaches to the promoter.
    d RNA polymerase attaches to the repressor. (pp. 234–36)

25 The *lac* operon is an example of
   a translational control.
   b posttranscriptional control.
   c replicational control.
   d transcriptional control.
   e positive control. (p. 236)

26 In the *lac* operon, RNA polymerase
   a binds to the promoter.
   b binds to the operator.
   c binds to the regulator gene.
   d is synthesized by the regulator gene.
   e binds to the regulator protein. (p. 235)

27 In bacteria, the structural genes can be turned off when the
   a end product of a reaction combines with the repressor protein and activates it.
   b substrate combines with a repressor protein and inactivates it.
   c substrate combines with the promoter.
   d substrate combines with its enzyme and produces a repressor molecule.
   e transcription factor binds to the promoter. (p. 235)

28 In the tryptophan operon of *E. coli*, the end product of the biochemical pathway, tryptophan, binds to the repressor protein, which then binds to the
   a promoter to inhibit transcription.
   b promoter to accelerate transcription.
   c operator to inhibit transcription.
   d operator to accelerate transcription.
   e repressor gene to accelerate transcription. (p. 235)

29 Nucleosomes
   a are composed of DNA and histones.
   b are small particles found in only the nuclei of plant cells.
   c disappear during transcription.
   d play a role in coiling and uncoiling the chromosomes.
   e are or do two of the above. (p. 238)

30 All the following are *true* about transcriptional control in eucaryotes *except* that
   a most control mechanisms involve activation of transcription rather than inhibition.
   b much of the mRNA transcript must be excised before translation.
   c the DNA wound around the histone-protein core must be unraveled before transcription can take place.
   d some base sequences are repeated thousands of times to form highly repetitive DNA.
   e almost 90 percent of the genome is transcribed and translated. (pp. 242–43)

31 Forms of posttranscriptional control in eucaryotes include all the following *except*
   a methylation of cytosine.
   b excision of introns.
   c movement of mRNA into the cytoplasm.
   d attachment of mRNA to ribosomes.
   e longevity of proteins. (pp. 243–44)

32 If an insect is given the hormone ecdysone, certain regions of the insect's chromosomes will soon exhibit puffing. What important process is taking place in the puffs?
   a mRNA synthesis
   b rRNA synthesis
   c DNA replication
   d enzyme synthesis
   e histone synthesis (pp. 240–41)

33 Regions of active gene transcription are found in
   a lampbrush chromosomes.
   b heterochromatic regions.
   c pseudogenes.
   d Barr bodies.
   e polyribosomes. (pp. 240–41)

34 All the following changes are involved in converting a normal cell into a cancer cell *except*
   a loss or reduction of contact inhibition.
   b loss of fixed-number-of-divisions control.
   c loss of anchorage dependence.
   d decrease in tissue-type affinity.
   e loss of vascularization. (pp. 244–46)

35 Cancer cells differ from normal cells in that cancer cells
   a have fixed-number-of-divisions control.
   b frequently possess extra chromosomes.
   c are anchorage-dependent.
   d show contact inhibition. (pp. 244–46)

36 Genes that cause cancer are referred to as
   a oncogenes.   d operator genes.
   b pseudogenes.   e promoter genes.
   c structural genes. (p. 251)

37 All the following may play a role in triggering cancer *except*
   a mutations.
   b viruses.
   c chromosomal puffing.
   d carcinogenic chemicals.
   e proto-oncogenes. (pp. 248–54)

For further thought

1. In what ways do aberrations of the cell surface contribute to the properties of cancer cells?

2. Describe the Jacob-Monod operon model for substrate induction. Include in your answer the role of the inducer, operator, regulator, promoter, repressor protein, and structural genes. Explain how end-product corepression differs from substrate induction.

3. Some states are trying to create electricity rate structures that reflect the environmental costs of different fuels used to create electricity. The carcinogenicity of the air pollutants given off by each type of fuel is an important part of the environmental cost. Explain why quantifying this cost is a difficult undertaking.

ANSWERS

Testing recall

1	yes	5	yes	9	C	13	N
2	yes	6	no	10	C	14	C
3	yes	7	C	11	N		
4	yes	8	N	12	C		

Testing knowledge and understanding

15	C	21	c	27	a	33	a
16	B	22	e	28	c	34	e
17	D	23	b	29	e	35	b
18	A	24	a	30	e	36	a
19	d	25	d	31	a	37	c
20	a	26	a	32	a		

Chapter 13

RECOMBINANT DNA TECHNOLOGY

A GENERAL GUIDE TO THE READING

Chapter 13 describes the important features of recombinant DNA technology. As you read this chapter, you will want to focus on the following topics.

1. Restriction enzymes. The discovery of these enzymes made the development of recombinant DNA technology possible. Figure 13.3 (p. 259) is useful for understanding how they work.

2. Vectors. Figures 13.5 (p. 261) and 13.6 (p. 262) explain how plasmids and bacteriophage can serve as vehicles for transporting foreign DNA into a host cell.

3. The identification of cells carrying the desired gene. The problems of identifying host cells carrying a gene of interest are discussed on pages 262 to 264. Studying Figures 13.7 (p. 263) and 13.8 (p. 263) will help you to understand two widely used identification techniques.

4. Complementary DNA (cDNA). In studying this section (pp. 264–65), focus on how the cDNA technique differs from the shotgun approach described earlier in the chapter and on why it is frequently preferable.

5. Transferring DNA to eucaryotic cells. It is important to realize that the vectors used to transfer genes into bacteria cannot normally enter eucaryotic cells. Study pages 265 to 268 to learn how gene transfer to plant and animal cells is accomplished.

6. Polymerase Chain Reaction (PCR). This new and powerful technique for amplifying minute amounts of DNA has many applications, such as "DNA fingerprinting," described in the Exploring Further (p. 269)

7. Applications of recombinant DNA technology. Agricultural and medical uses for this technology are growing rapidly. You will no doubt see reports in magazines and newspapers of exciting new advances.

KEY CONCEPTS

1. The aim of recombinant DNA technology is to impart some new characteristic or function to an organism by transferring the DNA of a different organism into its cells. (pp. 257–58)

2. Four basic steps are involved in recombinant DNA technology: (1) DNA from the donor organism is cut into small, manageable pieces; (2) the DNA fragments are joined to a vector, or transporting agent, that carries the donor DNA into the host cell; (3) the host cells and their vectors are allowed to multiply to produce large numbers of descendants; and (4) the descendants of these cells are screened for the gene of interest. (pp. 258–64)

3. Functional mRNA coding for a desired gene product can be used to make through reverse transcription complementary DNA (cDNA), which can then be used to clone the gene. (pp. 264–65)

4. Special procedures must be used to get donor DNA into eucaryotic cells. (pp. 265–68)

5. The polymerase chain reaction (PCR) can be used to amplify sequences of DNA. (p. 268)

4 Gene cloning makes it possible to use host cells as chemical factories to produce substances of medical importance, and also to study the sequencing and activity of genes from eucaryotic cells. Many other agricultural and medical applications of recombinant DNA technology are being developed. (pp. 271–74)

CROSSLINKING CONCEPTS

1 Transformation, the process by which recombinant plasmids are introduced into host bacterial cells, was first discovered through Griffith's experiments on pneumococci, described in Chapter 10.

2 The differences in the structure and control mechanisms of eucaryotic and procaryotic genes were described in Chapters 11 and 12. This chapter shows how these differences have resulted in problems getting eucaryotic genes to be expressed in bacterial cells.

3 The action of reverse transcriptase, which catalyzes the synthesis of cDNA from RNA, represents an exception to the rule that the normal flow of information in a cell is from DNA to RNA to protein, as described in Chapters 1, 10, and 11. This enzyme is unique to retroviruses, whose reproduction will be discussed in Chapter 37.

OBJECTIVES

1 Describe the objective of recombinant DNA technology and outline the four basic steps of the process. (pp. 257–58)

2 Describe the normal function of restriction enzymes in bacteria and their role in recombinant DNA procedures. In doing so, explain the terms palindrome and "sticky ends." (pp. 258–59)

3 Using Figure 13.5 (p. 261), explain what plasmids are and how they are used to produce recombinant DNA molecules. Then describe how the recombinant plasmids can be inserted into a host cell. Include the terms replication origin, vector, ligase, and transformation in your answer. (pp. 259–60)

4 Using Figure 13.6 (p. 262), differentiate transduction from transformation and explain the role of transduction in recombinant DNA technology. (pp. 261–62)

5 Define a gene library (clone bank), and using Figures 13.7 and 13.8 (p. 263), describe two methods of identifying a clone of bacterial cells carrying a desired foreign gene. (pp. 262–64)

6 Explain how complementary DNA (cDNA) is produced and how it is used in recombinant DNA technology. (pp. 264–65)

7 Describe four techniques that have been developed to transfer donor genes into eucaryotic cells. (pp. 265–68)

8 Describe the polymerase chain reaction (PCR) procedure and list three of its possible applications. (p. 268)

9 Explain how methods developed as spin-offs of recombinant DNA technology have been used for mapping genes. Describe the purpose of the Human Genome Project (pp. 268–71)

10 List five actual or potential agricultural and medical applications of recombinant DNA procedures. (pp. 271–74)

KEY TERMS

recombinant DNA technology (p. 257)
vector (p. 258)
restriction enzymes (p. 258)
"sticky ends" (p. 258)
plasmid (p. 259)
replication origin (p. 259)
ligase (p. 260)
transformation (p. 260)
bacteriophage (p. 261)
transduction (p. 261)
clone (p. 262)
gene library, or clone bank (p. 262)
probe (p. 263)
antibodies (p. 263)
reverse transcriptase (p. 264)
complementary DNA, or cDNA (p. 264)
interferon (p. 265)
polymerase chain reaction, or PCR (p. 268)
Human Genome Project (p. 271)
antisense compounds (p. 274)

SUMMARY

Recombinant DNA technology is the process by which DNA from one organism is isolated and transferred into the cells of another in order to impart some new characteristic or function to the host organism.

First, DNA from the donor organism is cut into small manageable pieces using *restriction enzymes.* Next, the DNA pieces are joined to a *vector,* which is most often a *plasmid*—a small, circular DNA molecule that replicates autonomously in the cytoplasm of bacteria. The foreign DNA pieces join to the plasmid DNA by complementary base pairing between their "sticky ends." After sealing with a *ligase,* the modified plasmids can be picked up by bacterial cells, which thereby acquire the foreign genes. The last step involves selecting the cells that have acquired the desired gene, using a radioactive probe or specific antibody. The selected clone is stimulated to reproduce repeatedly, so it produces many copies of the donor gene. The aim is to get the donor gene to be transcribed and translated in its new host cell.

Bacteriophage can also be used as vectors to transport genes into a bacterial cell. The donor gene is first added to the viral DNA, using restriction enzymes and ligases, and phage proteins are then added to complete the bacterial viruses. The bacteriophages attack bacteria and inject the viral DNA and the donor DNA into the cell.

An alternative procedure is to isolate functional mRNA from cells specialized to produce the desired product. *Reverse transcriptase* is used to synthesize a cDNA copy of the mRNA. DNA polymerases then replicate the cDNA strand, the new gene is inserted into a plasmid, and the plasmid is introduced into a bacterial host. The transformed bacterial cells grow and divide, and produce large numbers of bacteria that may synthesize the desired product.

Many different techniques are used to get DNA into eucaryotic cells. Two common methods are microinjection, shooting tiny DNA-coated tungsten pellets into the cells and disrupting the cellular membrane, and the use of viruses as delivery systems.

The *polymerase chain reaction* can be used to make millions of copies of DNA sequence to facilitate analysis.

The substantial technology that has grown up around recombinant DNA makes it possible to use bacterial cells as chemical factories to produce substances of commercial, agricultural, and medical importance: besides being used to introduce new genes into plants and animals, these techniques are also being used to map human chromosomes, and may soon be used for human gene therapy.

Recombinant DNA technology carries with it some risks. The public has questioned the safety of the procedures, and many individuals, scientists included, are concerned that some of these genetically engineered organisms may pose a threat to the environment.

QUESTIONS

Testing recall

Decide whether each of the following questions is true or false; if false, correct the statement.

1 During transduction, a plasmid is passed from one bacterium to another. (p. 261)

2 Ligases are enzymes used to cut DNA at specific sites, and so create sticky ends. (p. 260)

3 Gene cloning, using radioactive cDNA, can be used to map genes. (pp. 264–65)

4 The enzyme reverse transcriptase is used to catalyze RNA replication. (p. 265)

Testing knowledge and understanding

Choose the one best answer.

5 A plasmid is

   a a fragment of a bacterial chromosome.
   b a portion of a bacteriophage.
   c a self-replicating nonchromosomal circle of DNA.
   d a virus that infects bacteria.
   e formed from the DNA of a lysogenic virus.   (p. 259)

6 The process by which a virus transfers genes from one bacterial cell to another is called

   a transformation.
   b translation.
   c transcription.
   d transduction
   e translocation   (p. 261)

7 Restriction endonucleases are useful in recombinant DNA technique because they

   a cut DNA at specific sites.
   b restrict the number of nucleotides that can be removed at one time.
   c restore the bonds in the DNA backbone.
   d synthesize cDNA from mRNA.
   e can be used to locate genes for mapping.   (p. 258)

8 A genetic engineer prepares DNA fragments from two species and mixes them together. Two of the many fragments are shown below. Which one of the following statements is correct?

                           TTCC
              ATGC

   a No sticky ends were produced.
   b The two fragments shown above will join by complementary base pairing.
   c The two fragments were prepared by two different restriction endonucleases.
   d A single restriction endonuclease was used to cut at different locations in the two types of DNA.
   (pp. 258–59)

9 One problem with using the artificial transformation technique to insert genes from one organism into cells of another kind of organism such as bacteria is that

   a the two organisms often use different genetic codes.
   b the foreign genes generally contain introns that bacteria are unable to remove.
   c restriction endonucleases do not cleave plasmid DNA in the same way they cleave eucaryotic DNA.
   d ligases often remove bases at one end of the gene.
   (p. 264)

10 If you were to arrange the following steps of gene cloning in order, which step would be third?

   a A cDNA copy of mRNA is made using reverse transcriptase and the DNA strand is replicated.
   b The plasmid is inserted into the bacterial host cell.
   c mRNA molecules coding for the desired product are isolated.
   d Restriction endonucleases are used to cut DNA and plasmid, and so produce sticky ends.
   e The plasmid DNA and the cloned gene are joined and sealed by ligase. (pp. 258–65)

11 Suppose a pharmaceutical company planned to clone the gene for human phosphoribosyl transferase, an important enzyme. Which of the following steps would *not* be involved in this process?

   a joining of sticky ends by the enzyme ligase
   b breaking human DNA by a restriction endonuclease
   c breaking human DNA into individual nucleotides
   d transforming an E. coli with recombinant plasmid/human DNA
   e joining human DNA to a vector such as a plasmid (pp. 258–62)

12 The process by which a virus transports genes into a bacterial cell is

   a transformation   *d* transcription
   b translocation   *e* transduction
   c translation   (p. 261)

For further thought

1 In recombinant DNA technology, why is it necessary to attach a foreign gene to a plasmid? Why not just introduce the foreign gene directly into the bacterium?

2 Discuss some of the benefits and potential dangers of recombinant DNA technology.

3 Assume that you are working for a large pharmaceutical company in their research division. Your goal is to produce interferon in large quantity by genetic engineering techniques. You are provided with a DNA fragment, containing the interferon gene, which had been isolated using restriction endonuclease A. You know that the interferon gene is flanked on both ends by recognition sites for B, another restriction endonuclease. An *E. coli* plasmid with a single B recognition site is available. What would you do?

4 Describe some possibilities in applying recombinant DNA technology to plants and animals should our environment undergo warming, drying, and/or increased salinity.

ANSWERS

Testing recall

1 false—a virus carries bacterial genes from one bacterium to another
2 false—restriction endonucleases cut the DNA; ligases join DNA
3 true
4 false—reverse transcriptase catalyzes the synthesis of cDNA from RNA

Testing knowledge and understanding

5 *c*
6 *d*
7 *a*
8 *c*
9 *b*
10 *d*
11 *c*
12 *e*

Chapter 14

INHERITANCE

A GENERAL GUIDE TO THE READING

This chapter is concerned with the transmission of genetic information from generation to generation. Both the chapter in the text and the corresponding material in the *Study Guide* include a series of genetics problems that will give you an opportunity to become proficient in predicting genetic inheritance. You will want to do all of these problems, since the best way to gain an understanding of genetics is to work with it. Suggestions on how to approach such problems effectively are presented in "Study Guide for Genetics Problems," on pages 102 to 105 of this *Study Guide*, and answers to the genetics problems in the text are on pages 105 to 107. You will want to leave extra time for this chapter because of its length, the complexity of the information, and the problems. As you read Chapter 14 in your text, you will find it useful to concentrate on the following topics.

1. Mendel's work. You will want to read the material on Mendel carefully (pp. 277–79), since it provided the foundation for our present understanding of heredity.

2. Monohybrid and dihybrid crosses. Study the material on pages 279 to 284 carefully, since you will need to know it well in order to do the genetics problems at the end of the chapter.

3. Quantitative inheritance and modifier genes. These sections (pp. 284–86) discuss how genes interact with one another in various ways. The principle of quantitative (polygenic) inheritance is particularly important for understanding many traits of human beings.

4. Blood types. You will want to master this material (pp. 286–88), which illustrates the more general topic of multiple alleles. Table 14.2 (p. 288) is helpful.

5. Mutation. You will gain an understanding of the significance of mutations (first introduced in Chapter 10) to inheritance and evolution.

6. Sex and inheritance. Students often have difficulty with this topic. If you remember that "sex-linked" means "X-linked"—carried on the X chromosome—you will find it easier to understand sex-*linked* characters, as defined on page 295.

7. Linkage. Note that this concept underlies chromosomal mapping work, which is important in the much-publicized human genome project.

8. Chromosomal alterations. The section on this interesting topic (pp. 301–303) includes a box discussing human syndromes that result from one of these alterations.

KEY CONCEPTS

1. Somatic cells have pairs of homologous chromosomes; each pair consists of one chromosome from each parent. Each gene is found in two copies, one on each chromosome of the homologous pair, at corresponding loci (p. 279).

2. In meiosis, each gamete receives a copy of only one chromosome from each homologous pair, and hence only one of the two alleles (p. 279).

3. Genes do not alter one another; they remain distinct and segregate unchanged when meiosis occurs (Mendel's Law of Segregation) (p. 279).

4   When two or more pairs of genes located on different chromosomes are involved in a cross, the members of one pair are inherited independently of the other (p. 283).

5   Characteristics are often determined by many genes acting together (p. 284).

6   The expression of a gene depends both on the other genes present and on the physical environment. All organisms are products of their inheritance and their environment (p. 285).

7   A gene may exist in any number of allelic forms in a population, but a given individual can possess no more than two alleles because he or she possesses only two chromosomes (p. 286).

8   Mutational changes in genes occur at random; most of the mutations that have a phenotypic effect are deleterious (p. 288).

9   Natural selection can act against a deleterious allele only if it causes some change in the organism's phenotype. Deleterious alleles that are recessive may be retained in the population for a long time (p. 292).

10 Some alleles that are harmful when the organism is homozygous for them are beneficial when it is heterozygous (p. 293).

11 The inheritance patterns for characteristics controlled by genes on the X chromosomes are quite different from those for characteristics controlled by autosomal genes (p. 295).

12 Genes located on the same chromosome are linked together; they ordinarily remain together during meiosis. However, crossing over can occur between homologous chromosomes during synapsis, so that new linkages result. One way to remember is that generally genes are carried on different chromosomes and are therefore *independently assorted*. The exception to independent assortment is *linkage*. The exception to linkage is *crossing over* (p. 299).

13 Structural alterations to chromosomes sometimes occur, with a resulting rearrangement of the genetic material; also changes in chromosomal number may occur when the separation of chromosomes in cell division does not proceed normally (p. 301).

CROSSLINKING CONCEPTS

1   The text puts an evolutionary spin on the study of inheritance. As noted in Chapter 9 and this chapter, the independent assortment of chromosomes and crossing over increase the variety of genetic combinations a cross can produce. Natural selection acts on this variation to bring about evolution. Recall from your study of DNA (Chapter 10) how mutations serve as additional sources of variation. This conceptual base will be important when you study not just inheritance among *individuals* over generations, but the genetics of whole *populations* over generations (Chapter 31).

2   Mendel's work is an excellent example of the scientific method first introduced in Chapter 1.

3   An understanding of proteins and enzymes (Chapters 3 and 4) and the deleterious effects of mutations (Chapter 10) provides insight into the relationship between dominant alleles that produce functional structural protein or active enzyme, or recessive alleles that produce nonfunctional structural protein or inactive enzymes.

4   An understanding of meiosis (Chapter 9), particularly synapsis, is crucial for an understanding of independent assortment and crossing over.

5   An understanding of the genetics of blood type will contribute to your later study of the circulatory and immune systems (Chapters 20 and 21).

OBJECTIVES

1   Discuss Mendel's conclusions and relate them to the chromosomal theory of inheritance. Be sure to describe segregation and independent assortment (pp. 277–84).

2   Differentiate between the two terms in each of the following pairs: dominant–recessive, allele–gene, F_1–F_2, homozygous–heterozygous, phenotype–genotype, and monohybrid–dihybrid (pp. 277–84).

3   Use a Punnett square to find the expected genotypic ratios in a cross involving a single character (monohybrid). Compare the results when one allele is dominant over another with the results when there is incomplete dominance. [Problems 1 to 6 in your text (p. 304) cover this material.] (p. 282)

4   Explain how incomplete dominance differs from complete dominance (p. 282).

5   Use a Punnett square or the multiplication of probabilities (as shown on p. 284) to find the characteristic phenotypic ratios in the F_2 of a cross involving two characters (dihybrid) in which the two genes are independent. [Problems 7 to 12 in your text (p. 305) cover this material.] (p. 283)

6   Explain how human height and other traits that show continuous variation within a population are inherited. (p. 285)

7   Using the symbols shown in Table 14.2 (p. 288), list the possible genotypes of persons of each blood type (A, B, AB, and O) and explain the mode of inheritance. [Problems 13 to 15 in your text (p. 305) cover this material.] (p. 286)

8   Explain why most mutations are deleterious and how it is possible for a harmful recessive allele, even a lethal one,

to persist in a population. Explain why difficulties may arise from inbreeding (p. 288–93).

9 Differentiate between the terms sex chromosome and autosome, and explain how sex is determined genetically in human beings. Then discuss the pattern of inheritance of sex-linked characters and why they are expressed more often in males than in females (pp. 295–97).

10 Use a Punnett square or the multiplication of probabilities to do problems 16 to 18 (p. 305) in your text (p. 296).

11 Explain the concept of linkage. Show how crossover frequencies are calculated and how they can be used to make chromosomal maps (pp. 299–300).

12 Explain the process of nondisjunction and show how it affects the chromosomal composition of a cell. Give examples of trisomy and polyploidy (pp. 301–303).

13 Do all the problems on pages 304 to 306 of your text. Answers may be found on pages 105 to 106 of this *Study Guide*.

KEY TERMS

F_1 (p. 277)
P (p. 277)
F (p. 277)
dominant character (p. 278)
recessive character (p. 278)
Law of Segregation (p. 279)
locus (p. 279)
allele (p. 279)
homozygous (p. 279)
heterozygous (p. 279)
genotype (p. 280)
phenotype (p. 280)
monohybrid cross (p. 280)
test cross (p. 281)
incomplete dominance (p. 282)
codominance (p. 282)
dihybrid cross (p. 283)
independent (p. 283)
Principle of Independent Assortment (p. 284)
quantitative inheritance (p. 285)
modifier gene (p. 285)
lethal (p. 292)
sickle-cell anemia (p. 293)
pleiotropic (p. 294)
sex chromosome (p. 294)
autosome (p. 294)
X chromosome (p. 294)

Y chromosome (p. 294)
sex-linked character (p. 295)
reciprocal cross (p. 295)
Barr body (p. 295)
holandric (p. 297)
linkage (p. 299)
parentals (p. 299)
recombinants (p. 299)
chromosomal mapping (p. 300)
translocation (p. 301)
duplication (p. 301)
deletion (p. 301)
nondisjunction (p. 302)
trisomy (p. 302)
polyploidy (p. 304)

SUMMARY

Monohybrid inheritance From his breeding experiments on the garden pea the Austrian monk Gregor Mendel concluded that each pea plant possesses two hereditary factors for each character and that when gametes are formed, the two factors segregate into separate gametes (*Law of Segregation*). Each new plant thus receives one factor for each character from each parent. The hereditary factors exist as distinct entities within the cell; they do not blend or alter each other, and they segregate unchanged when gametes are formed. This theory is now supported by the discovery of chromosomal segregation during meiosis.

Mendel was working with two different forms (*alleles*) of the gene for flower color. When both alleles were present, the gene for red flowers was expressed (*dominant*) while the gene for white was masked (*recessive*). The dominant allele is customarily represented by an uppercase letter, the recessive by a lowercase letter. A diploid cell may be *homozygous* (have two of the same allele: *C/C, c/c*) or *heterozygous* (have one each of two different alleles: *C/c*). The term *genotype* refers to the allelic composition of an organism, *phenotype* to its expressed traits. An easy way to figure out the possible genotypes produced in a cross is to construct a Punnett square.

Extensive investigation has demonstrated that Mendel's results have general validity. Whenever a monohybrid cross is made between two contrasting homozygous individuals, the expected genotype ratio in the second generation of offspring (the F_2) is 1 : 2 : 1. When dominance is involved, the expected phenotypic ratio is 3 : 1.

In a test cross, the unknown is crossed with a homozygous recessive individual. If all the progeny show the dominant phenotype, the unknown genotype is probably homozygous dominant; if any of the progeny show the recessive genotype, the unknown is heterozygous.

One allele is not always completely dominant over the other; in some cases heterozygous individuals show the effects of both alleles and are different from both homozygotea. If the heterozygote's phenotype is intermediate between the parents it is *incomplete* or *partial dominance*; if both alleles are expressed independently in the heterozygote it is *codominance*. At the molecular level, most alleles probably show incomplete or codominance.

Dihybrid and polygenic inheritance Mendel also made crosses between strains that differed distinctly in two characteristics (a *dihybrid* cross). In his experiments Mendel found that the offspring consistently conformed to a 9 : 3 : 3 : 1 phenotypic ratio. From these results Mendel formulated the *Principle of Independent Assortment*, which states that during gamete formation, when two or more genes are involved in a cross and are located on nonhomologous chromosomes, the alleles of one gene are inherited independently of the alleles for the other gene.

Probably no inherited characteristic is controlled by only one gene pair. In *polygenic inheritance* many genes interact in additive fashion to affect a particular characteristic. Even when only one principal gene is involved, other genes act as *modifiers* to influence its expression. The action of any gene can be fully understood only in terms of the overall genetic makeup of the individual organism. The expression of a gene depends on the other genes present and on the physical environment.

Multiple alleles Genes may exist in a number of allelic forms (*multiple alleles*). Each individual has only two alleles for a given trait, but other alleles may be present in the population. An example of multiple alleles in humans is the A-B-O blood series, which involves three alleles: I^A, I^B, and i antigen.

Mutations and deleterious genes A variety of influences can cause changes or *mutations* in the chemical structure of genes. Mutations occur constantly; most are deleterious. Natural selection can act against a deleterious gene only if it is expressed phenotypically. Harmful dominant genes can be eliminated rapidly by natural selection, but recessive genes are not expressed in the heterozygous state and thus cannot be easily eliminated.

An allele whose phenotype, when expressed, results in the death of the organism is called a *lethal*. Some alleles that are harmful when the organism is homozygous for them are beneficial when it is heterozygous. For example, individuals homozygous for the sickle-cell anemia gene suffer from this fatal disease. Heterozygous individuals suffer from a mild anemia, but they have much higher than normal resistance to malaria. Thus the gene is beneficial in areas where malaria is common. Genes like this, with more than one effect, are said to be *pleiotropic*.

Inbreeding is considered dangerous because it increases the chance of offspring who are homozygous for deleterious traits.

Sex and inheritance In most higher organisms where the sexes are separate, the chromosomal endowments of males and females are different. One chromosomal pair, the *sex chromosomes*, differs in size and shape and determines the sex of the individual. All other chromosomes are called *autosomes*. In *Drosophila* and in human beings the females have two large *X chromosomes* whereas the males have an X and a smaller *Y chromosome*. The egg cells produced by meiosis are alike in chromosomal content, but the sperm cells are of two different types, one bearing an X chromosome, and the other a Y. The sex is determined at the time of fertilization by the type of sperm fertilizing the egg.

The genes on the X chromosome are said to be *sex-linked*. Because females have two X chromosomes, they will always have two alleles for any sex-linked characteristic, whereas males will have only one (one on the X, none on the Y). Consequently, recessive sex-linked genes are always expressed phenotypically in the male but may be masked by the dominant allele in the female. In the male, all sex-linked characteristics are inherited from the mother. The Y chromosome has only a few genes; these genes are said to be *holandric*.

Although females have two X chromosomes in each cell, only one is active; the other coils up into a tiny dark *Barr body* whose genes are inactive.

Linkage Genes that are located on different chromosomes can segregate independently of one another during meiosis. Genes located on the same chromosome are *linked* together; they ordinarily remain together during meiosis. However, crossing over can occur between homologous chromosomes during synapsis in meiosis; this results in new linkages.

The frequency of crossing over between any two linked genes will be proportional to the distance between them. The percentage of crossing over can be used to map gene locations. By convention, one unit of map distance is the distance within which crossing over occurs 1 percent of the time. Crossing over is the classical test of whether two characters are controlled by one gene or by two separate genes.

Chromosomal alterations Crossing over is one kind of chromosomal rearrangement; other alterations occur as well. Some involve *translocations* in which a segment is transferred from one chromosome to another nonhomologous chromosome; others involve alterations within a single chromosome, such as *deletions* or *duplications* of chromosomal material.

Separation of the chromosomes during meiosis does not always occur normally; sometimes both members of one homologous pair move to the same pole. The result of this *nondisjunction* may be the production of a cell with an extra chromosome (*trisomy*). Occasionally all the chromosomes move to the same pole. If this cell then unites with another during fertilization, the resulting zygote has more than two sets of chromosomes and is said to be *polyploid*.

CONCEPT MAPS

Map 14.1

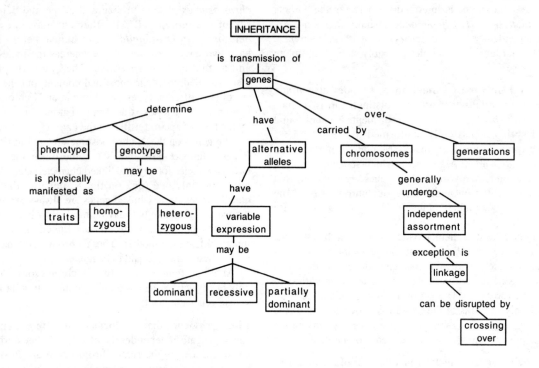

STUDY GUIDE FOR GENETICS PROBLEMS

The best way to gain an understanding of genetics is to work with it. The problems found in the study questions at the end of Chapter 14 in your text illustrate the various patterns of inheritance the chapter discusses. The following information is intended to get you started on the problems. Answers are provided in this section, but you are strongly advised to work the problems *first* and check your answers later. Genetics problems are considerably easier when you start with the answers and work backward, but you will not learn how to do problems that way.

General Rules

1. Know your terms: dominant (capital letter), recessive (small letter); P, F_1, F_2; homozygous, heterozygous; and genotype, phenotype. It is very important that you have a clear conceptual picture of what is happening in the genes on the chromosomes of the individuals being studied when you solve genetics problems.

2. Read the problem and assign appropriate symbols to the characteristics involved.

3. Set up a Punnett square correctly: female gametes go along the left side, male gametes across the top. Be sure to include all possible gametes.

	♂ gametes	
♀ gametes		

4. Know how to use the multiplication of probabilities instead of a Punnett square to solve problems. The basic principle is that the chance that a number of independent events will occur together equals the chance of the first event times the chance of each following event.

Example: The probability of getting heads when flipping a coin is 1/2 (or 0.5). If two pennies are flipped at the same time, the chance of getting two heads is the chance of the first's being heads (1/2) times the chance of the second's (1/2), or $1/2 \times 1/2 = 1/4$. The chance of getting five heads in a row is $1/2 \times 1/2 \times 1/2 \times 1/2 \times 1/2$, or 1/32. (See problem 12 in your text.)

Types of problems

1 **Monohybrid cross:** cross involving one character

   Example: If pole bean plants (P) are dominant over bush beans (p), what are the expected genotypes and phenotypes when two heterozygous individuals are mated?

	female	×	male
parents:	P/p		P/p
gametes:	P, p		P, p

   Punnett square:

	P	p
P	P/P	P/p
p	P/p	p/p

   genotypic ratio: 1 P/P : 2 P/p : 1 p/p
   phenotypic ratio: 3 pole : 1 bush

   Now try problems 1 to 5 in your text and 1 to 6 in the *Study Guide*.

2 **Test cross:** unknown organism crossed with a homozygous recessive

   Example: If the genotype of a pole bean plant is in doubt ($P/–$), it is crossed with a bush bean plant.

   parents: $P/– \times p/p$

   a If – is dominant (P) the cross is $P/P \times p/p$ and all offspring are pole.
   b If – is recessive (p) the cross is $P/p \times p/p$ and the offspring are half pole and half bush.

parents:	P/p	×	p/p
gametes:	P, p		p

   Punnett square:

	♂ p	p
♀ P	P/p	P/p
p	p/p	p/p

   genotypic ratio: 1 P/p : 1 p/p
   phenotypic ratio: 1 pole : 1 bush

3 **Dihybrid cross:** cross involving two characters

   Example: Red pole beans heterozygous for color and shape were crossed with each other. The genes for color (R = red, r = white) and for shape (P = pole, p = bush) are on different chromosomes.

   a Solving the problem using a Punnett square

	female		male
parents:	$R/r \; P/p$	×	$R/r \; P/p$
gametes:	RP, Rp, rP, rp	×	RP, Rp, rP, rp

   Punnett square:

	RP	Rp	rP	rp
RP	$R/R \; P/P$	$R/R \; P/p$		
Rp				
rP				
rp				

   Complete the Punnett square and determine the genotypic and phenotypic ratios. (Now try problems 7 to 10 in the *Study Guide*.)

   b Solving the problem using probabilities

   It is very laborious to make Punnett squares for two or more pairs of genes, so multiplication of probabilities is often used instead. A convenient and rapid way to do problems where you are asked to calculate all the genotypic and phenotypic ratios is to use probabilities and the fork-line method to do the crosses.

   According to Mendel's second law, the members of one pair of alleles are inherited independently of the members of another. Consequently, the probability of each of the various genotypic combinations in the offspring is equal to the product of the genotypic probabilities of each pair of alleles.

   To use the fork-line method, (i) calculate the genotypic probabilities of each pair of alleles separately, and then (ii) find all possible genotypic combinations and multiply their separate probabilities.

   Example: $R/r \; P/p \times R/r \; P/p$

   (i) Consider each pair of alleles separately.

   $Rr \times Rr$ gives you 1/4 R/R, 2/4 R/r, 1/4 r/r
   (3 possible genotypes)

   $P/p \times P/p$ gives you 1/4 P/P, 2/4 P/p, 1/4 p/p
   (3 possible genotypes)

   (3 × 3 = 9 total possible genotypic combinations)

   (ii) Now find all possible combinations and multiply their probabilities.

   Genotypes of offspring

   1/4 R/R ⟨ 1/4 P/P = 1/16 $R/R \; P/P$
            2/4 P/p = 2/16 $R/R \; P/p$
            1/4 p/p = 1/16 $R/R \; p/p$

   2/4 R/r ⟨ 1/4 P/P = 2/16 $R/r \; P/P$
            2/4 P/p = 4/16 $R/r \; P/p$
            1/4 p/p = 2/16 $R/r \; p/p$

   1/4 r/r ⟨ 1/4 P/P = 1/16 $r/r \; P/P$
            2/4 P/p = 2/16 $r/r \; P/p$
            1/4 p/p = 1/16 $r/r \; p/p$

   If you make a cross involving three pairs of genes, you can simply add another column of "forks." (Start with a big piece of paper!)

Probabilities can be used to solve many genetics problems that would be tedious to solve with a Punnett square. (See next example.)

Example: A heterozygous red pole bean plant with smooth seeds (*R/r P/p S/s*) is crossed with a red pole bean plant with wrinkled seeds (*R/r P/P s/s*). What fraction of the offspring will have the genotype *R/r P/P S/s*?

The chance of *R/r* from the cross *R/r* × *R/r* is 1/2.

The chance of *P/P* from the cross *P/p* × *P/P* is 1/2.

The chance of *s/s* from the cross *S/s* × *s/s* is 1/2.

The chance of the combination *R/r P/P s/s* is equal to the product of their separate probabilities; 1/2 × 1/2 × 1/2 = 1/8.

4 Sex linkage

Genes for sex-linked characteristics are carried on the X chromosome. Because females have two X chromosomes they always have two alleles for a characteristic on that chromosome, whereas males have only one. Consequently, recessive sex-linked traits show up more often in males than in females.

Sometimes students find it helpful to use special symbols for sex-linked problems; X^C can represent a dominant sex-linked gene and X^c the recessive. The Y chromosome is designated Y.

Example: The disease hemophilia is sex-linked and recessive. A normal man marries a woman who is a carrier (heterozygous) for this trait. What phenotypes will the children probably have?

Let X^H = allele for normal blood
     X^h = allele for hemophilia

The cross is ♀ X^H/X^h × ♂ X^H/Y.

Results: 1/2 daughters will be normal.

1/2 will be carriers.

1/2 sons will be normal.

1/2 sons will have hemophilia.

(Now try problems 16 to 17 in your text and 14 to 15 in the *Study Guide*.)

5 Pedigrees

Problems involving human pedigrees provide good practice, since they tie together many different aspects of inheritance. The following symbols are commonly used:

```
            normal    affected
   male      □          ■
 female      ○          ●
```

A horizontal line between two symbols indicates marriage, and the progeny are displayed below the parents:

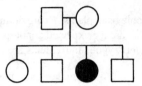

To do a pedigree problem, in which you are asked to determine the mode of inheritance of a given trait, first ascertain whether the trait in question is dominant or recessive. A dominant trait must occur in every generation, whereas a recessive trait can skip a generation. If you can find in the pedigree a family in which neither parent has the trait but one of the children does, you know the trait must be recessive. When a trait is identified as recessive, the next step is to decide whether it is autosomal or sex-linked, by analyzing the sex ratio of the affected individuals; remember that a recessive sex-linked trait is expressed more often in males than in females. Another clue is to look at the parents of the affected females. If a female offspring has the trait, it can be sex-linked only if the father expresses it. (For a female to express a sex-linked trait, she must get an allele from each parent; therefore the father would express the trait.) If a female has the trait and her father and mother do not have it, the trait is autosomal recessive. Hence, in the pedigree above, the trait in question is autosomal recessive. Now look at the following pedigree, in which the affected trait is myopia (nearsightedness), and determine the mode of inheritance.

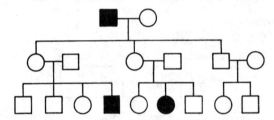

This trait too is autosomal recessive, as shown by the fact that a female has it even though neither parent expresses it. If the trait were sex-linked, her father would have expressed it.

(Now try problem 18 in your text and 16 to 18 in the *Study Guide*.)

6 Linkage

Genes are linked when they are on the same chromosome. In calculating crosses involving linked genes, special symbols are used. If the genes *A* and *B* are linked, the genotype of a heterozygous individual could be represented as *AB/ab*.

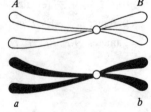

Linked genes are sometimes separated when homologous chromosomes change parts during synapsis in meiosis:

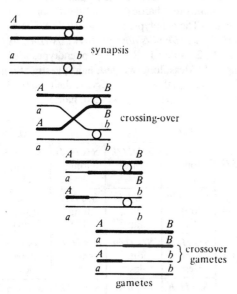

The normal and most common gametes are the non-crossovers (or "parentals"), *AB* and *ab*. The rare new gametes formed as a result of crossing over are the "recombinants"—*Ab* and *aB*. Linkage and crossing over can be detected by crossing with homozygous recessives:

AB/ab × *ab/ab*

	ab	# of offspring
Normal gamete *AB*	*Ab/ab*	45
Normal gamete *ab*	*ab/ab*	44
Crossover gamete *aB*	*aB/ab*	5
Crossover gamete *Ab*	*Ab/ab*	5

There are four phenotypes, but they are not present in equal proportions. (If *A* and *B* were not linked and were on different chromosomes, the ratios of the phenotypes would be 1 : 1 : 1 : 1.)

7 Chromosomal mapping

The distance between two genes can be calculated by means of the following procedure:

$$\frac{\text{number of crossover progeny}}{\text{total number of progeny}} \times 100 = \% \text{ crossing-over}$$

1% crossing-over = 1 map unit

From our linkage example above,

$$\frac{5+5}{100} \times 100 = 10\% = 10 \text{ map units}$$

(Now try problems 19 to 21 in your text and problems 22 to 26 in the *Study Guide*.)

Now do all the rest of the problems in your text (pp. 304–5). For additional practice, complete all the problems in the *Study Guide* (pp. 106–11).

Answers to genetics problems, in text (pp. 304–306)

1 *a* All heterozygous, white

	w	*w*
W	*W/w*	*W/w*
W	*W/w*	*W/w*

b Genotype ratio: *W/w* : *w/w* is 1 : 1
Phenotype ratio: 1 white : 1 yellow

	w	*w*
W	*W/w*	*W/w*
w	*w/w*	*w/w*

c Genotype ratio: 1 *W/W* : 2 *W/w* : 1 *w/w*
Phenotype ratio: 3 white : 1 yellow

	W	*w*
W	*W/W*	*W/w*
w	*W/w*	*w/w*

2 Let *B* be for brown eyes and *b* be for blue eyes. The man must be *b/b*, but the woman is *B/–*. However, since her father is *b/b*, she must have a *b* allele. Thus, she is *B/b*. The cross is *b/b* × *B/b*, and the Punnett square shows that 50 percent blue eyes are predicted.

	♂ *b*	*b*
♀ *B*	*B/b* brown	*B/b* brown
b	*b/b* blue	*b/b* blue

3 This time the man is *B/–*, and the woman is *b/b*. The cross is either *B/b* × *b/b* or *B/B* × *b/b*. The first cross would yield 50 percent blue-eyed children as in question 2, and the second cross would yield all brown-eyed children as shown in the figure. Note that every child must receive a dominant *B* allele from the father. In this case the father is probably homozygous, but one cannot be certain that this is true. Furthermore, an eleventh brown-eyed child makes no difference. It cannot be proved beyond a doubt that a character is homozygous dominant. In this case, if the man is heterozygous, there is still a 1 in 2^{11} chance of his having 11 consecutive brown-eyed children. It is also possible that some other factor may be prohibiting his *b* allele from being expressed.

	♂ *B*	*B*
♀ *b*	*B/b* brown	*B/b* brown
b	*B/b* brown	*B/b* brown

4 In this problem the brown-eyed man is *B/b* because his mother was *b/b*. His father was either *B/B* or *B/b*. His wife is *b/b*, so here brown-eyed parents must both have been heterozygous, *B/b*. The blue-eyed son is, of course, *b/b*.

5 There are probably a number of explanations, but it is best to assume only one genetic locus at first. This is

probably true since only tail length is involved. The ratio is 3 : 6 : 2, which by inspection is similar to 1 : 2 : 1. This ratio appears as the expected genotype ratio of a monohybrid cross. This would mean that the parental short-tailed cats are heterozygous, T^1/T^2. Assuming then that T^1/T^1 yields a long tail and T^2/T^2 yields no tail, the Punnett square confirms this possible explanation. This assumes no dominance, with short tails a hybrid between none and long alleles.

	♂ T^1	T^2
♀ T^1	T^1/T^1 long	T^1/T^2 short
T^2	T^1/T^2 short	T^2/T^2 none

6   a   This is a dihybrid cross, as in problems 7 and 8. The genotypic ratio, T/T S/S : T/T S/s : T/T s/s : T/t S/S : T/t S/s : T/t s/s : t/t S/S : t/t S/s : t/t s/s, is 1 : 2 : 1 : 2 : 4 : 2 : 1 : 2 : 1. The phenotypes are tall, smooth; tall, wrinkled; short, smooth; and short, wrinkled, in the ratio 9 : 3 : 3 : 1.

    b   The short, wrinkled parent is homozygous for both genes and hence can produce only one type of gamete, ts. The other parent can produce two types of gametes, Ts and ts. The phenotypic ratio, tall, wrinkled : short, wrinkled, is 1 : 1. The genotypic ratio, T/t s/s : t/t s/s, is also 1 : 1.

    c   In this case each parent has only two possible gametes, TS and ts, and ts and ts, respectively. The genotypic ratio, T/t S/s : T/t s/s : t/t S/s : t/t s/s, is 1 : 1 : 1 : 1. The phenotypes are tall, smooth; tall, wrinkled; short, smooth; and short, wrinkled, in the ratio 1 : 1 : 1 : 1.

7   Let B be belted, b be no belt, F be fused, and f be normal. The cross is b/b F/F × B/B f/f, and yields all double heterozygotes, B/b F/f, in the F_1 generation. The Punnett square shows the results of freely interbreeding these individuals, and the genotypic and phenotypic ratios are again, respectively, 1 : 2 : 1 : 2 : 4 : 2 : 1 : 2 : 1 and 9 : 3 : 3 : 1.

	♂ BF	Bf	bF	bf
♀ BF	B/B F/F belted fused	B/B F/f belted fused	B/b F/F belted fused	B/b F/f belted fused
Bf	B/B F/f belted fused	B/B f/f belted norm	B/b F/f belted fused	B/b f/f belted norm
bF	B/b F/F belted fused	B/b F/f belted fused	b/b F/F even fused	b/b F/f even fused
bf	B/b F/f belted fused	B/b f/f belted norm	b/b F/f even fused	b/b f/f even norm

8   Let S be long-winged, s be vestigial-winged, H be hairless, and h be hairy. The parental cross is s/s h/h × S/S H/H. Since the only gametes are sh and SH, the only F_1 offspring are S/s H/h, double heterozygotes. The Punnett square shows the results of the F_2 generation. (In the square "less" stands for "hairless," and "short" for "vestigial-winged.") The genotypic ratio, S/S H/H : S/S H/h : S/S h/h : S/s H/H : S/s H/h : S/s h/h : s/s H/H : s/s H/h : s/s h/h, is 1 : 2 : 1 : 2 : 4 : 2 : 1 : 2 : 1. The phenotypes are long-winged, hairless; long-winged, hairy; vestigial-winged, hairless; vestigial-winged, hairy, in the ratio 9 : 3 : 3 : 1. The F_1 phenotype is long-winged, hairless.

	♂ SH	Sh	sH	sh
♀ SH	S/S H/H long, less	S/S H/h long, less	S/s H/H long, less	S/s H/h long, less
Sh	S/S H/h long, less	S/S h/h long, hairy	S/s H/h long, less	S/s h/h long, hairy
sH	S/s H/H long, less	S/s H/h long, less	s/s H/H short, less	s/s H/h short, less
sh	S/s H/h long, less	S/s h/h long, hairy	s/s H/h short, less	s/s h/h short, hairy

9   The cross is $A/–$ $R/–$ × $A/–$ r/r. If the male's second black allele were R, there could be no yellow or cream offspring. Hence the allele is r. Similarly, if either second yellow allele were A, there coiuld be no black or cream offspring. Hence they are both a. The cross was A/a R/r × A/a r/r.

10   There are two possible crosses, $C/–$ i/i × $C/–$ $I/–$ and $C/–$ i/i × c/c $–/–$. At least half the offspring of the first cross would receive an I allele and be colorless, but all the offspring are colored. By the same argument, in the second cross neither inhibitory allele of the hen can be I. Finally, if the cock's second color allele were c, one would expect 50 percent white. Therefore, C/C i/i × c/c i/i is the most likely cross. The offspring are all C/c i/i.

11   The Punnett square shows the two gametes contributed by the parents. It shows that 75 percent of their offspring will be deaf. Without using a Punnett square, one could note that 50 percent of the offspring would have the necessary K allele for hearing, but only 50 percent of these, and thus 25 percent of the total, would have it without the M allele.

	♂ kM	km
♀ Km	K/k M/m deaf	K/k m/m normal
km	k/k M/m deaf	k/k m/m deaf

12   Assuming no linkage, each of the five characters assorts independently. The chances of any specific homozygous recessive or homozygous dominant appearing is 1/4. In addition, the probability of any one heterozygous combination appearing is 1/2. Thus, the chance of this particular offspring is (1/4)(1/4)(1/2)(1/4)(1/2), or 1/256.

13 The man is I^B/i since one of his parents was i/i. The cross is $I^B/i \times I^A/I^B$. Half the offspring will have an I^A allele from the mother and so will not be pure B type. The other half will get the I^B and will be B regardless of which allele they receive from the father.

14 Each of these parents has one unknown allele, so Shirley, who is i/i, may belong to either family. Jane, who is $I^B/-$, cannot be the daughter of the Joneses, in whom no I^B allele is present. Thus, a mixup did occur in her case, and she could belong to the Smiths. However, another possibility is that Shirley is the Smith's daughter and Jane came from a third family.

15 Let C^B be black, C^C be chcocolate, and C^Y be yellow. Both the black male and the chocolate female carry the recessive yellow allele, which each inherited from its yellow parent. Hence the cross is $C^B/C^Y \times C^C/C^Y$. The genotypic ratio of the offspring, $C^B/C^C : C^B/C^Y : C^C/C^Y : C^Y/C^Y$, is 1 : 1 : 1 : 1. So 1/2 the offspring would be expected to be black, 1/4 chocolate, and 1/4 yellow.

16 The cross is $X^B/Y \times X^B/X^b$. As a result half the male offspring will receive the lethal allele and die. All the females will receive the X^B from their father and survive. Therefore, there will be twice as many female as male children.

17 Let X^C be normal and X^c be color-blind. The cross is $X^C/Y \times X^c/X^c$. Thus all males receive an X^c from their mother and will be color-blind. All the females will receive the dominant X^C from their father and be normal.

18 Number the individuals in the diagram 1 through 15 from left to right and top to bottom.

   a The trait is not dominant autosomal because female 9 is deaf while neither parent is.
   b The trait could be autosomal recessive.
   c It is not sex-linked dominant for the same reason as a.
   d It is not sex-linked recessive, for all the sons of female 2 would have to be deaf and male 8 is not. Female 9 could not be deaf since her father, male 3, would have to be deaf for her to be homozygous recessive.
   e The trait is not holandric, for if it were, no female could have it.

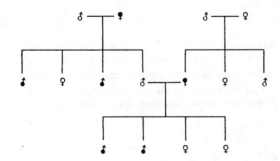

19 Let S be spotted, s be solid, L be short, and l be long. The cross—$Ss\ Ll \times ss\ ll$, a test cross—should segregate the characters into four equally occurring phenotypes if they are not linked. Since the phenotypes do not occur equally, the genes are linked.

$$\frac{14 + 10}{200} \times 100 = 12\% \text{ crossing over}$$

$$= 12 \text{ map units apart}$$

20 If the genes are 20 units apart, crossing-over would occur 20 percent of the time. Hence 20 percent of 1000 offspring would be expected to show the corssover phenotypes.

$$(1000 \times 0.20 = 200)$$

21 If A and B are 40 map units apart, and both A and B are 20 units from C, then C must lie halfway between them. Similarly, since both C and B are 10 units from D, D must lie halfway between them. The diagram below illustrates the conclusion, obtained through the incorporation of the various conditions in the order given in the text. The individual steps are lettered.

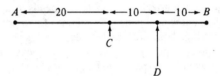

22 In humans, any individual with a Y chromosome will appear male:

Genotype	Sex	Number of Barr bodies
XXX	female	two
XYY	male	none
XXXX	female	three
XXXY	male	two
XXXXY	male	three

QUESTIONS

Choose the one best answer.

1 In peas, axial flowers are dominant over flowers borne terminally. What phenotypic ratios would you expect in offspring from a cross between a known heterozygous axial-flowered plant and one whose flowers are terminal?

   a 3 axial : 1 terminal   d all axial
   b 3 terminal : 1 axial   e none of the above
   c 1 axial : 1 terminal   (p. 280)

2 Cystic fibrosis is inherited as a simple recessive. Suppose a woman who carries the trait marries a normal man who does not carry it. What percent of their children would be expected to have the disease?

   a 0 percent   d 50 percent
   b 25 percent   e 75 percent
   c 33 percent   (p. 280)

3 The light color variation in the peppered moth is inherited as a simple recessive characteristic. If a light moth is crossed with a dark moth which had a light parent, what percent of their offspring will be light?

   a 25 percent   d 75 percent
   b 33 percent   e 100 percent
   c 50 percent   (p. 280)

4 A dominant gene W produces wire-haired texture in dogs; its recessive allele w produces smooth hair. A group of heterozygous wire-haired dogs are crossed and all their wire-haired progeny are then test-crossed. What is the expected phenotypic ratio among the test-cross progeny?

   a 3 wire-haired : 1 smooth
   b 2 wire-haired : 1 smooth
   c 1 smooth : 1 wire-haired
   d 100 percent wire-haired
   e none of the above   (p. 281)

5 In pigeons, the grizzle color pattern depends on a dominant autosomal gene G. A mating of two grizzle birds produced one nongrizzle youngster this year. If this pair of pigeons produces more youngsters next year, what percent would be expected to be grizzles?

   a 100 percent   d 25 percent
   b 75 percent   e 0 percent
   c 50 percent   (p. 280)

6 Among white human beings, when individuals with straight hair mate with those with curly hair, wavy-haired children are produced. If two individuals with wavy hair mate, what phenotypes and ratios would you predict among their offspring?

   a 3 curly : 1 wavy
   b 1 curly : 1 wavy : 1 straight
   c 1 straight : 1 curly : 2 wavy
   d 3 wavy : 1 straight
   e 1 straight : 2 curly : 1 wavy   (p. 280)

7 In cocker spaniels, black color is due to a dominant gene B, and red color to its recessive allele b. Solid color is dependent on a dominant gene S, and white spotting on its recessive allele s. A solid red male was mated to a black-and-white female. They had five puppies: one black, one red, one black-and-white, and two red-and-white. What were the genotypes of the parents?

   a male $b/b\ s/s$ and female $B/B\ s/s$
   b male $b/b\ S/s$ and female $B/b\ s/s$
   c male $b/b\ S/s$ and female $B/b\ S/s$
   d male $B/b\ S/s$ and female $B/b\ s/s$
   e male $B/b\ S/S$ and female $B/b\ s/s$ (p. 283)

8 Suppose a solid red male cocker spaniel whose mother was spotted red was mated with a solid black female whose mother was also spotted red. What phenotypes in what proportions might be expected from such a cross?

   a 3 solid black : 1 spotted black : 3 solid red : 1 spotted red
   b 9 solid black : 3 spotted black : 3 solid red : 1 spotted red
   c 1 solid black : 1 solid red : 1 spotted black : 1 solid red
   d 3 solid black : 1 spotted red
   e none of the above   (p. 283)

9 Ignoring modifier genes, we can think of brown eyes in human beings as determined by a dominant allele B, and blue eyes by a recessive allele b; free earlobes are determined by a dominant allele F, and attached earlobes by a recessive allele f. A brown-eyed man with attached earlobes (whose mother was blue-eyed) marries a blue-eyed woman with free earlobes (whose father had attached earlobes). What phenotypes may be expected among their children?

   a Both blue- and brown-eyed children may be expected, but all will have free earlobes.
   b All four possible combinations of eye color and earlobe condition may be expected, in roughly equal frequencies.
   c All brown-eyed children will have attached earlobes, and all blue-eyed children will have free earlobes.
   d Both blue- and brown-eyed children may be expected, but all will have attached earlobes.
   e All brown-eyed children will have free earlobes, and all blue-eyed children will have attached earlobes.
                                                (p. 283)

10 The allele for pea comb (P) in chickens is dominant to that for single comb (p), but the alleles black (B) and white (B') for feather color show partial dominance: B/B' individuals have "blue" feathers. If birds heterozygous for both alleles are mated, what proportion of the offspring are expected to be white-feathered and pea-combed?

   a 1/16   d 8/16
   b 3/16   e 9/16
   c 4/16   (p. 283)

11 In a P cross, an $A/A\ B/B\ C/C$ individual is paired with an $a/a\ b/b\ c/c$ individual. Assuming no linkage, what will be the expected frequency of $A/A\ b/b\ C/c$ individuals in the F_2 generation?

   a 16/64   d 2/64
   b 8/64   e 1/64
   c 4/64   (p. 283)

12 Knowledge of the blood-type genotypes of a certain couple leads us to say that if they were to have many children, the ratios of the children's blood types would be expected to approximate 1/2 type A and 1/2 type B. It follows that the blood types of the couple are

   a A and B
   b AB and AB.
   c AB and B.
   d AB and A.
   e AB and O.   (p. 287)

13 An adopted child (blood type O) has located her biological father and discovered that he has blood type B. Which blood type does she know her biological mother does *not* have?

   a A
   b B
   c AB
   d O
   e None of the blood types can be eliminated; the biological mother could have any of them.
   (p. 287)

14 A particular sex-linked recessive disease of human beings is usually fatal. Suppose that, by chance, a boy with the disease lives past puberty and marries a woman heterozygous for the trait. If they have a daughter, what is the probability that she will have the disease?

   a 0 percent
   b 25 percent
   c 50 percent
   d 75 percent
   e 100 percent
   (p. 296)

15 The rare trait ocular albinism (almost complete absence of eye pigment) is inherited as a sex-linked recessive. A man with ocular albinism marries a woman who neither has this condition nor is a carrier. Which one of the following is the best prediction concerning their offspring?

   a All their sons will have ocular albinism, and all their daughters will be carriers.
   b All their children of both sexes will have ocular albinism.
   c About 50 percent of their sons will have ocular albinism, and all their daughters will be carriers.
   d About 50 percent of their daughters will have ocular albinism, but all their sons will have normal eyes.
   e None of their children will have ocular albinism, but all their daughters will be carriers.
   (p. 296)

Questions 16 to 18 refer to the following pedigree for one type of deafness in human beings. Squares symbolize males, circles females; filled symbols designate deaf individuals; open symbols, individuals with normal hearing.

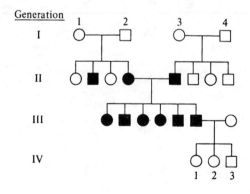

16 This type of deafness is probably inherited as

   a an autosomal dominant.
   b an autosomal recessive.
   c a sex-linked dominant.
   d a sex-linked recessive.
   e a holandric.
   (p. 280)

17 The genotype of individual 4 in generation I is probably

   a D/D.   *d* X^D/Y.
   b D/d.   *e* X^d/Y.
   c d/d.
   (p. 280)

18 The genotype of individual 3 in generation IV is probably

   a D/D.   *d* X^D/Y.
   b D/d.   *e* X^d/Y.
   c d/d.
   (p. 280)

19 A certain man has the genotype *AB/ab*; genes *A* and *B* are on one chromosome, and *a* and *b* are on the homologous chromosome. Suppose crossin- over occurs during a meiotic division in this man's testis. With regard to the two genes discussed here, how many genetically different types of sperm cells will result?

   a 1   *d* 8
   b 2   *e* 16
   c 4
   (p. 280)

20 Singer Arlo Guthrie has a 50 percent chance of dying prematurely from the same genetic disease that killed his father, Woody Guthrie. Neither Woody Guthrie's mother nor Arlo Guthrie's mother carries any allele for this disease (Huntington's chorea). What type of inheritance pattern does this disease have? (It can be figured out from the information given.)

   a autosomal dominant
   b sex-linked dominant
   c autosomal recessive
   d sex-linked recessive
   (p. 280)

21 Pleiotropism describes

   a a single gene's having multiple effects.
   b the gene interaction of multiple alleles.
   c a single trait's being influenced by several genes.
   d a trait that is not expressed for several generations.
   e polygenic inheritance.
   (p. 294)

22 Suppose alleles *M* and *n* are linked on one chromosome, and *m* and *N* are linked on the homologous chromosome. Individuals homozygous for *M* and *n* are mated with individuals homozygous for *m* and *N*. Their offspring are crossed with homozygous recessive individuals, and the following results are recorded:

Mn/mn 232
mN/mn 240
MN/mn 15
mn/mn 13

How many units apart are these genes on the chromosome?

- a 2.5
- b 5.1
- c 5.6
- d 12
- e 50   (p. 300)

23 The frequency of crossing over between linked genes A and B is 35 percent; between B and C, 10 percent; between C and D, 15 percent; between C and A, 25 percent; and between D and B, 25 percent. The sequence of the genes on the chromosome is

- a ACDB.
- b ACBD.
- c ABDC.
- d ABCD.
- e ADCB.   (p. 300)

24 In *Drosophila* a dominant allele P produces normal eye color, whereas the recessive allele p produces purple eye color. Another gene controls body color: the dominant allele C produces normal body color, whereas the recessive allele c produces black body color. A female fruit fly heterozygous for the traits of eye color and body color is crossed with a purple-eyed black-bodied male. The following offspring are obtained:

both normal traits	151
purple eye, normal body	8
normal eye, black body	10
purple eye, black body	131

How do you explain these data?

- a Genes C and P are linked and are 6 units apart on the chromosome.
- b Genes C and p are linked and are 6 units apart on the chromosome.
- c Genes C and P are on different chromosomes and assort independently of one another.
- d Both a and b are correct.
- e None of the above is correct.   (p. 300)

25 Consider two linked autosomal genes. The dominant allele C of the first gene causes cataracts of the eye, whereas its recessive allele c produces normal eyes. The dominant allele of the second gene P causes polydactyly (presence of an extra finger on each hand), whereas its recessive allele p produces normal hands. A man with cataracts and normal hands marries a woman with polydactyly and normal eyes. Their son has both cataracts and polydactyly. The son marries a woman with neither trait. Assuming no crossing-over, what is the probability that their first child will have both cataracts and polydactyly? (Hint: diagram the arrangement of the genes on the chromosomes of all the individuals mentioned in the question.)

- a 0 percent
- b 25 percent
- c 50 percent
- d 75 percent
- e 100 percent   (p. 283)

26 Gene A and gene B are known to be 10 units apart on the same chromosome. Individuals homozygous dominant for these genes are mated with homozygous recessives. The offspring are then mated with homozygous recessive individuals. If there are 1000 offspring from this cross, how many of the offspring would you expect to show the crossover phenotypes?

- a 10
- b 50
- c 100
- d 250
- e 500   (p. 300)

27 A person with XYY syndrome will have how many Barr bodies in his nuclei?

- a 0
- b 1
- c 2
- d 3
- e either 1 or 2   (p. 295)

28 The transfer of a segment of a chromosome to a nonhomologous chromosomes is called

- a trisomy.
- b translocation.
- c polyploidy.
- d transformation.
- e duplication.   (p. 301)

For further thought

Answers to these questions are given in the Answers section.

1 Blood typing is often used as evidence in paternity cases in court. In a series of paternity cases, the mothers and their respective children had the blood types listed in the table below. For each, indicate the blood type(s) which, if found in the accused man, would exonerate him as the father. (p. 286)

MOTHER	CHILD	MAN EXONERATED IF HE BELONGS TO GROUP
A	O	
B	AB	
O	A	
AB	A	
O	O	
B	B	
A	B	
AB	AB	
A	A	
A	AB	
B	A	
B	O	
AB	B	

2 In pea plants, the genes for flower color and pollen-grain shape are on the same chromosome. Purple flowers (*P*) are dominant over red (*p*) and long pollen grains (*L*) dominant over round (*l*). Plants heterozygous for purple flowers and long pollen (*PL/pl*) are test-crossed with the homozygous recessive (*pl/pl*). The following offspring are produced:

purple, long	118
red, round	122
purple, round	32
red, long	28

How far apart are the genes for color and pollen-grain shape on the chromosome? (p. 300)

3 A tomato breeder wants to find out whether the genes for plant height and plant color are on the same chromosome. The allele for tall plants (*T*) is dominant over the dwarf allele (*t*), and the green allele (*G*) is dominant over the light-green allele (*g*). She crosses plants homozygous for tallness and green color with homozygous recessive plants. The progeny are then test-crossed, and the following results are observed:

tall, green	434
dwarf, light green	466
tall, light green	51
dwarf, green	49

Are the genes for height and color on the same chromosome? If so, how far apart are they on the chromosome? (p. 300)

4 The frequency of crossing over between linked genes *A* and *B* is 6 percent; between *B* and *C*, 13 percent; between *C* and *D*, 18 percent; between *C* and *A*, 7 percent; and between *D* and *B*, 5 percent. What is the sequence of these genes on the chromosome? (p. 300)

5 In watermelons, green color is dominant over striped and short length is dominant over long. The dominant allele for green (*G*) is linked to the dominant allele for short (*S*). In a test cross of heterozygous green short watermelons, how many types of gametes can each parent produce, assuming there is no crossing-over? Give the genotypes of the progeny of such a cross. (p. 298)

6 Make a concept map of the interaction of evolution and genetics. Be sure to include the more general concepts of natural selection, phenotypic variation, and allelic frequencies. More specific concepts should be sources of variation, including the chromosomal alterations described in this chapter. (p. 301)

ANSWERS

1	*c*	9	*b*	17	*b*	25	*a*
2	*a*	10	*b*	18	*b*	26	*c*
3	*c*	11	*d*	19	*c*	27	*a*
4	*b*	12	*e*	20	*a*	28	*b*
5	*b*	13	*c*	21	*a*		
6	*c*	14	*c*	22	*c*		
7	*b*	15	*e*	23	*e*		
8	*a*	16	*b*	24	*a*		

For further thought

1

MOTHER	CHILD	MAN EXONERATED IF HE BELONGS TO GROUP
A	O	AB
B	AB	B, O
O	A	B, O
AB	A	—
O	O	AB
B	B	—
A	B	A, O
AB	AB	O
A	A	—
A	AB	A, O
B	A	B, O
B	O	AB
AB	B	—

2 20 map units apart
3 same chromosome, 10 map units apart
4 *CABD* or *DBAC*

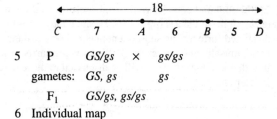

5 P   *GS/gs* × *gs/gs*
   gametes: *GS, gs*   *gs*
   F₁   *GS/gs, gs/gs*

6 Individual map

Chapter 15

MULTICELLULAR ORGANIZATION

A GENERAL GUIDE TO THE READING

This chapter introduces basic information about plant and animal tissues, which will serve as a foundation for material presented in other chapters. The chapter is well illustrated; you will find learning the material easier if you pay close attention to the photographs and drawings. As you read Chapter 15 in your text, you will want to concentrate on the following topics.

1. Plant tissues. The section on these tissues (pp. 308–13) provides an introduction for Chapters 16, 19, and 23; learning the material now will save you time later on.

2. Epithelium and connective tissue. Carefully read the material on these tissues (pp. 315–17), since they will not be discussed further. The subsequent material on cartilage, bone, muscle, and nerves (p. 319) serves primarily as an introduction; these tissues will be discussed in more detail in later chapters.

3. All plant tissues can be divided into two major categories: meristematic tissue and permanent tissue. (pp. 308–309)

4. All animal tissues are divided into four categories: epithelium, connective tissue, muscle, and nerve. Each of these is an assemblage of different subtypes. (pp. 314–15)

CROSSLINKING CONCEPTS

1. This chapter introduces the plant and animal tissues and organ systems and thus serves as a framework for Part III, which discusses each system in depth. Each chapter from Part III will draw on your knowledge of the cells and specialized tissues that form the organs of each system.

2. Surface-to-volume ratio, discussed with regard to cell size in this chapter, is a theme that will recur in later chapters in Part III.

KEY CONCEPTS

1. The need for efficient diffusion puts a strict limit on the surface-to-volume ratio of cell, and consequently limits cell size. (p. 307)

2. The bodies of most multicellular organisms are organized on the basis of specialized tissues, organs, and systems. (p. 308)

OBJECTIVES

1. Explain why single-celled organisms are limited in size. (p. 307)

2. Describe the hierarchical organization of most multicellular organisms and define tissue, organ, and organ system. (p. 308)

3. Using a diagram of a flowering-plant body (Fig. 15.1, p. 308) identify the root, shoot, vascular system, and apical meristems.

4. Describe meristematic cells and indicate where in the plant body such cells are found. Differentiate between the apical and lateral meristems and give the function of each. (pp. 308–309)

5. List the three subcategories of permanent tissue and give examples of each. (pp. 309–13)

6. Give the characteristics of epidermal cells, indicate where they are found in the plant body, specify the function of the cuticle, and describe the periderm. (p. 309)

7. Describe parenchyma cells; specify their location in the plant body and list some of their functions. Contrast collenchyma cells (Fig. 15.7, p. 312) and sclerenchyma cells (Fig. 15.8, p. 312) with respect to function, the thickness of cell walls, and whether or not the cells are living or dead at maturity. (pp. 311–13)

8. Give the function of xylem; indicate which types of cells are present in it; and explain how a complex tissue such as xylem differs from simple tissues such as parenchyma and collenchyma. Give the function of phloem, and indicate which types of cells are present in it. (p. 313)

9. Explain where epithelial tissues are found in the body and give their functions. Differentiate among squamous, cuboidal, and columnar cells. Explain how you would recognize simple epithelium and stratified epithelium. (pp. 315–17)

10. Using Figure 15.10 (p. 315), describe the structure and function of spot desmosomes, tight junctions, and gap junctions. (p. 315)

11. List the four main types of connective tissue. Explain how you would recognize dense fibrous connective tissue and loose connective tissue, and specify where in your body you might expect to find each. Describe cartilage ad bone, and give the function of each; in doing so, show how the matrix of cartilage differs from that of bone and how that difference is reflected in the respective functions of cartilage and bone. (pp. 317–19)

12. Give the function of muscle cells, and list the three principal types. (p. 319)

13. Give the function of a nerve cell and explain how its structure, as shown in FIgure 15.18 (p. 319), is adapted to its function.

14. Using Figure 15.19 (p. 321), point out the different categories of tissue and explain how the cells are integrated to form an organ. (pp. 315–20)

KEY TERMS

tissue (p. 307)
organ (p. 308)
system (p. 308)
root (p. 308)
shoot (p. 308)
simple (p. 308)
complex (p. 308)
meristematic tissue (p. 308)
permanent tissue (p. 308)
apical meristem (p. 309)
primary tissue (p. 309)
lateral meristem (p. 309)
secondary tissue (p. 309)
epidermis (p. 309)
periderm (p. 309)
cuticle (p. 309)
guard cells (p. 311)
root hairs (p. 311)
ground tissue (p. 311)
parenchyma (p. 311)
collenchyma (p. 312)
sclerenchyma (p. 312)
fiber (p. 312)
sclereid (p. 313)
vascular tissue (p. 313)
xylem (p. 313)
tracheid (p. 313)
vessel element (p. 313)
phloem (p. 313)
sieve element (p. 313)
epithelium (p. 315)
spot desmosome (p. 315)
tight junction (p. 315)
gap junction (p. 315)
squamous cell (p. 315)
cuboidal cell (p. 315)
columnar cell (p. 315)
simple epithelium (p. 317)
stratified epithelium (p. 317)
basement membrane (p. 317)
matrix (p. 318)
connective tissue (p. 318)
loose connective tissue (p. 318)
cartilage (p. 318)
bone (p. 318)
adipose tissue (p. 318)
blood (p. 319)
lymph (p. 319)
skeletal muscle (p. 319)
smooth muscle (p. 319)
cardiac muscle (p. 319)
nerve (p. 319)

SUMMARY

The bodies of multicellular organisms are organized on the basis of *tissues* that are groups of cells similar in structure and function, *organs* that are various tissues grouped together to form structural and functional units, and *systems* that are groups of interacting organs.

Plant tissues and organs All plant tissues can be divided into meristematic tissue (growth tissue—undifferentiated cells capable of dividing) and permanent tissue (mature differentiated cells). Regions of *meristematic tissue* are found at the tips of roots and stems (*apical meristems*) and, in many plants, in areas around the periphery of the roots and stems (*lateral meristems*). Apical meristems produce primary tissues; lateral meristems produce secondary tissues.

The permanent tissues fall into three subcategories; surface tissues, ground (fundamental) tissues, and vascular tissues. The surface tissues (*epidermis, periderm*) form the protective outer covering of the plant body.

There are three types of ground tissues: *parenchyma,* which are thin-walled, loosely packed cells found throughout the plant body; *collenchyma,* supportive tissue whose cell walls are irregularly thickened; and *sclerenchyma,* supportive tissue with very thick cell walls.

The vascular, or conductive, tissue is characteristic of vascular plants and consists of two principal types of complex tissue: xylem and phloem. *Xylem* supports the plant and transports water and dissolved minerals upward. *Phloem* conducts organic materials up and down the plant body.

The body of vascular land plants is divided into two major parts: the root and the shoot. The root functions in absorbing water and nutrients whereas the shoot supports the leaves, which carry out photosynthesis and synthesize the plant's food.

Animal tissues and organs Animal tissues are divided into epithelium, connective tissue, muscle, and nerve. *Epithelial tissue* covers or lines the internal and external surfaces of all free body surfaces. Epithelial cells are packed tightly together, with almost no intercellular spaces. Specialized intercellular junctions join the cells to one another. Epithelial tissues provide a continuous barrier protecting the underlying cells from the external medium.

Connective tissue, composed of cells and fibers embedded in an extensive intercellular matrix, connects, supports, or surrounds other tissues or organs; examples of connective tissue are blood and lymph, connective tissue proper, cartilage, and bone.

The three types of *muscle tissue* (skeletal or striated, smooth, and cardiac) consist of cells specialized for contraction and are responsible for most movement in complex animals.

Nervous tissue is highly specialized for the ability to respond to changes in the environment (stimuli). It carries information from the outside environment and from different body parts to allow the animal to respond to stimuli.

Organs are grouped into functional complexes called organ systems.

QUESTIONS

Testing recall

Match each of the following plant tissues with its characteristics.

- a collenchyma
- b epidermis
- c meristematic
- d parenchyma
- e phloem
- f sclerenchyma
- g xylem

1 thin-walled, loosely packed, unspecialized plant tissue (p. 311)

2 protective outer covering of the plant body (p. 309)

3 undifferentiated plant cells capable of dividing (p. 309)

4 conducts organic materials up and down the plant body (p. 313)

5 uniformly thick-walled supportive tissues (p. 312)

6 conducts water and dissolved materials upward in the plant body (p. 313)

Match each of the following animal tissues with its description below.

- a simple epithelium
- b stratified epithelium
- c vascular tissue
- d loose connective tissue
- e dense connective tissue
- f cartilage
- g bone
- h muscle
- i nerve

7 blood and lymph (p. 319)

8 specialized for stimulus reception and conduction (p. 319)

9 a specialized form of dense fibrous connective tissue with a rubbery intercellular matrix (p. 318)

10 contractile tissues (p. 319)

11 tissue with compact arrangement of many fibers, limited amount of ground substance, and relatively small number of cells (p. 318)

12 tissue with loose, irregular arrangement of fibers, a large amount of ground substance, and numerous cells of various types (p. 318)

13 covering or lining tissue two or more cells thick (p. 315)

14 covering or lining tissue one cell thick (p. 315)
15 tissue that has a hard relatively rigid matrix containing numerous collagen fibers and a large amount of water and that is impregnated with inorganic salts (p. 318)

Testing knowledge and understanding

Choose the one best answer.

16 Which of the following cell types would you *not* expect to find in a leaf?

    a epidermal cell   *d* tracheid
    b parenchyma cell   *e* collenchyma cell
    c periderm   (p. 309)

17 The region of tissue at the growing tips of roots and stems would be classified as

    a surface tissue.   *d* periderm.
    b apical meristem.   *e* parenchyma.
    c lateral meristem.   (p. 309)

18 The corky outer bark on trees is classified as

    a epidermis.   *c* sclerenchyma.
    b cuticle.   *d* periderm.   (p. 309)

19 All of the following are simple tissues *except*

    a parenchyma.   *c* sclerenchyma.
    b collenchyma.   *d* phloem.   (p. 313)

20 Which one of the following is *not* an example of connective tissue.

    a blood   *d* cartilage
    b bone   *e* muscle
    c lymph   (p. 319)

21 A certain tissue is made up of several varieties of cells that are embedded in an extensive matrix composed of ground substance and many irregularly arranged fibers. This tissue would be classified as

    a lymph.
    b loose connective tissue.
    c dense connective
    d cartilage.
    e bone.   (p. 318)

22 A specialized cellular connection that permits the movement of ions and some small molecules between animal cells is the

    a gap junction.   *d* tight junction.
    b spot desmosome.   *e* intercellular matrix.
    c basement membrane.   (p. 315)

23 Skin is an animal organ. Of which of the following tissue types is it composed?

    a connective tissue   *d* nerve
    b epithelium   *e* all the above
    c muscle   (p. 320)

ANSWERS

Testing recall

1	*d*	5	*f*	9	*f*	13	*b*
2	*b*	6	*g*	10	*h*	14	*a*
3	*c*	7	*c*	11	*e*	15	*g*
4	*e*	8	*i*	12	*d*		

Testing recall

| 16 | *c* | 18 | *d* | 20 | *e* | 22 | *a* |
| 17 | *b* | 19 | *d* | 21 | *b* | 23 | *e* |

Chapter 16

NUTRIENT PROCUREMENT AND GAS EXCHANGE IN PLANTS AND OTHER AUTOTROPHS

A GENERAL GUIDE TO THE READING

This chapter describes how plants obtain the nutrients they need in order to synthesize their own organic compounds and how they exchange gases with the environment. As you read Chapter 16 in your text, you will find it useful to focus on the following.

1. The introduction, which contrasts autotrophic and heterotrophic modes of nutrition. You should learn the terms autotrophic and heterotrophic (p. 324), since they are used frequently.

2. The sections on the structure and fuction of roots. These sections (pp. 329–36) require careful reading; particular attention should be paid to Figures 16.8–16.15 (pp. 330–36). Make sure you understand the terms apoplast and symplast, and the importance of the endodermis with its Casparian strip.

3. The problem of gas exchange. Factors that complicate the maintenance of an adequate respiratory surface in various organisms, especially terrestrial plants, are discussed on pages 338 to 339. You will find the material on gas exchange presented in this chapter and in Chapter 18 easier if you first learn the four basic requirements for gas-exchange systems (pp. 338–339). Then, when reading the remainder of the chapter and Chapter 18, try to understand how each gas-exchange mechanism meets these four requirements.

4. Gas exchange in terrestrial plants. Remember that the gas-exchange problem in plants (pp. 339–40) is different from that in animals. Like animal cells, all living plant cells require O_2 for cellular respiration, but those cells carrying on photosynthesis also require CO_2. Transpiration, a process resulting in the loss of water in plants, is mentioned in this chapter; it will be discussed in more detail in Chapter 19.

KEY CONCEPTS

1. Autotrophic organisms manufacture their own organic compounds from inorganic raw materials absorbed directly from the environment. (p. 324)

2. In order to carry out their life processes, all organisms require prefabricated high-energy organic compounds or the raw materials from which these compounds can be synthesized. (pp. 323–24)

3. Some of these essential compounds form structural components of plants and are needed in large quantities, whereas others serve as enzymes or coenzymes and are therefore needed in small quantities. (pp. 325–26)

4. Because of the presence of the Casparian strip, all materials absorbed by the root must pass through the living endodermal cells to reach the stele, and the plant can exercise some control over the movement of substances into the vascular tissue. Water and minerals move independently of each other. (pp. 334–36)

5 Some plants have evolved specialized adaptations to acquire nitrogen. (p. 337)

6 Plants must obtain oxygen as well as carbon dioxide. (p. 337)

7 Gas exchange between a living cell and its environment always takes place by diffusion across a thin, moist cell membrane. (p. 338)

8 All organisms require protected respiratory surfaces of adequate dimensions relative to their volumes. The larger the organism, the more surface area it requires. (p. 338)

CROSSLINKING CONCEPTS

1 Knowledge of how transport occurs across cell membranes (Chapter 5) is fundamental to an understanding of the transport of water and minerals in plants. Prior key concepts include diffusion, facilitated diffusion, osmosis, and active transport.

2 Your study of noncyclic photophosphorylation in Chapter 7 should help you understand the biochemical basis of gas exchange in leaves.

3 As an organism grows, its volume increases faster than its surface area. You will find many similarities between the adaptations of plants that increase their surface area for nutrient procurement and gas exchange and the adaptations of animals, which will be examined in the next two chapters.

4 The relationships of mycorrhizae and plant roots and of legumes and nitrogen-fixing bacteria are examples of mutualism, a concept to be examined more closely in Chapter 31.

5 In its description of the effort to develop crop plants that can fix nitorgen, this chapter illustrates how recombinant DNA technology (Chapter 13) is being applied to agriculture.

OBJECTIVES

1 Contrast the nutrient requirements of autotrophic and heterotrophic organisms. (p. 324)

2 For each of the nutrients in the following list, indicate whether it is required by plants and, if so, whether it is a macronutrient, needed in relatively large amounts, or a micronutrient, needed in minute amounts; then give one function of each nutrient. Describe the function most micronutrients perform.

nitrogen	potassium
protein	magnesium
carbon dioxide	water
copper	vitamins
phosphorus	fats
glucose	calcium (pp. 325–26)

3 Explain why most of the mass of a plant's body comes from air, not from the solid earth in which it grows. (p. 325)

4 Describe the process of nitrogen fixation in certain soil bacteria and the relationship of some nitrogen-fixing bacteria and legumes. Explain why nitrogen fixation is a useful process. (pp. 328–29)

5 Describe how plants obtain carbon dioxide from the air. (p. 329)

6 Explain how the root system is adapted to provide the extensive absorptive surface a plant needs in order to obtain sufficient nutrients to support its large volume. (p. 329)

7 Describe the mutualistic relationship between mycorrhizae and plant roots. (pp. 330–31)

8 Using a cross section of a root in Figures 16.10 (p. 332) and 16.14 (p. 335), identify the epidermis, root hairs, cortex, endodermis, pericycle, stele, xylem and phloem, and vascular tissues. (pp. 331–33)

9 Explain how water can move through the cortex of the plant root by flowing through the symplast or apoplast. (pp. 334–36)

10 Draw several endodermal cells with their Casparian strips, and relate their structure to the role of the endodermis in water and mineral absorption. (p. 333)

11 Explain how insectivorous plants supplement their diet, and name the most beneficial nutrient that is obtained from this process. (p. 337)

12 Discuss why terrestrial plants have evolved special respiratory surfaces and mechanisms, whereas small aquatic autotrophs and even large kelp have not evolved such adaptations. In doing so, explain how gas exchange is affected by the relationship of surface to volume, the shape of an organism, the permeability of the outer covering of an organism, and the water loss in terrestrial plants. (pp. 338–40)

13 List the three basic requirements for gas-exchange systems. (p. 338)

14 Indicate how the structure of leaves of a typical terrestrial plant meets the three basic requirements of gas-exchange systems, and indicate where the actual gas-exchange process takes place. In doing so, specify what two gases pass in and out the stomata. (pp. 338–39)

118 • CHAPTER 16

15 Using Figures 16.21 (p. 339) and 16.22 (p. 339), explain the structure and function of guard cells. Explain the role of potassium-ion concentration in changing the turgidity of guard cells, and the effect this has on the opening and closing of the stomata.

16 Discuss the process of gas exchange in stems and roots. In doing so, answer the following questions. What are lenticels, and what is their function? Do roots have specialized structures for gas exchange? How do gases enter and leave the roots? Do plants, like animals, have specialized structures for transporting gases? How are gases distributed effectively in plants? Why wouldn't the same method work in animals? (pp. 340–41)

KEY TERMS

nutrient (p. 324)
macronutient (p. 324)
micronutient (p. 324)
autotrophic (p. 324)
heterotrophic (p. 324)
nitrogen fixation (p. 328)
nodules (p. 328)
stomata (p. 329)
primary root (p. 330)
secondary root (p. 330)
fibrous root system (p. 330)
taproot system (p. 330)
root hair (p. 330)
mycorrhizae (p. 330)
mutualistic symbiosis (p. 331)
epidermis (p. 331)
cortex (p. 333)
periderm (p. 333)
endodermis (p. 333)
Casparian strip (p. 333)
stele (p. 333)
pericycle (p. 333)
xylem (p. 333)
phloem (p. 333)
pith (p. 334)
symplast (p. 334)
apoplast (p. 336)
guard cells (p. 339)
transpiration (p. 340)
lenticels (p. 340)

SUMMARY

All organisms require proteins, carbohydrates, lipids, and nucleic acids to carry out their life functions. Organisms capable of manufacturing their own organic nutrients from inorganic raw materials are *autotrophic*; those that cannot synthesize organic materials from inorganic raw materials and require prefabricated complex organic materials from the environment are *heterotrophic*. These two groups differ in both their nutrient requirements and the problems associated with nutrient procurement that they have.

Most autotrophs are *photosynthetic*; they use light energy to drive their synthesis of organic compounds. Green plants, the largest group of photosynthetic organisms, require carbon dioxide and water as raw materials. In addition, they need certain mineral nutrients. Macronutrients, or minerals that are required in substantial amounts, function as structural components of complex organic molecules; micronutrients, which are required in minute amounts, usually function as parts of enzymes or coenzymes.

Nutrient procurement Minerals are absorbed in ionic form from soil water. The availability of inorganic ions depends on the fertility and pH of the soil. Simple diffusion, facilitated diffusion, and active transport are all involved in the absorption process. Most ions are absorbed by active transport.

Leguminous plants form a mutualistic association with certain bacteria, which can convert nitrogen gas into ammonia that can then be used by the plant for synthesizing amino acids and other organic constituents. The plant in turn provides the bacteria with the organic compounds necessary for its nutrition and growth.

The carbon dioxide required by the plant is absorbed by diffusion into cells in the interior of the leaf. Tiny openings called *stomata* in the *epidermis*, or outer covering of the leaf, allow the carbon dioxide to penetrate to the interior of the leaf, where it can circulate in the intercellular spaces and is available to individual cells.

Most higher land plants take in water and mineral nutrients from the soil through their roots. To absorb enough of these needed materials, plants require a sufficiently large surface area. As an organism gets bigger, its volume increases much faster than its surface area. Therefore, a large multicellular plant must have an enormous absorptive surface area to support its large volume. A typical root is extensively branched and subdivided and has many *root hairs*, tiny hairlike extensions of the epidermal cells, which vastly increase the total absorptive surface. Mycorrhizal associations with vascular plant roots also help plants to absorb water and ions, and may enable the plant to survive in harsh environments. Roots also store organic compounds and anchor the plant to the soil.

Water from the soil may be absorbed directly into the epidermal cells and move from cell to cell through the *cortex* to the vascular cylinder, or *stele*, by flowing through the *symplast*. Alternatively, it may move across the epidermis and cortex without entering a cell by flowing along the cell walls and intercellular spaces that make up the *apoplast*. However, the *endodermis*, which separates the cortex from the stele, presents a barrier to movement. The side and end walls of the endodermal cells contain a *Casparian strip*, which forms a waterproof seal between the endodermal cells; consequently all materials must pass through the living endodermal cells to reach the stele, and the plant can exercise some control over the movement of substances into the vascular tissue.

A few photosynthetic plants, particularly those living in nutrient-poor environments such as bogs, supplement their inorganic diet with organic nitrogen compounds obtained by trapping and digesting insects and other small animals.

Gas exchange Since aerobic respiration is the chief method of respiration in both plants and animals, the vast majority of organisms must have some method of obtaining oxygen and getting rid of waste carbon dioxide.

Gas exchange between a living cell and its environment always takes place by *diffusion* across a cell membrane. Therefore, there must be a surface area of adequate dimensions, which must be kept moist and protected from mechanical damage. The cells of most small aquatic organisms are in close contact with the surrounding medium and carry out gas exchange across their whole body surface; no special respiratory devices are necessary.

Solutions in terrestrial plants In plants, gas exchange in the leaves takes place at a high rate. The loosely packed mesophyll cells in the interior of the leaf provide an adequate surface area, and the epidermis with its waxy cuticle protects the mesophyll from mechanical damage and water loss. The stomata allow the gases to penetrate to the interior, where they circulate through the numerous intercellular spaces.

The size of each *stoma* in the epidermis is regulated by two guard cells. When the guard cells are turgid, the stoma is open; when the guard cells lose water and are flaccid, the stoma is closed. When the stomata are open, gases can move into the interior of the leaf, but at the same time the plant loses water by *transpiration*. The stomata close at night, and so prevent excessive water loss.

Root cells obtain oxygen by simple diffusion across the moist membranes of the individual cells; air reaches the interior cells via the intercellular spaces of the apoplast system. Gas exchange in older stems and roots takes place through lenticels connected with the extensive apoplast system.

CONCEPT MAP

Map 16.1

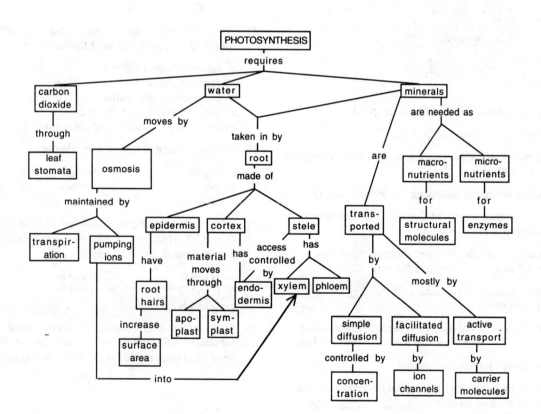

120 • CHAPTER 16

QUESTIONS

Testing recall

Match each nutrient below with the plant molecules it helps to form. (p. 326)

1 phosphorous
2 nitrogen
3 iron
4 magnesium
5 calcium
6 manganese

a micronutrient, found in electron transport molecules
b macronutrient, important for cell wall formation
c macronutrient, found in protein
d macronutrient, found in ATP and nucleic acids
e macronutrient, found in chlorophyll
f micronutrient, activates certain enzymes

Match each item below with the associated part of the root shown in cross section in the drawing. Answers may be used once, more than once, or not at all.

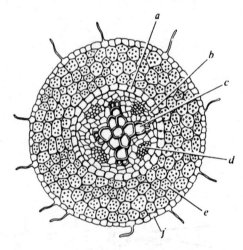

7 region that functions in starch storage (p. 333)
8 outer boundary of the stele (p. 333)
9 part of enormous surface area where much water absorption occurs (p. 330)
10 cells with a waterproof band along side and end walls (p. 333)
11 cells that conduct water and dissolved minerals to other parts of the plant body (p. 336)
12 area where water and dissolved minerals must enter a living cell to reach the vascular tissue (p. 331)
13 area that contains meristematic cells that may give rise to lateral roots (p. 333)
14 region where adjacent plant cells are interconnected and form a symplast through which water can flow (p. 334)

Complete the following statements.

15 To obtain enough energy to carry on life processes, most plants and animals require _____ for respiration. (p. 324)

16 Gas exchange between cells and the environment always takes place by _____ across _____ membranes. (p. 338)

17 Gas exchange in the leaf takes place across the cell membrane of the _____ cells. (p. 329)

18 Stomata are usually _____ when the guard cells are carrying out photosynthesis in bright sunlight. (p. 339)

19 In the leaf, excessive water loss by transpiration is prevented by _____. (p. 339)

20 Gaseous oxygen can reach living cells deep within the plant stem by entering the intercellular air space system via openings called _____. (p. 329)

Testing knowledge and understanding

Choose the one best answer.

21 Autotrophic organisms

a must digest their nutrients before taking them into the cell.
b synthesize their own organic materials from inorganic materials in the environment.
c require complex organic molecules already synthesized by other organisms.
d require no external energy source since they synthesize their own high-energy compounds. (p. 324)

22 Three nutrients needed by green plants and included in considerable quantity in most fertilizers are

a calcium, boron, and lead.
b carbon dioxide, water, and nitrogen.
c copper, zinc, and sodium.
d glucose, PGAL, and sucrose.
e nitrogen, phosphorus, and potassium. (p. 325)

23 Nitrogen is needed for the formation of

a sugars.
b fats.
c cellulose.
d proteins.
e starches. (p. 326)

24 Of the four most abundant elements in most plants (C, H, O, and N), which does a terrestrial green plant procure mainly through its roots from the soil?

a H and O
b C and O
c O and N
d H and N
e C and N (p. 327)

25 Plants rely on membrane selectivity to control which substances will enter the xylem for transport. The cell layer most responsible for this regulation is the

a cortex.
b endodermis.
c pericycle.
d xylem.
e pith. (p. 333)

26 Most of the mineral nutrients required by plants are absorbed by the root cells by the process of

   a osmosis.
   b passive diffusion.
   c active transport.
   d phagocytosis.
   e pinocytosis.   (p. 328)

27 The nutrient from which 93 percent of the total weight of glucose is synthesized is

   a water.
   b carbon dioxide.
   c potassium.
   d nitrogen.
   e phosphorus.   (p. 325)

28 Below are shown endodermal cells, each with its Casparian strip shaded in. Which one of the arrows represents a correct pathway for water? (p. 333)

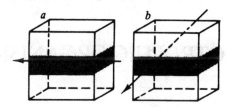

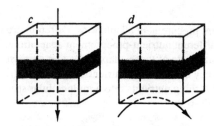

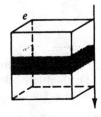

29 Minerals can be transported into roots by all the following processes *except*

   a diffusion.
   b osmosis.
   c facilitated diffusion.
   d active transport.   (p. 327)

30 The nitrogen source used by bacteria of the genus *Rhizobium* living in the root nodules of legume plants is

   a N_2.
   b NO_2^-.
   c NO_3^-.
   d NH_3.
   e NH_4^+.   (p. 328)

31 Insectivorous green plants supplement their diet by trapping and digesting insects. The most beneficial nutrient obtained from this process is

   a nitrogen.
   b glucose.
   c calcium.
   d phosphorus.
   e sulfur.   (p. 337)

32 Which of the following organisms require oxygen?

   a algae
   b maple tree
   c fish
   d dog
   e two of the above
   f all the above   (p. 337)

33 The principal site of gas exchange in the leaf is the

   a upper epidermis.
   b cuticle.
   c lenticel.
   d mesophyll.
   e lower epidermis.   (p. 338)

34 When the guard cells are turgid,

   a the stomata are open.
   b the stomata are closed. *e*lower epidermis.
   c gases can move in through the lenticels.
   d they have a relatively low concentration of potassium ions.
   e two of the above are true.   (p. 339)

For further thought

1 Contrast the nutrient requirements of animals and green plants. Why are animals ultimately dependent on green plants? (pp. 323–24)

2 Describe the processes involved in the uptake of water and minerals by a plant root. Explain why such factors as soil acidity may affect the absorption of ions. (pp. 326–28)

3 Describe a variety of plant adaptations useful in obtaining nitrogen. If you were a genetic engineer, what genes would you try to introduce into crop plants? (pp. 328–29)

ANSWERS

Testing recall

1 d	5 b	9 f	12 a
2 c	6 f	10 a	13 b
3 a	7 e	11 c	14 e
4 e	8 a		

15 oxygen
16 diffusion, moist
17 mesophyll
18 open
19 the cuticle and closing the stomata
20 lenticels

Testing knowledge and understanding

21 b	25 b	29 b	32 f
22 e	26 c	30 a	33 d
23 d	27 b	31 a	34 a
24 d	28 c		

Chapter 17

NUTRIENT PROCUREMENT IN HETEROTROPHIC ORGANISMS

A GENERAL GUIDE TO THE READING

This chapter describes the way in which heterotrophs—nonphotosynthetic bacteria, fungi, nonphotosynthetic protozoans, and animals—procure the complex organic nutrients necessary for their activities. The following are some of the highlights of Chapter 17.

1. Definitions of saprophytic, parasitic, herbivore, carnivore, and omnivore. You will want to learn these terms, introduced on page 343, as they are used frequently.

2. Nutrient requirements. The material on essential amino acids (pp. 345–47) is particularly relevant, since vegetarianism is becoming increasingly common.

3. Nutrient procurement by protozoans and animals. When you read this material (pp. 348–59) pay special attention to the differences between intracellular and extracellular digestion. Figures 17.10 to 17.15 (pp. 354–57) will be particularly helpful.

4. The digestive system of vertebrates. You will want to become thoroughly familiar with this material, which focuses on the human digestive tract. (pp. 359–65)

5. Enzymatic digestion in humans. Notice that in most cases digestion is a two-step process: one type of enzyme breaks down the complex food molecule into smaller units by hydrolyzing internal bonds in the molecule, and then other types of enzymes complete its hydrolysis to building-block units by removing the units one by one from the ends of the molecule. The summary diagrams on page 125 of this *Study Guide* provide a helpful summary of enzymatic digestion. (pp. 365–68)

6. Nutrient procurement adaptations. Different organisms have different nutrient-procurement adaptations. Take particular note of those adaptations that increase the efficiency of obtaining nutrients, digesting nutrients, and absorbing nutrients into a multicellular structure. (pp. 352–63)

KEY CONCEPTS

1. Heterotrophic organisms must obtain high-energy organic compounds already synthesized; ultimately heterotrophs depend on nutrients synthesized by autotrophs. (p. 343)

2. Much of the diversity among living things is a result of adaptations toward one of the three major nutritive modes: photosynthetic, absorptive, or ingestive. (p. 343)

3. Heterotrophs must mechanically and chemically digest their food; they must break it down into smaller molecules, before their cells can absorb and metabolize it. (p. 352)

4. Extracellular digestion is an adaptation for utilizing larger pieces of food than could be eaten with only intracellular digestion; it is the general rule in multicellular animals. (p. 356)

5. The presence of specialized sections for different functions in a digestive tract produces an efficient digestive system. (p. 357)

6. Organisms have adopted a variety of strategies to increase mechanically the surface area of the food to be digested,

as well as ways to increase the surface areas of surface areas of sections of the digestive tract for nutrient absorption. (pp. 358–63)

7 Many enzymes work together to break down large nutrient molecules into their building blocks. (pp. 365–70)

CROSSLINKING CONCEPTS

1 In living organisms, energy is transferred from the sun to photosynthetic autotrophs to heterotrophs. This principle will be particularly important when you study ecosystems in Chapter 34. (p. 343)

2 Evolution (Chapter 31) is a theme that links together all the special adaptations that organisms have undergone to procure food. Evolutionary adaptations can involve the loss, acquisition, or functional conversion of structures or systems.

3 Your study of transport across the cell membrane (Chapter 5) gives you a deeper appreciation of phagocytosis in protozoans and the active transport of digested nutrients across the intestine. Absorption across the intestinal wall is analogous to the absorption of nutrients across the root hairs in plants, studied in the previous chapter.

4 Like the previous chapter, this one supplies examples of mutualistic nutritional relationships, such as that between herbivores and the cellulose-digesting microorganisms that inhabit their caeca.

OBJECTIVES

1 Define saprophyte, parasite, herbivore, carnivore, and omnivore. Differentiate between absorptive and ingestive heterotrophs and give an example of each. (p. 343)

2 List the nutrients required by heterotrophs; specify for each nutrient at least one of its roles in metabolism or physiology. (pp. 344–48)

3 Define "essential amino acid" and explain why it is important to include a variety of different proteins in the diet. (p. 345)

4 Expalin what a vitamin is, and indicate the general function of almost all water-soluble vitamins. (p. 348)

5 Discuss the role of various minerals in the human body and indicate why many are needed in only small quantities. (p. 348)

6 Explain what digestion is and why heterotrophs need to digest complex foods before they can be absorbed. (pp. 348–49)

7 Briefly describe how fungi procure nutrients, and indicate whether their digestion is intracellular or extracellular. Describe a way in which fungi are similar to animals, and a way in which they are different. (pp. 348–49)

8 For *Paramecium*, hydra, planarian, and earthworm, explain how each organism procures its food, indicate for each whether digestion is primarily intracellular or extracellular, and explain how each digests and absorbs its food. (pp. 352–57)

9 Explain what is meant by a complete digestive tract and describe some of its advantages. In doing so, be sure to discuss the importance of the mechanical breakup of bulk food in animals; give three examples of adaptations for this process. Also, explain the adaptive significance of a storage chamber in the digestive tract. (pp. 357–58)

10 Trace the route food follows in a human from the time it enters the mouth until the undigestible residue leaves the anus; indicate where most digestion and absorption occur. Explain how the lining of the small intestine is adapted to increase the absorptive surface area. (pp. 359–64)

11 Compare the food type, tooth type, and length of the small intestine of herbivores, carnivores, and omnivores. (pp. 360–62)

12 Explain how organisms such as cows, rabbits, and horses can derive some nutrients from the cellulose they eat, even though mammals cannot digest cellulose. (p. 362)

13 Assume you have just eaten a hamburger with pickle on a bun. Describe the steps in your digestion that will break down the starch in the bun and the protein and fat in the meat. In doing so, use Figure 17.28 (p. 365) to show how starch is hydrolyzed to maltose and how maltose is hydrolyzed to glucose; specify what enzymes are involved in these reactions and where they are produced. Explain how proteins are broken down to amino acids; specify what enzymes are involved in these reactions and where they are produced. Explain how fat can be digested into fatty acids and glycerol; specify the role bile plays in fat digestion, and indicate whether fat must be hydrolyzed prior to absorption. (pp. 361–69)

KEY TERMS

absorptive heterotroph (p. 343)
saprophytic (p. 343)
parasitic (p. 343)
ingestive heterotroph (p. 343)
herbivore (p. 343)
carnivore (p. 343)
omnivore (p. 343)
essential amino acid (p. 345)
protein-energy malnutrition (PEM) (p. 346)
kwashiorkor (p. 346)
essential fatty acid (p. 347)
water-soluble vitamin (p. 348)
fat-soluble vitamin (p. 348)

digestion (p. 348)
extracellular digestion (p. 348)
intracellular digestion (p. 352)
food vacuole (p. 353)
oral groove (p. 353)
cytopharynx (p. 353)
gastrovascular cavity (p. 354)
nematocyst (p. 354)
pharynx (p. 356)
intestine (p. 356)
complete digestive tract (p. 357)
mouth (p. 357)
anus (p. 357)
esophagus (p. 357)
crop (p. 357)
gizzard (p. 357)
filter feeder (p. 359)
oral cavity (p. 359)
amylase (p. 361)
peristalsis (p. 361)
sphincter (p. 362)
pyloric sphincter (p. 362)
duodenum (p. 362)
villi (p. 363)
microvilli (p. 363)
caecum (p. 364)
appendix (p. 364)
colon (p. 364)
rectum (p. 365)
pancreas (p. 365)
pancreatic duct (p. 365)
pancreatic amylase (p. 365)
maltase (p. 365)
sucrase (p. 365)
lactase (p. 365)
pepsin (p. 367)
trypsin (p. 367)
chymotrypsin (p. 367)
endopeptidase (p. 367)
exopeptidase (p. 367)
liver (p. 368)
gall bladder (p. 368)
bile (p. 368)
lipase (p. 369)

SUMMARY

Heterotrophic nutrient procurement All organisms require high-energy organic compounds to carry out their life functions. Since many of the organic molecules found in nature are too large to be absorbed unaltered through cell membranes, they must first be hydrolyzed (i.e., digested) by enzymes into their constituent building-block molecules.

There are four main groups of heterotrophic organisms: nonphotosynthetic bacteria, fungi, protozoans, and animals.

Bacteria and fungi are *absorptive heterotrophs*; they lack internal digestive systems and depend mainly on absorption as their mode of feeding. In contrast, protozoans and animals are *ingestive heterotrophs*; they take in particulate or bulk food and digest it inside their body.

Nutrient requirements Many bacteria and fungi thrive on a diet containing only carbohydrates because they can synthesize other organic compounds from them. However, most heterotrophic organisms require carbohydrates, fats, and proteins in bulk. They also require certain vitamins and minerals in small quantities.

Nine *amino acids* are *essential* in the diet of most animals because the animals cannot synthesize them. All the essential amino acids must be present simultaneously and in the correct relative amounts if effective protein synthesis is to take place.

Some animals require no fat in their diet; others cannot synthesize enough linoleic acid for their needs, so th1s *essential fatty acid* must be included in their diet.

Vitamins are organic compounds necessary in small quantities to given organisms that cannot synthesize them. Most vitamins function as coenzymes or parts of coenzymes. A prolonged vitamin deficiency impairs metabolic processes within the cell, and often produces symptoms of a deficiency disease in the organism. Vitamins are classified into two groups on the basis of solubility: those of the B complex and vitamin C are water-soluble, and vitamins A, D, E, and K are fat-soluble.

Extracellular and intracellular digestion *Extracellular digestion* takes place in the environment (as in fungi) or in specialized digestive structures (as in animals). In fungi, digestive enzymes are secreted directly into the food supply and the products of digestion are then absorbed from the environment. Many fungi are *saprophytic* (living on dead material); others are *parasitic* (living on or in other living organisms).

Animals ingest their food and digest it in digestive structures before the nutrients cross cell membranes and become useful to the organisms. In protozoans and some animals, digestion is *intracellular*; the food is ingested directly into the cell by endocytosis, then hydrolyzed in a food vacuole, and the products of digestion are absorbed.

Gastrovascular cavities The unicellular protozoans carry on only intracellular digestion. The radially symmetrical cnidarians have a saclike body containing a *gastrovascular cavity*. The specialized cells lining the cavity carry on both extracellular and intracellular digestion. Extracellular digestion allows the organism to eat larger pieces of food than it could handle using only intracellular digestion; it is the general rule in multicellular animals.

The free-living flatworms are bilaterally symmetrical, elongated animals with a gastrovascular cavity. They carry out some extracellular digestion, but most of the food is digested intracellularly.

Complete digestive tracts Animals more complex than cnidarians and flatworms have a *complete digestive tract*, each section of which can be specialized for a different function—

mechanical breakup of bulk food, temporary storage, enzymatic digestion, absorption of products, reabsorption of water, storage of wastes, etc. For example, the earthworm has a muscular *pharynx* for sucking in food, a *crop* for temporary storage, a *gizzard* for mechanical breakup, and an *intestine* with a large surface for extracellular enzymatic digestion and absorption.

The digestive tract of mammals consists of a series of chambers specialized for different functions. The first chamber is the *oral cavity*, where the teeth break up the food mechanically. The teeth of different vertebrates are specialized in a variety of ways and may differ in number, structure, and arrangement. Sharp pointed teeth, poorly adapted for chewing, characterize *carnivores* (meat eaters), whereas broad flat teeth, well adapted for chewing, characterize *herbivores* (animals that eat plant materials). Humans, who eat both plant and animal material, are *omnivores*; their teeth are rather unspecialized.

The muscular tongue manipulates the food during chewing and mixes it with *saliva*. The lubricated food is pushed backward through the *pharynx* into the *esophagus*. The food moves quickly through the long esophagus, pushed along by waves of muscular contraction called *peristalsis*, into the *stomach*, where it is stored and further broken up by the squeezing and churning action of the stomach muscles. *Pepsin* secreted in the *gastric juice* in the stomach begins protein digestion. The food moves from the stomach into the *duodenum*, which connects to ducts carrying secretions from the liver and pancreas. The rest of the *small intestine* is long and coiled. Most of the digestion and the absorption of the products of digestion occurs in the small intestine, where several structural adaptations increase the absorptive surface area. The length of the small intestine in various animal species is proportional to the amount of plant material in their diets; herbivores ordinarily have longer small intestines than carnivores, because the cellulose cell walls of plants are difficult to break up and tend to interfere with digestion and absorption.

Indigestible material, water, and unabsorbed substances move on into the *large intestine*, or *colon*. A *caecum* projects from the junction of the small and large intestine. Two important functions of the large intestine are the reabsorption of water and the excretion of certain salts when their concentration in the blood is too high. The last portion of the large intestine, the *rectum*, functions as a storage chamber for the feces until release.

Enzymatic digestion Enzymatic digestion, the complete hydrolysis of organic nutrients into their building-block molecules, occurs mainly in the mouth, stomach, and small intestine. It is generally a two-step process. First the complex nutrients are hydrolyzed enzymatically into smaller fragments; then other enzymes complete the hydrolysis into the building-block compounds. Absorption of the simple sugars and amino acids into the blood involves active transport. A summary of the action of the various enzymes (with their sites of production shown in parentheses) follows.

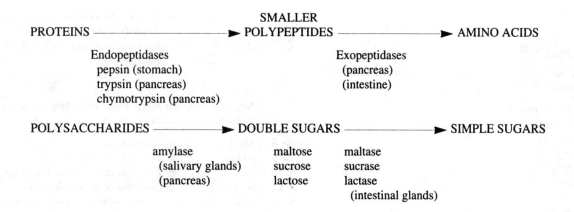

The digestion and absorption of fat follows a different pattern. Much fat is absorbed directly since it is lipid-soluble and can cross the cell membrane. Some fat is digested; the liver produces *bile* that acts to emulsify the fats so the principal fat-digesting enzyme, *lipase*, can hydrolyze them more effectively.

```
                    emulsification
    FATS ─────────────────────► EMULSIFIED FAT ─────────────► FATTY ACIDS AND GLYCEROL
          bile salts (liver)                    lipase (pancreas)
```

126 • CHAPTER 17

CONCEPT MAP

Map 17.1

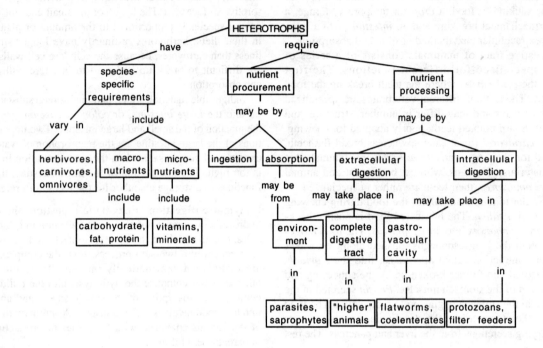

QUESTIONS

Testing recall

1 Which of the following substances are nutrients required by animals?

    a carbon dioxide
    b amino acids
    c vitamins
    d polysaccharides
    e phosphates
    f fats
    g water
    h sugars
    i proteins
    j nitrates   (pp. 344–45)

Match the vitamins listed below with their descriptions.

a 2 vitamin A   a fat-soluble, prevents xerophthalmia, important for vision
f 3 vitamin K   b water-soluble, prevents scurvy, important for collagen fibers
d 4 vitamin D   c water-soluble, coenzymes in cellular respiration, formation of red blood cells
b 5 vitamin C   d fat-soluble, prevents rickets, important for calcium absorption

c 6 vitamin B   e fat-soluble, muscle/nerve condition, detoxifies oxygen radicals
e 7 vitamin E   f fat-soluble, important for blood clotting
    (p. 349)

Match the minerals listed below with their functions. Choices may be used once, twice, or not at all.

c 8 iodine   a bones, teeth
e 9 sodium   b component of hemoglobin
b 10 iron   c component of thyroid hormone
d 11 phosphorous   d component of nucleic acids
a 12 calcium   e affects osmotic balance, component of body fluid
    (p. 348)

13 Distinguish between (among) the following terms:

    a autotroph—heterotroph
    b absorptive heterotroph—ingestive heterotroph
    c saprophyte—parasite
    d herbivore—omnivore—carnivore
    e extracellular digestion—intracellular digestion
    f incomplete digestive system—complete digestive system
    (pp. 343–57)

Match each statement below with the associated part of the digestive system in the drawing below. Answers may be used once, more than once, or not at all.

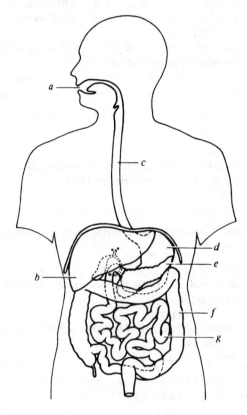

14 Carbohydrate digestion begins here. (pp. 361, 365)

15 Protein digestion begins here. (p. 367)

16 Enzymes that digest fat, protein, and carbohydrates are produced here. (p. 365)

17 Most water reabsorption occurs here. (p. 364)

18 Most fat digestion occurs here. (p. 368)

19 Most products of digestion are absorbed here. (p. 362)

20 Fat-digesting enzymes are synthesized here. (p. 369)

21 This organ produces a substance that emulsifies fat. (p. 369)

22 This organ secretes enzymes active at an extremely low pH. (p. 362)

23 This structure is lined by villi. (p. 363)

Testing knowledge and understanding

Choose the one best answer.

24 Nutritionally, a lion could best be described as

  a an autotroph.
  b an absorptive heterotroph.
  c an ingestive heterotroph.
  d a herbivore.
  e an omnivore. (p. 360)

25 Which one of the following statements most accurately describes a vitamin?

  a It can be obtained in the diet only from animal products, such as meat, eggs, milk, and cheese.
  b It is essential for all animals but can be synthesized only by green plants.
  c Though it is a useful addition to the diet, it is not essential.
  d It is a universal building block for the body's proteins.
  e It is an organic compound that is required in small quantities in the diet and that cannot be synthesized by the body. (p. 348)

26 Essential amino acids

  a can be synthesized from other amino acids within the body.
  b can be obtained only from animal protein.
  c must all be present at the same time in the proper proportions for proper utilization.
  d can be supplied by eating large quantities of a single plant protein. (p. 345)

27 Which one of the following statements most accurately describes an essential amino acid?

  a It can be obtained in the diet only from animal products such as meat, eggs, milk, and cheese.
  b It is a universal building block for the body's proteins and is found in all the proteins our bodies synthesize.
  c Though it is necessary for life, the organism in question cannot synthesize it and hence must obtain it in the diet.
  d It functions as a coenzyme in a biochemical pathway that is essential for life.
  e It is essential for all animals but can be synthesized only by green plants. (p. 345)

28 A deficiency of vitamin C (ascorbic acid) in the human diet results in the disease called scurvy, characterized by bleeding gums, loosening of teeth, delayed healing of wounds, and painful and swollen joints. These deficiency symptoms are understandable because ascorbic acid plays a major role in

  a forming red blood cells.
  b maintaining good night vision.
  c facilitating calcium absorption.
  d promoting blood clotting.
  e forming collagen fibers in connective tissue. (p. 349)

29 Beriberi can be prevented or cured if the diet contains sufficient

  a vitamin A.   d vitamin D.
  b vitamin B_1. e vitamin K.
  c vitamin C.   (p. 349)

30 Which one of the following organisms obtains its nourishment only by extracellular digestion?

  a hydra   d fungus
  b amoeba   e green plant
  c planarian   (p. 348)

31 Gastrovascular cavities

   a are found only in unicellular organisms such as *Paramecium*.
   b have only one opening to the exterior, which functions as both mouth and anus.
   c generally include a gizzard instead of teeth to break up bulk food.
   d are commonly found in herbivorous animals such as sheep and goats.
   e usually have several chambers, each specialized for a different function. (p. 354)

32 All the following are functions of the stomach in humans *except*

   a temporary food storage.
   b mechanical breakup of food.
   c digestion of proteins.
   d fermentation by microorganisms.
   e secretion of acid. (p. 362)

33 What do fungi and animal nutrient procurement have in common?

   a Both are heterotrophs.
   b Both are autotrophs.
   c Both have extracellular digestion.
   d Both have intracellular digestion.
   e *a* and *c* are true. (p. 348)

34 Which would have the longest intestine for its body size?

   a fungi   d mosquito
   b human   e cow
   c dog   (p. 362)

35 All the following are adaptations to a herbivorous diet *except*

   a sharp teeth.
   b pouches, such as the caecum, for microbial digestion.
   c a long small intestine.
   d partial dependence on bacterial synthesis and digestion. (pp. 360–64)

36 Which one of the following animals has the least subdivision of its digestive system into separate chambers specialized for different operations on food?

   a earthworm
   b cow
   c human being
   d hydra
   e chicken (pp. 354–57)

37 Which one of the following animals would you expect to have the most developed grinding surface on the molar teeth?

   a dog
   b cat
   c human
   d sheep
   e lion (p. 360)

38 Which one of the following statements concerning the digestive system of most higher land animals is *false*?

   a It begins with a mouth and ends with an anus.
   b It frequently depends on microorganisms to perform vital functions.
   c The surface area is reduced as much as possible to permit easy passage of food.
   d It has a mechanism for chewing and grinding the food.
   e It includes a chamber for food storage. (p. 357)

39 All the following increase surface area of food to be digested *except*

   a molars.
   b muscles for churning action in the stomach.
   c the gizzard in earthworms and birds.
   d bile.
   e zymogens in the lysosomes of ciliates. (p. 357)

40 All the following increase surface area for absorption *except*

   a villi and microvilli in the small intestine.
   b thick muscle tissue in the wall of the gizzard of earthworms.
   c root hairs in plants.
   d extensive branching of the gastrovascular cavity in flatworms.
   e the typhlosole in earthworms. (p. 357)

41 Which one of the following digestive enzymes would be most likely to catalyze the reaction

   $C_{12}H_{22}O_{11} + H_2O \rightarrow C_6H_{12}O_6 + C_6H_{12}O_6$

   a amylase   d endopeptidase
   b sucrase   e exopeptidase
   c lipase   (p. 365)

Questions 42 to 44 refer to the following situation.

A hungry student eats a ham and cheese sandwich and drinks a glass of milk for lunch, and has a piece of apple pie for dessert.

42 Digestion of the starch in this meal is carried out primarily in the

   a oral cavity and stomach.
   b stomach and small intestine.
   c small intestine and large intestine.
   d oral cavity and small intestine.
   e stomach and large intestine. (pp. 361, 365)

43 The products of digestion of the protein in the student's meal are absorbed in the

   a stomach.   d pancreas.
   b small intestine.   e large intestine.
   c liver.   (p. 362)

44 Enzymatic hydrolysis in the mouth will primarily affect the

   a ham.   d milk sugar.
   b cheese.   e milk fats.
   c bread.   (p. 361)

45 The digestion of fats takes place almost entirely in the

   a mouth.
   b esophagus.
   c stomach.
   d small intestine.
   e large intestine. (p. 368)

46 The enzyme that catalyzes hydrolysis of fat is

   a amylase.
   b pepsin.
   c lipase.
   d trypsin.
   e bile. (p. 369)

47 Bile aids in fat digestion by

   a hydrolyzing the bonds between glycerol and fatty acids.
   b breaking peptide bonds.
   c converting unsaturated fats to saturated fats.
   d emulsifying fat droplets so that more surface area is exposed to the action of digestive enzymes.
   e stimulating increased release of fat-digesting enzymes from the pancreas. (p. 369)

48 Which one of the following organs is part of the digestive system but does *not* secrete digestive enzymes?

   a pancreas
   b large intestine
   c small intestine
   d stomach
   e salivary gland (p. 364)

49 Which one of the following can be absorbed in the digestive tract without any hydrolysis?

   a fat molecule
   b sucrose
   c polypeptide
   d starch
   e dipeptide (p. 368)

50 Which of the following pairings of enzyme and function is *incorrect*?

   a salivary amylase—starches hydrolyzed to disaccharides
   b pepsin—peptides hydrolyzed to amino acids
   c lipase—fats hydrolyzed to glycerol and fatty acids
   d maltase—disaccharides hydrolyzed to monosaccharides
   e trypsin—internal peptide bonds (p. 367)

51 The *best* reason that the stomach is not digested by proteolytic enzymes is that

   a there are no proteolytic enzymes in the stomach.
   b only exopeptidases are found in the stomach.
   c the stomach has a mucus lining.
   d the acidity of the gastric juice inhibits enzymatic activity.
   e two of the above answers are correct. (p. 367)

For further thought

1 What is the general function of the digestive enzymes? Name some factors in the digestive tract that influence the function of the enzymes and discuss their effect. Describe some specific ways enzymes may "cut up" a protein molecule. (pp. 365–69)

2 Discuss and compare adaptations for nutrient procurement and processing in hydra, earthworm, cow, and human being. (pp. 354–60; Chapter 3, p. 57)

3 Why is folic acid often given as a supplement to pregnant women? (p. 349)

4 What effect can taking antibiotics have on your digestive system? Explain. (p. 364)

ANSWERS

Testing recall

1 *b, c, d, e, f, g, h, i*
2 *a*   5 *b*   8 *c*   11 *d*
3 *f*   6 *c*   9 *e*   12 *a*
4 *d*   7 *e*   10 *b*

13 *a* An *autotroph* can synthesize its own high-energy organic compounds. A *heterotroph* must obtain its high-energy compounds from the environment.
   b An *absorptive heterotroph* lacks an internal digestive system and relies on absorption for obtaining nutrients. An *ingestive heterotroph* can take in particulate or bulk food and digest it in an internal digestive system.
   c A *saprophyte* obtains its nutrients from dead organisms. A *parasite* obtains its nutrients from living organisms.
   d A *herbivore* eats primarily plants. A *carnivore* eats primarily meat. An *omnivore* eats both plants and animals.
   e *Extracellular digestion* occurs outside body cells, either in the environment or in a specialized digestive structure. *Intracellular digestion* occurs directly in the cells of the organism.
   f An *incomplete digestive system* has one opening to bring in nutrients and to rid the organism of undigested material. A *complete digestive system* has two separate openings and allows for more specialization of organs and an assembly-line approach to digestion.

14 *a*   17 *f*   20 *e*   22 *d*
15 *d*   18 *g*   21 *b*   23 *g*
16 *e*   19 *g*

Testing knowledge and understanding

24 *c*   31 *b*   38 *c*   45 *d*
25 *e*   32 *d*   39 *e*   46 *c*
26 *c*   33 *e*   40 *b*   47 *d*
27 *c*   34 *e*   41 *b*   48 *b*
28 *e*   35 *a*   42 *d*   49 *a*
29 *b*   36 *d*   43 *b*   50 *b*
30 *d*   37 *d*   44 *c*   51 *c*

Chapter 18

GAS EXCHANGE IN ANIMALS

A GENERAL GUIDE TO THE READING

This chapter is designed to give you an understanding of gas exchange and of how various animals have solved it. As you read Chapter 18 in your text, you will want to concentrate on the following aspects of the topic.

1. The nature of gas exchange. This topic was first discussed in Chapter 16, with reference to gas exchange in plants. This chapter reviews the three basic requirements of gas-exchange systems mentioned in the earlier chapter and describes a fourth particularly applicable to animals: a method of transporting gases between the gas-exchange surface and internal cells (p. 374). In reading the rest of the chapter, focus on how each gas-exchange mechanism meets these requirements.

2. The gills of a fish. These are described in some detail (pp. 376–78) to give you a good understanding of the problems associated with gas exchange in water and of how fish have solved these problems. Figures 18.8 (p. 378) and 18.9 (p. 379) will help you understand how the gills of a fish function.

3. Tracheal systems of arthropods. You should learn how the insect tracheal system functions. Focus on how this type of system is adaptive for arthropods but would be ineffective for birds or mammals. (pp. 379–80)

4. Human lungs. The human gas-exchange system is discussed on pages 381 to 385. Make sure you understand negative-pressure breathing and how it differs from positive-pressure breathing. (p. 386).

KEY CONCEPTS

1. A basic problem for most living organisms, both plant and animal, is procuring sufficient oxygen for respiration and eliminating carbon dioxide. (p. 373)

2. Gas exchange between a living cell and its environment always takes place by diffusion across a thin, moist cell membrane. (pp. 373–74)

3. Every organism requires a protected respiratory surface of adequate dimensions relative to its volume. The larger the organism, the more surface it requires. (pp. 373–74)

4. Many animals need a transport system to carry the gases from the respiratory surface to the internal cells. (pp. 374–75)

5. Two basic forms of respiratory systems have evolved: invaginated (primarily in terrestrial animals) and evaginated (primarily in aquatic animals). (p. 375)

6. Countercurrent exchange systems are very efficient. (pp. 376–78)

CROSSLINKING CONCEPTS

1. Surface–area-to-volume ratio. The gills of fish, tracheal tubes of arthropods, and lungs of terrestrial vertebrates all possess large surface areas for gas exchange. Earlier we saw the importance of a high surface–area-to-volume ratio in the microvilli of the small intestine (Chapter 17) and the root hairs of plants (Chapter 16). Mammals and

birds have evolved lungs with an extremely high respiratory surface–area-to-volume ratio; because they are warm-blooded animals, they have a high demand for oxygen.

2 The mammalian gas-exchange system and circulatory system (detailed in Chapter 20) are closely interrelated. The respiratory system provides the thin moist surface for gas exchange. The circulatory system transports the gases to locations where they are needed.

OBJECTIVES

1 Discuss the problem of gas exchange faced by nearly all living organisms. In doing so, consider the reason why most organisms require molecular oxygen. (p. 373)

2 Explain how gas exchange is affected by the relationship of surface to volume, the shape of an organism, and the permeability of the outer covering of an organism. (pp. 373–74)

3 List the four basic requirements for gas-exchange systems. (pp. 373–74)

4 Give three reasons why obtaining oxygen is a greater problem for aquatic animals than for air breathers. (p. 375)

5 Explain how the sea star, the segmented marine worm, the squid, and the fish meet the four basic requirements for gas-exchange systems. Indicate what features the gas-exchange mechanisms of these organisms have in common. (pp. 376–78)

6 Using Figures 18.83 (p. 378) and 18.9 (p. 379), show how the flow of water across the gills of a fish maximizes the amount of oxygen picked up. Describe what is meant by a countercurrent exchange system. (pp. 377–78)

7 Distinguish between an invaginated and an evaginated respiratory surface, and give an example of each. Give three reasons why most land animals have evolved invaginated rather than evaginated gas-exchange systems. (pp. 378–79)

8 Describe the tracheal system of an insect, and identify two fundamental ways in which this invaginated system differs from the respiratory system of the land vertebrtes Indicate whether a tracheal system would be sufficient for larger aniamls, and why or why not. (pp. 379–80)

9 Trace the route a molecule of air follows in your body from inhalation to your lung. Then, explain how the structure of your lung satisfies the four basic requirements for a gas-exchange system. (pp. 381–84)

10 Describe the process of breathing in your body, and explain how this differs from the breathing process of a frog. (pp. 385–86)

11 Explain why birds and mammals have such high oxygen requirements and how birds obtain oxygen from the air more efficiently than mammals. (p. 385)

KEY TERMS

invaginated (p. 375)
evaginated (p. 375)
gill (p. 376)
parapodia (p. 376)
countercurrent exchange system (p. 378)
tracheae (p. 378)
spiracle (p. 380)
lung (p. 378)
external nares (p. 381)
nasal cavity (p. 381)
pharynx (p. 381)
glottis (p. 382)
epiglottis (p. 382)
larynx (p. 382)
trachea (p. 382)
bronchus (p. 382)
broncioles (p. 382)
alveolus (p. 382)
rib cage (p. 385)
diaphragm (p. 385)
negative-pressure breathing (p. 386)
positive-pressure breathing (p. 386)

SUMMARY

The problem of gas exchange Since aerobic respiration is the chief method of respiration in both plants and animals, the vast majority of organisms must have some method of obtaining oxygen and getting rid of waste carbon dioxide.

Gas exchange between a living cell and its environment always takes place by *diffusion* across a moist cell membrane. With the evolution of large three-dimensional organisms came the necessity for a corresponding evolution of specialized respiratory surfaces to meet the increased oxygen requirements and the necessity of getting oxygen to interior cells. A wide variety of structures for gas exchange have evolved in animals, but each respiratory system must meet four basic requirements:

1 a respiratory surface of adequate dimensions
2 protection for the fragile respiratory surface
3 a moist surface
4 in most animals, a method of transporting the gases to the cells

Solutions in animals Most large multicellular animals have evolved some type of special gas-exchange surface. These may be outward-oriented (*evaginated*) or inward-oriented (*invaginated*) extensions of the body surface.

Aquatic animals The cells of most small aquatic organisms are in close contact with the surrounding medium and carry out gas exchange across their whole body surface; no special respiratory devices are necessary. Most multicellular aquatic animals utilize evaginated gas-exchange surfaces called *gills*, which have finely subdivided surfaces that expose an immense exchange surface to the water. Most gills contain a rich supply of blood vessels, which carry special carrier pigments that transport oxygen to the individual cells.

In fish, the water flows over the gills in the opposite direction to that of the blood flow; this *countercurrent exchange system* maximizes the amount of oxygen the blood can pick up from the water.

Because of oxygen's low solubility and slow diffusion rate in water, most aquatic organisms must constantly ventilate, or move water across, the exchange surface.

Terrestrial animals Most terrestrial animals have evolved invaginated respiratory systems, either *lungs* or *tracheae*.

The tracheal system typical of land arthropods consists of many small air ducts called *tracheae* that run from openings (*spiracles*) in the body wall and carry air directly to the individual cells. Because osygen is delivered directly to each cell by the system, there is no significant transport of oxygen by the blood. This system limits the size of the organism.

Lungs are invaginated gas-exchange organs supplied with blood vessels. The evolution of the lung in higher vertebrates has tended toward increased surface area and an increased blood supply to the exchange surface.

In human beings, air is drawn through the *external nares* into the *nasal cavities*, then moves into the throat, or *pharynx*, a common passageway for food and air. After leaving the pharynx through the *glottis*, the air enters the *larynx*. During swallowing, the *epiglottis* closes the glottis. The air then moves into the *trachea*, which divides at its lower end into the two *bronchi*, which lead into the two lungs. Each bronchus branches repeatedly and forms *bronchioles*, which branch into smaller ducts that terminate in the *alveoli*. Each alveolus is surrounded by a dense bed of blood capillaries; the diffusion of oxygen and carbon dioxide takes place here.

Air is drawn into and pushed out of the lungs during the breathing process. In humans, inhalation occurs when the rib muscles contract and draw the rib cage up and out, and the *diaphragm* contracts and moves downward, and thus reduces the air pressure within the cavity below the atmospheric pressure and draws air into the lungs. Normal exhalation is a passive process; the muscles relax and the rib cage and diaphragm return to their resting positions, and thus increase the air pressure in the lungs, which forces air out.

In birds, the special arrangement of the lungs and their associated *air sacs* permits a continuous one-way flow of air through the lungs during both inhalation and exhalation. Because of the one-way flow and the countercurrent arrangement of blood vessels, birds extract oxygen from air more efficiently than mammals.

When air is drawn into the lungs as a result of an enlarged body cavity, the process is termed *negative-pressure breathing*; when air is forced into the lungs, it is termed *positive-pressure breathing*. Birds and mammals utilize negative-pressure breathing. Frogs use a combination of positive- and negative-pressure breathing.

CONCEPT MAP

Map 18.1

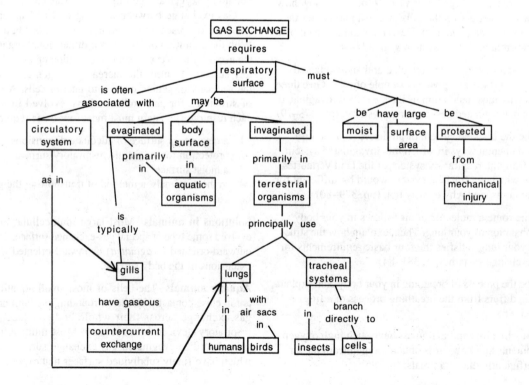

QUESTIONS

Testing recall

Complete the following statements.

1. To obtain enough energy to carry on life processes, most plants and animals require _____ for respiration. (p. 373)

2. Many small aquatic organisms such as Hydra have no special respiratory mechanism, since dissolved gases can quickly move a distance of up to _____ mm by simple diffusion. (p. 376)

3. Gas exchange between cells and the environment always takes place by _____ across _____ membranes. (p. 373)

4. Gills are an example of an _____ respiratory surface whereas lungs are an example of an _____ surface. (pp. 376–78)

5. In the gills of fish, the countercurrent flow of water and blood greatly _____ the pickup of O_2 by the blood. (p. 378)

6. The rate of diffusion is _____ in air than in water. (p. 375)

7. Most land organisms utilize an _____ type of respiratory system, of which there are two principal types, _____ and _____. (p. 378)

8. _____ are the gas-exchange openings on the body surface of insects. They lead into many smaller air ducts called _____, which conduct air to internal cells. (p. 380)

9. Gas exchange in the human takes place by the process of _____ across the thin, moist membranes of the _____, which are surrounded by a dense network of _____. (p. 383)

10. In humans, incoming air is warmed, moistened, and filtered in _____. (p. 381)

11. Frogs force air into their lungs in a process called _____-pressure breathing whereas birds and mammals _____ air into their lungs using _____-pressure breathing. (p. 386)

12. When the diaphragm contracts, the air pressure in the lungs _____. (p. 385)

13. List the four basic requirements for a specialized gas-exchange surface.

    _____   _____
    _____   _____
    (pp. 373–74)

Testing knowledge and understanding

Choose the one best answer.

14. All the following organisms use gills as a gas-exchange system *except* the
    a squid.
    b sea star.
    c fish.
    d marine semgent worm.
    e frog. (p. 386)

15. Water flowing over the gills of the fish
    a flows in the same direction as the blood.
    b flows in a direction opposite to that of the blood.
    c exchanges O_2 and CO_2 with the blood in the capillaries of the gills.
    d does *a* and *c*
    e does *b* and *c* (p. 378)

16. The passageway common to the respiratory and digestive systems of humans is the
    a nasal cavity.   d larynx.
    b trachea.   e esophagus.
    c pharynx.   (p. 381)

Questions 17 to 19 refer to the following graph, which shows pressure changes in the thoracic (chest) cavity during breathing in a human.

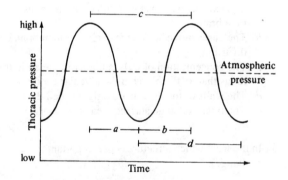

17. Which segment of the curve (*a*, *b*, *c*, or *d*) covers the period of inhalation? (p. 385)

18. Which segment of the curve covers the period of exhalation? (p. 385)

19. Which segment of the curve covers the period when the diaphragm muscle is relaxing? (p. 385)

20. Which of the following features do *all* gas-exchange systems have in common?
    a The exchange surfaces are moist.
    b They are enclosed in a special chamber.
    c They are maintained at a constant temperature.
    d They are exposed to air.
    e They are found only in vertebrates. (p. 373)

21 Which one of the following statements is *false* for gills operating in water?

    a Water can support the delicate gill filaments.
    b Most fish must actively pump water across their gills.
    c Keeping membranes moist is no problem.
    d Water carries more oxygen than air. (p. 375)

22 Insects carry out respiratory gas exchange

    a in specialized external gills.
    b in specialized internal gills.
    c in the alveoli of their lungs.
    d directly across the membranes of internal cells.
    e across thier thin moist skin. (p. 379)

23 The human respiratory system is *most* like that of

    a land snails.   *d* house flies.
    b squid.   *e* spiders.
    c fish. (p. 381)

24 The normal air pathway in the human respiratory system is

    a trachea → pharynx → alveolus → bronchus.
    b bronchus → bronchiole → pharynx → trachea.
    c alveolus → bronchus → bronchiole → pharynx.
    d trachea → bronchus → alveolus → pharynx.
    e pharynx Æ trachea Æ bronchus Æ alveolus. (pp. 381–82)

25 Which one of these statements about human lungs is *false*?

    a Gas exchange always takes place across moist membranes.
    b The gases move across the exchange membranes by diffusion.
    c The concentration of CO_2 is higher in the air in the lungs than in the blood in the alveolar capillaries.
    d The walls of the alveoli are only one cell thick.
    e The total exchange surface area is very large. (p. 383)

26 In the frog, which structure is *not* important in gas exchange?

    a nostrils   *d* diaphragm
    b mouth   *e* skin
    c lungs (p. 386)

27 Gas exchange in plant stomata and intercellular spaces is most similar to gas exchange

    a across thin moist skin.
    b in the lungs.
    c in the gills.
    d in the spiracles and tracheae.
    e in water bubbles. (p. 380)

For further thought

1 Discuss basic requirements for gas-exchange mechanisms, and show how these are fulfilled in a fish, a human, an insect, and a large tree. (pp. 378–83; see Chapter 16, pp. 337–41)

2 Explain why gills are not suitable for life on land. (p. 378)

3 The second largest invertebrate group is the phylum Mollusca. A familiar example of the mollusc is the marine clam. In the marine clam, gas exchange is performed by two double gills, which hang in the mantle cavity. The gills consist of many interconnected thin plates, or lamellae. Each lamella is supplied with small blood vessels, which bring blood to the gill and return it to the heart. Cilia cover the gills and function to produce a flow of water through the gills.

    a For gas exchange, give one advantage and one disadvantage of a marine environment.
    b Discuss how the clam meets the basic requirements of a gas-exchange system. (p. 375)

4 Compare and contrast the respiratory systems of a human being and a terrestrial insect. (pp. 379–84)

5 Cigarette smoke has been found to have the following effects on the respiratory system:

    a destruction of many of the cilia that line parts of the respiratory tract
    b thickening of the walls of the bronchioles, and thus reduction of the interior diameter of the tubes
    c rupturing of the walls of some of the alveoli.

For each of these effects, indicate how the *normal* functioning of the respiratory tract is altered by the use of cigarettes. (pp. 383–84)

6 You are probably aware that your exhaled breath contains water vapor. How do you account for this? (p. 381)

7 Why do professional football teams generally keep oxygen masks on the sidelines when playing games against the Denver Broncos in Mile High Stadium, located in the foothills of the Rocky Mountains?

ANSWERS

Testing recall

1. oxygen
2. one
3. diffusion, moist
4. evaginated, invaginated
5. facilitates
6. faster
7. invaginated, lungs, tracheae
8. spiracles, tracheae
9. diffusion, alveoli, capillaries
10. nasal cavities
11. positive, draw, negative
12. decreases
13. large surface area, method of transport, protection for fragile surface, moist surface

Testing knowledge and understanding

14	e	18	b	22	d	25	c
15	e	19	b	23	a	26	d
16	c	20	a	24	e	27	d
17	a	21	d				

Chapter 19

INTERNAL TRANSPORT IN PLANTS

A GENERAL GUIDE TO THE READING

This chapter, the first of two on the topic of internal transport, discusses transport in higher plants; the next considers transport in animals. As you read Chapter 19 in your text, you will want to give special attention to the following aspects of the topic.

1. Organisms without special transport systems. Though this chapter deals primarily with transport in higher plants, you need to realize that many organisms do not have special transport mechanisms and rely on diffusion for hte transportation of materials. The section on these organisms (pp. 389-90) makes this point clear.

2. Stems. Focus your attention on the three types of plant stems: monocots, herbaceous dicots, and woody dicots. Notice the similarities and differences. You will find learning about stems easier if you carefully study the photographs and diagrams on pages 391 to 396 (Figs. 19.2-19.11).

3. Structure of xylem and phloem cells. Study the material on the different types of conducting cellscarefully: notice in particular the differences between xylem and phloem cells. Figures 19.12 to 19.15 (pp. 397-98) are helpful in learning this material.

4. Hypotheses about the mechanisms of transport. You will want to understand the transpiration theory of xylem transport (p. 400) and the pressure-flow model of phloem transport (pp. 403-404). Figure 19.20 (p. 404) is crucial for understanding the latter.

KEY CONCEPTS

1. Diffusion plays an important role in the movement of materials, but it is a very slow process. Consequently most organisms have evolved some sort of specialized transport mechanism to distribute substances more rapidly. (p. 389)

2. The successful exploitation of the land environment by plants was dependent on the evolution of specialized conducting tissues that would transport water and nutrients from one part of the plant to another. (p. 390)

3. All tissues produced by the apical meristem are primary tissues; they contribute to the plant's growth in length. All tissues derived from the vascular cambium are secondary tissues; they contribute to growth in diameter. (pp. 391-93)

4. The two main types of plant vascular tissue are xylem, which conducts water and inorganic ions upward from the roots to the aerial parts of the plant, and phloem, which conducts water and organic solutes (and some inorganic ions) from one part of the plant to another, both upward and downward. (pp. 402-403)

5. According to the transpiration-adhesion-tension-cohesion (TATC) theory, water lost by transpiration from the aerial parts of plants is replaced by withdrawal of water from the water column in the xylem, and in the process the whole column is pulled upward. It requires an interaction of the forces of transpiration, adhesion, tension, and cohesion. (pp. 400-401)

6 According to the pressure-flow hypothesis, there is a mass flow of solutes under pressure through the sieve tubes of phloem from an area of high solute concentration to an area of low solute concentration. (p. 403)

CROSSLINKING CONCEPTS

1 Xylem and phloem were introduced earlier as plant tissues (Chapter 15). The structure and function of these tissues are discussed more thoroughly in this chapter.

2 Although this chapter focuses on the *stems* of plants, you should remember that vascular tissue is continuous from the roots through the stems to the leaves (Chapter 15).

3 Plasmodesmata were introduced in Chapter 5 as a means of transport between plant cells. In this chapter, you study a type of plasmodesmata called pits.

4 An understanding of the chemical structure of water, particularly hydrogen bonds (Chapter 2), underlies the transpiration theory of sap movement in the xylem.

OBJECTIVES

1 Explain why the unicellular and multicellular algae can get along without a specialized internal-transport system, whereas most higher plants cannot. (pp. 389–90)

2 Give two reasons why the evolution of specialized transport tissue was necessary for the higher plants to exploit the land environment fully. (p. 390)

3 Using cross sections in Figures 19.2–19.4 (pp. 391–92), point out the xylem and phloem, and where present, the vascular cambium, pith, epidermis, and cortex. Explain how you can distinguish among herbaceous dicot, woody dicot, and monocot stems. (pp. 391–95)

4 Compare the structure and arrangement of tissues in a typical dicot stem (as shown in Fig. 19.6, p. 393) with those of a typical dicot root (see Fig. 16.10, p. 395). Do the same for a monocot stem (as shown in Fig. 19.4, p. 392) and a monocot root (see Fig. 16.14, p. 398). Explain how you can tell whether the cross section you are looking at is a root or a stem.

5 Explain the differences between primary and secondary tissue in a dicot, and describe the process of secondary growth in a dicot stem. Using the cross sections in Figure 19.8 (p. 394), identify the primary xylem, secondary xylem, vascular cambium, primary phloem, secondary phloem, and periderm. (pp. 391–95)

6 Describe the pattern of growth that leads to the rings seen in cross sections of trees; explain the difference between heartwood and sapwood. Explain why phloem does not form growth rings. (pp. 393–94)

7 Draw diagrams showing how the structure of a typical 25-year-old tree trunk (as seen in cross section) would differ from that of a stem of the same species during its first season of growth. (p. 394)

8 Contrast the structure of xylem tissue with that of phloem with respect to the types of cells present, the structure of the conducting cells, and whether the conducting cells are living or dead. (pp. 397, 401)

9 Describe the potential mechanisms by which sap moves upward in xylem tissue. In doing so, consider the force necessary to raise water to the tops of the tallest trees. Describe the role of root pressure in the ascent of sap. Describe the transpiration (TATC) theory, considering the role of transpiration, hydrogen bonding (in both adhesion and cohesion), and the tension produced by osmotic of water loss in the leaves. (pp. 397–401)

10 Using Figure 19.20 (p. 404), describe the mechanism of phloem transport according to the pressure-flow hypothesis. Define the "source" and the "sink." Indicate whether the movement through sieve tubes would be predominantly upward or downward on a sunny day in summer, and in the same plant at night. (pp. 403–404)

KEY TERMS

cytoplasmic streaming (p. 389)
nonvascular plants (p. 390)
vascular plants (p. 390)
xylem (p. 390)
phloem (p. 390)
herbaceous (p. 390)
woody (p. 390)
annual (p. 390)
perennials (p. 390)
bundle sheath (p. 391)
apical meristem (p. 391)
primary tissue (p. 392)
vascular cambium (p. 392)
secondary phloem (p. 393)
secondary xylem (p. 393)
secondary vascular tissue (p. 393)
heartwood (p. 394)
sapwood (p. 394)

cork cambium (p. 395)
cork cell (p. 395)
periderm (p. 395)
outer bark (p. 395)
inner bark (p. 395)
tracheid (p. 397)
pit (p. 397)
vessel element (p. 397)
vessel (p. 397)
vascular ray (p. 397)
sap (p. 399)
guttation (p. 399)
root pressure (p. 400)
cohesion (p. 400)
tension (p. 400)
adhesion (p. 400)
TATC theory (p. 400)
transpiration (p. 400)
sieve element (p. 401)
sieve tube (p. 402)
sieve plate (p. 402)
companion cell (p. 402)
translocation (p. 402)
source (p. 403)
sink (p. 403)
pressure-flow hypothesis (p. 403)

SUMMARY

Unicellular organisms In general, only very small organisms can rely exclusively on the processes of diffusion and intracellular transport for the movement of substances. Diffusion is a very slow process. *Cytoplasmic streaming*, which is considerably faster than diffusion, creates definite currents along the cell vacuole's surface. Most higher organisms, both plant and animal, require a specialized transport system to deliver required materials to the cells and remove the waste products.

Vascular tissue structure The large multicellular land plants, the *vascular plants*, have evolved two types of specialized conducting tissue: *xylem* and *phloem*. These tissues form continuous pathways running through the roots, stems, and leaves and serve to transport materials from one part of the body to another.

Stems function as organs of transport and support. In young stems, the tissues are produced by the *apical meristem* as the stem grows in length; these tissues are *primary tissues*. If the vascular cambium becomes active, it produces *secondary tissues*, which contribute to growth in diameter.

There are two types of herbaceous stems: monocot and dicot. In a monocot stem, the vascular tissue is in discrete bundles scattered throughout the stem. In an herbaceous dicot, the vascular tissue is arranged in a ring of discrete bundles separating the central pith from the cortex. In the latter, the vascular cambium, which arises between the xylem and phloem, may become active and produce *secondary xylem* cells to the inside and *secondary phloem* cells to the outside. This increases the diameter of the stem.

In woody stems, the cambium is active and each year new secondary xylem and phloem cells are produced. Since xylem cells produced early in the growing season are larger than cells produced later, a series of concentric annual rings is formed. Usually only the newer, outer rings, or *sapwood*, are functional in transport; the cells of the older rings, the *heartwood*, have become plugged and function only in support. As woody stems grow in diameter, a layer of cells outside the phloem, the *cork cambium*, begins to produce cork cells, which form the outer bark, or *periderm*.

The sequence of tissues (moving from outside toward the center) in an old woody stem is cork tissue and cork cambium (outer bark), primary phloem and secondary phloem (inner bark), vascular cambium, secondary xylem (wood), primary xylem, and pith.

Xylem contains two types of conductive cells: *tracheids* and *vessel elements*. Both are dead at functional maturity; the cellular contents disintegrate and leave thick, lignified cell walls that form hollow tubes within which vertical movement of materials can take place. Numerous *pits* in the wall allow the lateral movement of materials from cell to cell. Xylem also contains *fiber* and *parenchyma* cells. The thick-walled fiber cells are supportive elements; the parenchyma cells are often arranged to form *rays*, which function as pathways for the lateral movement of materials.

Phloem also contains both fiber and parenchyma cells in addition to those unique to it: *sieve elements* and *companion cells*. The elongate sieve elements with their perforated end walls—the *sieve plates*—are arranged end to end and form a *sieve tube* through which vertical transport takes place. At maturity the nucleus of the sieve element disintegrates, but the cytoplasm remains. Companion cells, which retain both their nuclei and their cytoplasm, are intimately associated with the sieve elements in most advanced plants.

Xylem transport: TATC theory Water and inorganic ions absorbed by the roots (sap) move upward through the plant body in the tracheids and vessels of the xylem. According to the *transpiration-adhesion-tension-cohesion (TATC) theory*, water lost by *transpiration* from the aerial parts of the plant is replaced by withdrawal of water from the water column in the xylem. The water in the xylem forms a continuous chain from the aerial parts of the plant to the roots. Water molecules *adhere* to the hydrophilic walls of vessels and tracheids. The chain of water is under *tension* created by transpiration from the leaves. And these water molecules exhibit *cohesion* because of hydrogen bonding between the water molecules. Thus the withdrawal of water molecules from the top of a water column will pull the whole chain of water molecules upward.

In some plants, especially in early spring, *root pressure* may contribute to the upward movement of sap in the xylem. Root pressure is created when ions are actively transported into the stele, and water follows passively in such quantity as to build up pressure in the xylem and push the sap slowly upward.

Phloem transport: pressure-flow hypothesis Inorganic ions are transported, or *translocated*, upward from the roots to the leaves primarily through the xylem; however, the downward transport of these ions is through the phloem. Most organic solutes (carbohydrates and organic nitrogen compounds) are transported, both up and down, through the phloem.

Unlike the xylem transport, the movement of substances in the phloem (i.e., *translocation*) is through living cells that retain their cytoplasm. The end walls are penetrated only by the tiny pores of the sieve plate. The most widely supported hypothesis of phloem function is the *pressure-flow* hypothesis, according to which there is a mass flow of water and solutes under pressure through the sieve tubes from an area of high turgor pressure (the "source") to an area of low turgor pressure (the "sink"). The source is any place where more food is being produced than is used (e.g., photosynthesizing leaves during the day). The sink is any place where more food is being used up than produced (e.g., a storage cell in a root or stem).

CONCEPT MAP

Map 19.1

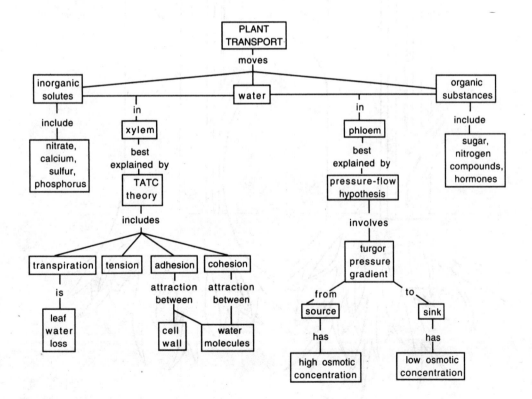

140 • CHAPTER 19

KEY DIAGRAM

In the following diagrams of monocot and woody dicot plants, color in the following using a different color for each structure. First color the names, then the appropriate structures. (Reference: Figs. 19.3, p. 391 and 19.11, p. 396)

Which plant is the monocot and which the dicot?

How can you tell the difference between the two?

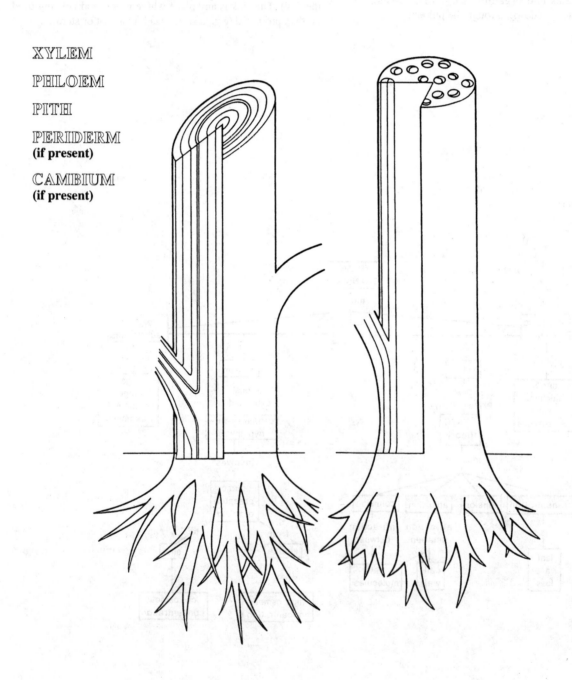

XYLEM

PHLOEM

PITH

PERIDERM
(if present)

CAMBIUM
(if present)

QUESTIONS

Testing recall

1 Complete the following chart; it will be a useful study aid. (pp. 397–404)

	Xylem	Phloem
Conducting cells	vessels trach.	sieve tubes
Are cells living or dead at functional maturity?	dead	living
Materials transported	inorg.	organic
Direction of movement	↑	↑ and ↓
Probable mechanism of transport	TACT theory	pressure flow the

Match each item below with the associated part of the plant stem shown in cross section in the diagram. Then refer to the diagram for questions 10 and 11. Answers may be used once, more than once, or not at all.

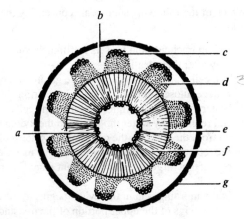

2 primary phloem (p. 394) c
3 secondary phloem (p. 394) d
4 vascular cambium (p. 394) e
5 primary xylem (p. 394) a
6 secondary xylem (p. 394) f
7 tissues produced by the vascular cambium (p. 392)
8 cork (p. 395) b
9 wood (p. 394)
10 As the stem grows, tissue c — a
  a will be pushed outward.
  b will be pushed inward.
  c will remain in the same position. (p. 393)

11 The stem shown is
  a a monocot.
  b a herbaceous dicot.
  c a woody dicot. (p. 394)

Decide whether the following statements are true or false, and correct the false statements.

12 In a stem, the secondary phloem is located outside the layer of primary phloem. (p. 393) F

13 As the stem of a woody plant grows through successive years, the cortex and pith become less and less important and are often virtually obliterated. (p. 395) T

14 Phloem is the tissue of which wood is principally composed. (p. 394) F

15 Mature functioning vessel cells are dead, but mature functioning tracheids are alive. (p. 397) F

16 Mature sieve elements lack a nucleus but retain cytoplasm. (p. 402) T

17 Mature xylem tracheid cells are metabolically more active than mature phloem sieve cells. (p. 397) F

18 At present, most biologists believe transpiration pull is more important than root pressure in the upward movement of water in xylem. (p. 400) T

19 The phloem of a tree can transport organic materials upward only. (p. 405) F

20 Some mineral nutrients are much more mobile in the phloem than others. (p. 402) T

21 Current evidence on the mechanism of transport in the phloem favors the pressure-flow hypothesis. (p. 403) T

Testing knowledge and understanding

Choose the one best answer.

22 Which of the following statements about transport is *false*?
  a Transport systems help organisms get nutrients to cells.
  b Transport systems help organisms maintain moist respiratory surfaces for gas exchange.
  c Transport systems help organisms rid themselves of wastes.
  d Transport systems are most important to multicellular organisms.
  e Transport is not a concern to unicellular organisms. (p. 389)

23 Stems can function in all the following except
  a holding the plant firmly attached in the soil.
  b photosynthesis in chloroplasts.
  c nutrient storage in pith.
  d transport of nutrients.
  e support of the plant. (p. 390)

24 Which one of the following parts of the trunk of a large tree is the most important in transport of sucrose from the leaves to the roots?

   a heartwood   d outer bark
   b sapwood   e cambium
   c inner bark   (p. 395)

25 In a tree trunk, the tissue just outside the vascular cambium is the

   a cork.   d primary xylem.
   b primary phloem.   e secondary xylem.
   c secondary phloem.   (p. 393)

26 The plant tissue responsible for the growth in height of a vascular plant is the

   a vascular cambium. d cork cambium.
   b apical meristem.   e phloem.
   c xylem.   (p. 392)

27 In a 25-year-old tree, most of the "wood" is

   a pith.   d secondary phloem.
   b primary xylem.   e cork cells.
   c secondary xylem.   (p. 395)

28 Baseball bats are made primarily from

   a xylem.   d periderm.
   b phloem.   e stele.
   c cambium.   (p. 395)

29 Heartwood functions primarily in

   a transporting nutrients.
   b transporting gases.
   c transporting minerals.
   d transporting wastes.
   e support.   (p. 394)

30 One cell type in a plant stem is particularly hard to study because its contents are under pressure and the cell bursts when cut. This cell is a

   a tracheid.   d epidermal cell.
   b sieve element.   e vessel element.
   c vascular cambial cell.   (p. 403)

31 A nail is driven into the trunk of a young 6-foot-tall tree, 3 feet above the ground. Ten years later the tree is 30 feet tall. The nail will be how many feet above the ground?

   a 3 feet   d 15 feet
   b 6 feet   e 27 feet
   c 10 feet   (pp. 391–92)

32 A beaver gnawed completely around a birch tree but did not proceed further to fell the tree. It was noticed that the leaves retained their normal appearance for several weeks, which were hot and sunny, but that the tree eventually died. It can be concluded that the most peripheral region it is *certain* that the beaver left functional was

   a the phloem.   d the bark.
   b the cortex.   e the pith.
   c the xylem.   (p. 396)

33 Which one of the following mature cells in a plant would be least active metabolically?

   a root endodermal cell   d phloem companion cell
   b stem cortex cell   e xylem vessel element
   c mesophyll cell   (p. 397)

34 Which of the following retains *both* cytoplasm and nucleus when mature?

   a tracheid   c sieve element
   b vessel   d companion cell
                                        (p. 402)

35 Which one of the following statements concerning a tracheid cell in mature secondary xylem in a large tree is *false*?

   a It was produced by cell division in the cambium.
   b It once contained a nucleus, but no longer has one.
   c Its cytoplasm is under the control of companion cells.
   d Its thick cellulose walls are impregnated with lignin.
   e Its walls are interrupted by numerous pits.
                                        (p. 397)

36 Which of the following statements concerning the xylem is *false*?

   a Xylem tracheids and vessels fulfill their vital function only after their death.
   b The cell walls of the tracheids are greatly strengthened with cellulose fibrils forming thickened rings or spirals.
   c Water molecules are transpired from the cells of the leaf and replaced by water molecules in the xylem pulled up from the roots by the cohesion of water molecules.
   d Movement of materials is by mass flow; materials move owing to a turgor-pressure gradient from "source" to "sink."
   e In the morning, sap in the xylem begins to move first in the twigs of the upper portion of the tree and later in the lower trunk.   (p. 403)

37 Which of the following mechanisms could raise water to the greatest height in the xylem vessels of a tall tree?

   a transpiration from the leaves combined with cohesion of water molecules
   b root pressure
   c cytoplasmic streaming
   d active transport of water
   e mass flow along a turgor-pressure gradient
                                        (p. 400)

38 Which of the following does *not* play a role in xylem transport?

   a pressure–flow hypothesis
   b transpiration
   c root pressure
   d adhesion and cohesion   (p. 403)

39 Which of the following organic compounds may be transported in the xylem?

    a fatty acids   *c* glycerol
    b amino acids   *d* starch   (p. 403)

40 Root pressure

    a is not sufficient to raise water above ground level in any plant.
    b is negative in all but the tallest trees.
    c can push water to the top of a 10-foot corn plant but not to the top of a 200-foot tree.
    d is the best explanation known for the transport of water to the tops of the tallest trees.
    e is the driving force for the mass flow of sugar. (p. 400)

41 What is the principal pathway by which sugar travels in a plant?

    a xylem vessels   *d* sieve elements
    b companion cells   *e* capillaries
    c parenchyma cells   (p. 403)

42 According to the pressure-flow hypothesis, materials are moved through the conducting cells as a result of

    a a turgor-pressure gradient along the sieve tubes.
    b a pressure gradient along the sieve tubes produced by cytoplasmic streaming.
    c active transport between adjacent sieve elements.
    d a flow along membranous interfaces within the sieve elements.
    e a pull exerted by transpiration from the aerial parts of the plant. (p. 403)

43 Which one of the following statements concerning the flow of sap in the xylem of trees is *correct*?

    a In the morning sap begins to flow first in the twigs and later in the trunk.
    b Flow makes the trunk increase in diameter during the day and decrease at night.
    c Flow is driven by the high concentration of sugar in vessel elements.
    d Rapid flow of water puts the xylem under pressure much greater than atmospheric pressure.
    e Flow from the roots to the twigs would be accelerated if the leaves were removed. (p. 400)

44 Arrange the following five events in an order that explains the mass flow of materials in the phloem.

    1 Water diffuses into the sieve elements.
    2 Leaf cells produce sugar by photosynthesis.
    3 Solutes are actively transported into sieve elements.
    4 Sugar is transported from cell to cell in the leaf.
    5 Sugar moves down the stem.

    a 2-1-4-3-5   *d* 1-2-3-4-5
    b 2-4-3-1-5   *e* 4-2-1-3-5
    c 2-4-1-3-5   (p. 403)

45 The "pressure" in the pressure-flow hypothesis comes from

    a root pressure.
    b the osmotic uptake of water by sieve elements in the leaf.
    c the accumulation of minerals and water by the stele in the root.
    d transpiration in the leaves. (p. 403)

For further thought

1 As plants evolved to land forms, they were confronted with the problem of support, since they no longer had water to support their weight. How was this problem solved? (p. 390)

2 Trace the route a molecule of water would follow from the place where it enters a vascular plant until it is lost from a leaf by transpiration. (pp. 400–401)

3 Explain how the carbohydrates synthesized in the leaves get from the leaves to the root and other parts of the growing plant. (p. 403)

4 Hypothesize which plant tissue is tapped into for maple sugaring. Explain your choice. (p. 399)

5 Describe two research experiments that contributed to our knowledge of how transport occurs within the xylem and the phloem. (pp. 399–404)

ANSWERS

Testing recall

1

	Xylem	Phloem
Conducting cells	tracheids vessels	sieve elements
Are cells living or dead at functional maturity?	dead	living
Materials transported	water, inorganic nutrients	water, organic materials—sucrose, amino acids, etc.
Direction of movement	up only	up and down
Probable mechanism of transport	transpiration, cohesion	pressure-flow

2 *c*
3 *d*
4 *e*
5 *a*
6 *f*
7 *d, f*
8 *g*
9 *a, f*
10 *a*
11 *c*
12 false—inside
13 true
14 false—xylem
15 false—both are dead
16 true
17 false—less active (they are dead)
18 true
19 false—upward and downward
20 true
21 true

Testing knowledge and understanding

22	*e*	28	*a*	34	*d*	40	*c*
23	*a*	29	*e*	35	*c*	41	*d*
24	*c*	30	*b*	36	*d*	42	*a*
25	*c*	31	*a*	37	*a*	43	*a*
26	*b*	32	*c*	38	*a*	44	*b*
27	*c*	33	*e*	39	*b*	45	*b*

Chapter 20

INTERNAL TRANSPORT IN ANIMALS

A GENERAL GUIDE TO THE READING

This chapter briefly describes the circulatory system in a variety of animals and then focuses on the human circulatory system, a subject most students find fascinating. As you read Chapter 20 in your text, you will want to concentrate on the following material.

1. The introductory discussion. This part of the chapter (pp. 407–409) covers a number of important topics, including the concept of open and closed circulatory systems.

2. The parts of the human heart and the main blood vessels. You will want to learn these, as shown in Figures 20.4 and 20.5 (pp. 410–11).

3. The exchange of materials in the capillaries. This process, one of the more difficult subjects presented in the chapter, is discussed on pages 418 to 423. Figure 20.17 (p. 422) is the key to understanding it.

4. Thromboembolic and hypertensive disease in human beings. The box on this subject (pp. 419–20) is particularly interesting.

5. The lymphatic system. It is necessary to understand the importance of this system; read pages 424 to 425 with care.

6. The function of the red blood cells and hemoglobin. The discussion of this topic (pp. 428–30) refers to the material in Chapter 3 on the structure of the hemoglobin molecule. Make sure you understand the role of hemoglobin in the transport of O_2 and CO_2.

KEY CONCEPTS

1. The active way of life and rapid metabolism of animals require that each cell be provided with a continuous, abundant supply of oxygen and nutrients to metabolize for energy. Therefore many complex animals have evolved a true circulatory system to transport materials to and from the body cells more efficiently. (p. 407)

2. The high metabolic rate to support the active way of life of the endothermic birds and mammals is made possible by the four-chambered heart and the evolution of separate pulmonary and systematic circulations. (p. 412)

3. Nearly all the exchange of materials between the blood and the tissue fluid occurs in the capillaries; this exchange is governed by a delicate balance between the blood pressure (hydrostatic pressure) and osmotic pressure. (pp. 421–22)

4. The cells of higher animals require a fairly uniform, or stable, chemical environment (homeostasis) in order to maintain normal function. The circulatory system plays an important role in maintaining stability within the body by delivering nutrients and oxygen and by removing carbon dioxide and nitrogenous wastes. (p. 407)

CROSSLINKING CONCEPTS

1. The blood was briefly introduced as a type of connective tissue in your study of animal tissues in Chapter 15. The components of the blood are studied in more detail in this chapter.

2. This chapter describes the importance of nodal tissue in controlling the heartbeat. Later, you will study how the involuntary nervous system contributes to this control.

3. This chapter explores the connections between the fluids of the circulatory and lymphatic systems. In the next chapter, which covers immunology, you will learn more about the functions of the lymphatic system in immunology.

4. In Chapter 8, you studied cellular respiration from a biochemical point of view. In the discussion of the gas-exchange system (Chapter 18), you learned more about the exchange of carbon dioxide and oxygen in the lung. In this chapter, you will learn in more detail how these gases are transported by the blood. You will see how the circulatory system aids in temperature regulation.

5. In this chapter, you will study how the circulatory system transports nitrogen wastes, ions, and hormones. The excretion of nitrogen wastes and maintenance of ionic balance will be presented in more detail when you study the kidney (Chapter 22). The potassium and calcium ions carried in the plasma play a significant role in transmission of nerve impulses (Chapter 28) and muscle contraction (Chapter 30). Hormones will be described in considerable detail when you study the endocrine system (Chapter 25).

6. Your study of protein structure (Chapter 3) is crucial to an understanding of the structure and function of the hemoglobin molecule.

OBJECTIVES

1. Explain why many very small animals, such as hydra and planarian, can get along without a specialized circulatory system, whereas most higher animals cannot. Give one major way in which the circulatory system of plants differs from that of animals. (pp. 407–409)

2. Differentiate between an open and a closed circulatory system, and compare them with respect to speed and efficiency. Cite the advantages of a closed circulatory system over an open one. Explain why insects can function so successfully with an open circulatory system, whereas animals such as earthworms must utilize a closed system. (p. 409)

3. Describe the principal parts of your heart, and the main blood vessels. Using Figure 20.5 (p. 411), name the parts of the heart as are appropriate, and trace the flow of blood from the posterior vena cava, through the heart, to the lungs, back to the heart, and out the aorta. Indicate which portion of this pathway constitutes the pulmonary circuit.

4. Using Figures 20.7 (p. 411) and 20.8 (p. 413), compare the circulatory systems of modern fish, amphibians, reptiles, and mammals and explain why the mammalian system is advantageous to an endotherm with a high metabolic rate.

5. Discuss the events that take place during one cycle of contraction and relaxation of the heart. In doing so, trace the path of an electric impulse from its point of origin to the other parts of the heart, indicate whether the heart requires stimulation from the nervous system to contract, and explain the terms diastole and systole. (pp. 410–14)

6. Discuss the structure and function of arteries, veins, and capillaries. In doing so, explain why the blood pressure is considerably lower in capillaries and veins than in arteries and why the blood pressure is constant rather than fluctuating. Specify the mechanisms other than blood pressure that function in moving blood through the veins. Indicate whether blood flows more slowly or more quickly through capillaries than through arteries, and explain how this difference in speed is correlated with capillary function. Give two mechanisms by which water and solutes move out of the capillaries. (pp. 411, 417–22)

7. Using Figure 20.17 (p. 422), explain how water and dissolved materials are forced out of the capillaries at the arteriole end of the capillary bed and into the capillaries at the venule end. (p. 422)

8. Describe the structure of the lymphatic system and discuss three important ways in which this system helps maintain the normal functions of the body. (p. 424)

9. Name the substances that make up blood—the constituents of the plasma, the cells, and the solid components. In doing so, list the three major groups of cells and solid components and give one function of each; then list the solutes found in the plasma and give a major function of each. (p. 426)

10. Draw a diagram that shows the major steps in the formation of a blood clot. Be sure to include the damaged cells, thrombin, fibrin, platelets, prothrombin, fibrinogen, thromboplastin, and Ca^{++} in your drawing. (pp. 426–28)

11. Review Figure 3.28 (p. 69), which shows the quaternary structure of hemoglobin. Point out where the O_2 molecules bind, and specify how many O_2 molecules can bind to each hemoglobin molecule. (Chapter 3, p. 69; p. 429)

14. Explain how most CO_2 is transported in the blood. (p. 430)

KEY TERMS

homeostasis (p. 407)
heart (p. 408)
closed circulatory system (p. 409)
open circulatory system (p. 409)
artery (p. 409)
vein (p. 409)
capillary (p. 410)
arterioles (p. 410)

venules (p. 410)
right atrium (p. 410)
right ventricle (p. 410)
pulmonary trunk (p. 410)
pulmonary artery (p. 410)
pulmonary vein (p. 410)
left atrium (p. 410)
left ventricle (p. 410)
aorta (p. 410)
anterior vena cava (p. 410)
posterior vena cava (p. 410)
pulmonary circulation (p. 410)
systemic circulation (p. 412)
autorhythmic (p. 412)
sino-atrial (S-A) node (p. 412)
nodal tissue (p. 412)
atrio-ventricular (A-V) node (p. 412)
bundle of His (p. 412)
Purkinje fibers (p. 412)
systole (p. 414)
diastole (p. 414)
blood pressure (p. 416)
hydrostatic pressure (p. 421)
lymph (p. 424)
plasma (p. 426)
erythrocyte (p. 426)
leukocyte (p. 426)
platelet (p. 426)
stem cells (p. 426)
fibrinogen (p. 426)
fibrin (p. 426)
thromboplastin (p. 427)
prothrombin (p. 427)
thrombin (p. 427)
erythropoietin (p. 428)
hemoglobin (p. 429)
heme (p. 429)
oxyhemoglobin (p. 429)
anemia (p. 429)

SUMMARY

Most animals, with their active way of life and high metabolic rates, rapidly consume energy and therefore require an almost continuous supply of nutrients and oxygen. One of the important functions of a transport system is to help maintain *homeostasis* of the cells and extracellular fluids of the body by delivering materials to the tissues and removing the accumulated waste products.

Circulatory systems Most higher animals have a true circulatory system in which a pumping device, the *heart*, generally forces blood around the body along a fairly definite path. The system is an *open circulatory system* if blood leaves the vessels and flows through large open spaces within the body; it is a *closed circulatory system* if the blood is always contained within well-defined vessels.

Open circulatory systems are characteristic of most molluscs and all arthropods. The open circulatory system of insects consists of a dorsal longitudinal vessel which contracts and forces the blood out at the anterior end into the head. The blood then flows backward through large sinuses and finally reenters the dorsal vessel. Although this system seems inefficient for highly active organisms, recall that oxygen delivery in insects is dependent on a tracheal system, not on a circulatory system.

Closed circulatory systems are characteristic of earthworms and all vertebrates. The system characteristic of vertebrates consists of a heart and numerous vessels: *arteries*, which carry blood away from the heart; *veins*, which carry blood to the heart; and *capillaries*, tiny vessels that connect the arteries to the veins. The actual exchange of materials between the blood and tissues takes place in the capillaries.

The human heart The human heart is a double pump; each side of the heart is divided into two chambers, an upper chamber, or *atrium*, that receives the blood and pumps it into a lower chamber, or *ventricle*, which then pumps the blood away from the heart.

Deoxygenated blood from all over the body enters the right atrium and then moves into the right ventricle. The right ventricle then pumps it through the *pulmonary arteries* to the lungs, where the blood picks up O_2 and gives up CO_2. The oxygenated blood then enters the left atrium by the *pulmonary veins*. This portion of the circulatory system is called the *pulmonary circulation*.

The left ventricle pumps the blood into the *aorta* and its numerous branches, from which the blood moves into capillaries, where the exchange of materials takes place, then into veins, and finally back via the *anterior* or *posterior vena cava* to the right atrium of the heart. This portion of the circulatory system is called the *systemic circulation*.

The four-chambered heart, with complete separation of the right and left sides, is characteristic of the endothermic birds and mammals. Complete heart separation and the double pump to maintain high blood pressure makes this the most efficient transport system. Amphibian and reptile hearts without complete separation have some mixing of oxygenated and deoxygenated blood.

The heartbeat is initiated when a wave of contraction spreads out from the *S-A node* across the atria to the *A-V node*, which sends excitatory impulses down the fibers of the bundle of His, to the *Purkinje fibers*, stimulating both ventricles to contract. When the ventricles contract (*systole*), the blood is forced out of the heart and into the arteries under high pressure. During relaxation (*diastole*), the blood pressure in the arteries falls. Friction causes the blood pressure to decrease as the blood moves farther away from the heart. One-way valves, skeletal muscle action, and the motions of the chest during breathing aid in moving blood in the veins.

Capillary exchange The enormous numbers, extensive branching, and small diameters of the individual capillaries ensure that all tissues are supplied with a large capillary surface area for the exchange of materials. Exchange may be accomplished by diffusion through the capillary cell walls, by transport in endocytotic vesicles, and by filtration between capillary cells.

The movement of materials into and out of the capillaries is governed by the balance between the hydrostatic blood pressure and osmotic pressure. At the arteriole end of a capillary the hydrostatic pressure usually exceeds the osmotic pressure, and water is forced out of the capillary; at the venule end the reverse is true, and water moves into the capillary.

Temperature regulation The circulatory system plays an important role in the regulation of both heating and cooling. Temperature control is particularly important in endotherms, which must maintain an optimum temperature for proper enzyme function. that are cold often experience shivering, which heats the blood passing through the capillaries in the muscles. The warmth is carried by the circulatory system to the rest of the body. In addition, a system of countercurrent exchange aids in conservation of heat: arteries carrying warm blood out from the body to the extremities lie against veins returning with cold blood from the extremities.

To cool themselves animals evaporate water; for humans, this means sweating. Blood is shunted to the skin, where it can radiate heat directly into the air and simultaneously benefit from evaporative cooling.

Lymphatic system The *lymphatic system* helps maintain the osmotic balance of the body fluids by returning excess tissue fluid and proteins to the blood. Lymph nodes located along the major veins act to filter out particles and are sites of formation of some white blood cells.

Blood Blood consists of the *plasma*, which is the liquid portion of the blood, and the three major types of cells or solid components: *red blood cells*, *white blood cells*, and *platelets*. The water in the plasma is the solvent for the inorganic ions, plasma proteins, organic nutrients, nitrogenous waste products, hormones, and dissolved gases that are transported by the blood.

Blood clotting is initiated when the surfaces of the damaged vessel release *thromboplastin* which helps convert the plasma protein *prothrombin* into *thrombin* with the aid of calcium ions. The thrombin then converts another plasma protein, *fibrinogen*, into *fibrin*, which forms the clot.

White blood cells, or *leukocytes*, defend the body against disease and infection.

Red blood corpuscles, or *erythrocytes*, contain the oxygen-carrying pigment *hemoglobin*. Each hemoglobin molecule can combine loosely with four molecules of oxygen to form oxyhemoglobin. When the concentration of oxygen is relatively high (as in the lungs), the hemoglobin picks up oxygen; when the concentration is low (as in the tissues), the oxyhemoglobin releases oxygen. The blood also transports carbon dioxide from the tissues to the lungs. Most of the CO_2 is carried in the form of the bicarbonate ion (HCO_3^-). The excess H^+ ions are bound to hemoglobin and plasma proteins.

CONCEPT MAP

Map 20.1

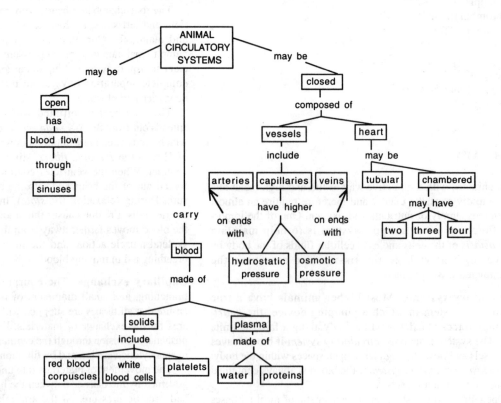

QUESTIONS

Testing recall

Decide whether the following statements are true or false, and correct the false statements.

1. Insects have blood capillaries but no red blood cells. (p. 409) F

2. A weak aorta is more likely to rupture during systole than during diastole of the cardiac cycle. (p. 416) T

3. At the arteriole end of a capillary, the osmotic pressure of the blood is greater than the blood pressure. (p. 422) F

4. The lymphatic system returns excess tissue fluid to the blood. (p. 424) T

5. Blood clotting begins as soon as the platelets are exposed to air. (p. 426) T

6. Carbon monoxide is a dangerous poison because it has a strong tendency to bind to hemoglobin. (p. 429) T

7. Iron is an essential component of the hemoglobin molecule. (p. 429) T

8. Carbon dioxide is transported in the blood plasma principally as dissolved gas (CO_2). (p. 430) F

9. The pacemaker (S-A node) of the heart is located in the brain. (p. 412) F

10. Earthworms, like insects, have an open circulatory system. (p. 409) F

11. The open circulatory system is a disadvantage to insects because it does not efficiently deliver oxygen to the cells. (p. 409) T

12. Conditions caused wholly or partly by atherosclerosis are the leading cause of death in the United States. (pp. 419–20) T

Questions 13 to 22 refer to the following diagram of the adult heart.

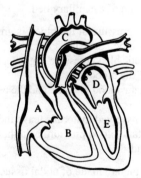

13. A correct sequence for the flow of blood through the heart would be

   a A-B-C-E-D.
   b A-B-D-E-C.
   c D-E-C-B-A.
   d B-A-D-E-C.
   e C-E-D-B-A.   (p. 410)

14. The heartbeat originates in the wall of what part of the heart? (p. 412) A

15. Blood going to the lungs leaves from what part of the heart? (p. 410) B

16. Blood from all over the body is returned to what part of the heart? (p. 410) A

17. Which chamber pumps blood that will be distributed to all parts of the body? (p. 410) E

18. The mitral or bicuspid valve is between which two chambers of the heart? (p. 414) DE

19. Which two chambers contract as a unit? (p. 414) AB BE

20. Which chambers pump oxygenated blood? (p. 414) DE

21. The pulmonary circulation connects which two chambers of the heart? (p. 414) BD

22. The pacemaker (S-A node) is located in which part of the heart? (p. 412) A

For statements 23 and 24, tell whether the relationship is true or false, and correct the false relationships.

23. The greater the total cross-sectional area of a set of blood vessels, the slower the rate of flow of the blood. (p. 418) T

24. Hemoglobin unloads oxygen when the concentration of oxygen in the surrounding medium is high. (p. 429) F

Testing knowledge and understanding

Choose the one best answer.

25. Both open and closed circulatory systems

   a usually have one-way valves.
   b have capillary beds.
   c carry respiratory gases.
   d are low-pressure systems.   (p. 409)

26. Which one of the following blood vessels carries deoxygenated blood in a mammal?

   a aorta
   b artery to the kidney
   c pulmonary artery
   d pulmonary vein
   e coronary artery   (p. 410)

27. In which one of the following blood vessels of a person would you expect to find the highest systolic blood pressure?

   a pulmonary artery
   b pulmonary vein
   c arteriole in the leg
   d capillary in the brain
   e anterior vena cava   (pp. 413–14)

28. In which one of the above vessels would there be the greatest difference between systolic and diastolic pressure? (pp. 413–14)

29 A labeled red blood corpuscle is released into the arterial circulation in the left leg. It is recaptured 30 seconds later in the left lung. What is the minimum number of chambers of the heart it must have passed through?

 a 0
 b 1
 c 2
 d 3
 e 4   (p. 410)

30 A red blood corpuscle leaves the left ventricle of a human and travels into an artery leading to the right arm. How many capillary beds will it pass through before it returns to the left ventricle?

 a 0
 b 1
 c 2
 d 3
 e 4   (p. 410)

31 A correct continuous sequence for blood flow in the adult human heart is

 a pulmonary vein, left atrium, left ventricle.
 b posterior vena cava, left atrium, left ventricle.
 c right atrium, right ventricle, aorta.
 d left atrium, right atrium, aorta.
 e left atrium, left ventricle, pulmonary artery.   (p. 410)

32 The driving force that moves blood back to the heart in the veins is

 a active transport.
 b beating of the heart.
 c closing of one-way valves.
 d skeletal-muscle contraction.
 e stretching of the walls of the veins.   (pp. 417–18)

Questions 33 to 38 refer to the following diagram of two mammalian blood vessels (A and B) connected by a capillary bed. Blood pressure is higher in B than in A. The arrows indicate the direction of net diffusion for O_2 and CO_2. Choose the one best answer.

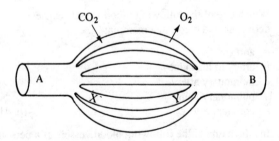

33 This capillary bed is probably part of

 a the systemic circuit.
 b the pulmonary circuit.
 c either the systemic or the pulmonary circuit.   (p. 410)

34 The hydrostatic blood pressure exceeds the osmotic pressure at

 a X.
 b Y.   (p. 422)

35 Which blood vessel is a vein?

 a A
 b B   (p. 410)

36 Blood flows from

 a A to B.
 b B to A.   (p. 410)

37 The concentration of O_2 in the blood will be highest at

 a X.
 b Y.   (p. 410)

38 Which chemical reaction is occurring in these capillaries (Hb = hemoglobin)?

 a $Hb + 4O_2 \rightarrow Hb(O_2)4$
 b $Hb(O_2)4 \rightarrow Hb + 4O_2$ (p. 429)

39 The following diagram of a capillary shows the hydrostatic blood pressure measured at three points along this particular capillary. Net outward movement of water at the arteriole end and net inward movement at the venule end would be facilitated if the osmotic pressure of the blood (relative to the tissue fluid) were

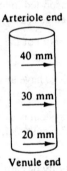

 a 15 mm Hg.
 b 20 mm Hg.
 c 30 mm Hg.
 d 40 mm Hg.
 e 50 mm Hg.   (p. 422)

40 At the venule end of a capillary bed,

 a the osmotic pressure of the blood is greater than the hydrostatic blood pressure.
 b the hydrostatic blood pressure is greater than the osmotic pressure of the blood.
 c the hydrostatic blood pressure and the osmotic pressure are equal.
 d water and dissolved materials leave the capillary.
 e the hydrostatic blood pressure is higher than it is at the arteriole end.   (p. 422)

41 All the following are made of nodal tissue *except*

 a pacemaker.
 b Purkinje fibers.
 c bundle of His.
 d A-V node.
 e mitral valve.   (p. 412)

42 During capillary filtration, loss of fluid from the arteriole end is mostly counterbalanced at the venule end by

   a osmotic pressure.
   b diffusion.
   c endocytosis.
   d exocytosis.
   e hydrostatic pressure. (p. 422)

43 Which one of the following statements concerning the regulation of the human heartbeat is *correct*?

   a The heart immediately stops beating when nervous connections are severed.
   b The rate of heartbeat is not influenced by nerves or chemical factors from outside the heart.
   c Heartbeat is initiated and coordinated by nodal tissue.
   d The right half of the heart beats before the left half.
   e The left half of the heart beats before the right half. (p. 412)

44 All the following are parts of or contained in the lymphatic system *except*

   a capillary networks.
   b arterioles.
   c valves.
   d leukocytes.
   e nodes. (p. 424)

45 Which one of the following would you *least* expect to find in lymph?

   a salt
   b fat
   c water
   d protein
   e hemoglobin (p. 424)

46 The lymphatic system performs all the following functions *except* to

   a absorb fat from the intestine.
   b filter out dead cells.
   c store dust or soot in the nodes.
   d store excess tissue fluid in the nodes.
   e monitor for infection. (p. 424)

47 The circulatory system performs all the following functions *except* to

   a carry nutrients to the body cells.
   b control body responses to stimuli.
   c carry oxygen to the body cells.
   d carry wastes away from the cells.
   e regulate body temperature. (pp. 418–24)

48 Which one of the following statements concerning blood clotting is *false*?

   a For clotting to occur, the blood must be exposed to molecular oxygen.
   b When platelets are mechanically damaged, they release a substance that initiates clotting.
   c For clotting to occur, calcium ions must be present.
   d The plasma protein fibrinogen is necessary for blood clotting.
   e The plasma protein prothrombin is necessary for blood clotting. (pp. 426–28)

49 During clotting, surfaces of damaged vessels release

   a thromboplastin.
   b thrombin.
   c fibrinogen.
   d calcium ions.
   e prothrombin. (p. 427)

50 Which one of the following statements concerning mammalian red blood cells is *false*?

   a They are spherical in shape.
   b They lack a nucleus when mature.
   c They contain hemoglobin.
   d They transport O_2.
   e They are destroyed by the liver and the spleen. (p. 428)

51 CO_2 released by a mammalian cell would probably be found moving in the blood primarily as

   a carboxyhemoglobin.
   b bicarbonate ions.
   c carbonic acid.
   d acid hemoglobin.
   e oxyhemoglobin. (p. 430)

One of the best ways to learn the human circulatory system is to trace the route a red corpuscle follows as it moves through the circulatory system, naming, in order, all the vessels and chambers of the heart that the red corpuscle would pass through. For example,

trace the path of a corpuscle moving by the most direct route possible from the capillaries in the lung to a vein in the brain: capillaries in lung, pulmonary vein, left atrium, left ventricle, aorta, artery leading to brain, arterioles, capillaries in brain, venules, and vein in brain.

Now try questions 52 and 53.

52 Trace the path of a corpuscle moving by the most direct route possible from an artery in the leg to an artery in the arm. (p. 410)

53 Trace the path of a corpuscle moving by the most direct route possible from the anterior vena cava to an artery leading to the kidney. (p. 410)

For further thought

1 Distinguish between open and closed circulatory systems. What are the advantages of a closed system? (p. 409)

2 If a person touched a high-voltage electric power line, what would be the effect on her heart? Explain. (p. 412)

3 A man goes to his doctor and finds that his blood pressure is 190/110. What do these numbers refer to? What is the effect of high blood pressure on the exchange of materials in the capillaries? What symptoms may result over a long period of time if this condition remains untreated? (pp. 414–22)

4 John Jones was admitted to the hospital with a swollen

and painful infected finger. There was a red streak up his arm, and several large, tender lumps in his armpit.

a What caused the lumps under his arm?
b State the factors that contributed to the swelling.
c Discuss the ways in which the human body protects itself against invading bacteria and viruses. (You will learn much more about this topic in Chapter 21.)

(pp. 424–28)

ANSWERS

Testing recall

1 false—no blood capillaries and no corpuscles
2 true
3 false—less
4 true
5 false—clotting initiated only by damaged tissue
6 true
7 true
8 false—as HCO_3^-
9 false—in the right atrium
10 false—earthworms have a closed circulatory system
11 false—is not a disadvantage because tracheae deliver oxygen

12	true	16	A	19	B, E	22	A
13	b	17	E	20	D, E	23	true
14	A	18	D, E	21	B, D	24	false—loads
15	B						

Testing knowledge and understanding

25	a	32	d	39	c	46	d
26	c	33	a	40	a	47	b
27	a	34	b	41	e	48	a
28	a	35	a	42	a	49	a
29	c	36	b	43	c	50	a
30	c	37	b	44	b	51	b
31	a	38	b	45	e		

52 artery in leg, arterioles, capillaries in leg, venules, veins in leg, posterior vena cava, right atrium, right ventricle, pulmonary artery, arterioles, capillaries in lung, venules, pulmonary vein, left atrium, left ventricle, aorta, artery in arm

53 anterior vena cava, right atrium, right ventricle, pulmonary artery, arterioles, capillaries in lung, venules, pulmonary vein, left atrium, left ventricle, aorta, artery leading to kidney

Chapter 21

DEFENSE OF THE HUMAN BODY: THE IMMUNE SYSTEM

A GENERAL GUIDE TO THE READING

This chapter describes how the human immune response protects the body against potential pathogens. You will find Chapter 21 easier if you proceed slowly and carefully and allot plenty of time to studying it. The payoff for concentrated study is great; you will emerge from this reading with a sound basis for understanding the immune response to many human pathogens. A box on AIDS describes how this debilitating disease affects the immune systems. You will want to concentrate on the following topics.

1. The nonspecific immune response. This section explains the important roles of epithelial tissues and the inflammatory response in protecting the body from pathogens. Figure 21.3 (p. 435) is a useful summary of this material.

2. The specific immune response. Notice that although T lymphocytes and B lymphocytes are both important for the immune response, they function in quite different ways. You should focus on the two different immune responses: humoral and cell-mediated. Also note the fundamental mechanism of clonal selection (pp. 440–41), which functions in both responses. Figure 21.15 (p. 443) is central to an understanding of this area.

3. AIDS. This box describes the nature of the AIDS virus, how the virus attacks the immune system, and how the disease is spread. (pp. 444–46)

4. Self-recognition and autoimmune disease. You will want to focus on how the immune system distinguishes between "self" and "nonself." It is important for the immune system to be able to attack foreign invaders without attacking its own cells. Focus on the role of MHC molecules. (pp. 447–50)

KEY CONCEPTS

1. The nonspecific immune response nonselectively defends the body against foreign invaders: it acts as the first line of defense against infectious agents and most potential pathogens. (pp. 433–35)

2. An important defense against disease in vertebrate animals is the ability to produce cells and antibodies that can inactivate or destroy foreign substances. (p. 433)

3. When stimulated, the B lymphocytes differentiate and then manufacture and secrete antibodies in the process called the humoral immune response, which is particularly effective against bacterial cells, viruses, and toxins. (pp. 436–38)

4. Antibodies are proteins that consist of four polypeptide chains: two identical "heavy" chains and two identical shorter "light" chains; these chains are linked by disulfide bonds. Most of the antibody molecule is constant in its amino acid sequence and structure; the two binding sites for the antigens are at the ends of the variable portion of the molecule. (p. 436)

5. Antibodies can bind to two antigen molecules; the result is agglutination of the antigens. The agglutinated clumps can trigger three antigen-destroying responses: macrophage ingestion, natural killer-cell activity, and complement-mediated lysis. (pp. 437–38)

6. Stimulated T lymphocytes are responsible for the cell-mediated immune response, which primarily defends

against pathogens such as viruses that are living inside cells. (p. 438)

7 T-cell receptors are always membrane-mounted molecules consisting of two polypeptide chains. T cells cannot respond to the antigen directly; the antigen must be first joined to an MHC molecule and displayed on the surface of a cell. (p. 438)

8 The balance among the three types of T cells—cytotoxic T cells, helper T cells, and suppressor T cells—determines the overall level of the immune response. (pp. 438–40)

9 According to the clonal selection hypothesis, the human body contains an enormous number of lymphocytes, each of which, whether of the B or T type, has only one type of membrane-mounted antibody or receptor. An antigen can bind only to those lymphocytes whose receptor molecules have binding sites for it. A lymphocyte that has been activated by the binding of the appropriate antigen begins to grow and divide, and produces a clone of cells. (p. 440)

10 The specific immune response begins when a circulating phagocytic white blood cell such as a macrophage engulfs a pathogen, processes its antigens, and displays them in combination with MHC-II protein onits cell surface. The macrophage activates the rest of the immune system. (pp. 440–42)

11 Immunity in which the individual actively forms antibodies and memory cells in response to contact with an antigen is called active immunity. Passive immunity is acquired when and individual receives antibodies that were previously synthesized by another immune individual. (pp. 442–43)

12 Recognition of "self" originates during fetal development, when no foreign antigens are present. Rejection of grafts and transplanted organs introduced into the adult organism occurs because the foreign tissue bears alien MHC proteins. (p. 447)

13 When the immune system has difficulty distinguishing "self" from "nonself," the immune cells may attack the body's own proteins and thus bring on autoimmune diseases such as multiple sclerosis, rheumatoid arthritis, and insulin-dependent diabetes. (pp. 447–48)

CROSSLINKING CONCEPTS

1 The relationship of humans to the normal microbial flora that help to protect us from pathogens is an example of a mutualistic symbiosis, a topic which is discussed in detail in Chapter 31.

2 A thorough understanding of enzyme-substrate binding (Chapter 4) will enhance your understanding of antibody-antigen binding.

3 The control of the immune response by means of helper and suppressor T cells that have opposite effects illustrates a regulatory principle that you will encounter again in your study of the endocrine system (Chapter 25) and the nervous system (Chapter 28).

4 This chapter discusses acetylcholine and its receptors in relation to myasthenia gravis, an autoimmune disease. you will learn more about this important neurotransmitter in Chapters 28 and 30.

OBJECTIVES

1 Describe the nonspecific immune response, including the role of the skin and other epithelial tissues, and the functions of the inflammatory response. (pp. 433–35)

2 Differentiate between T lymphocytes and B lymphocytes with respect to the site of maturation (thymus vs. bone marrow), the type of immune response elicited (humoral vs. cell-mediated), and the type of antigens to which each responds. (pp. 435–36)

3 Explain what happens when an organism is exposed to an antigen such as a bacterial cell and is stimulated to produce antibodies, making clear the role of the B lymphocytes, plasma cells, memory cells, antibodies, macrophages, natural killer cells, and the complement cascade. Figures 21.7 (p. 437) and 21.9 (p. 438) may be helpful in meeting this objective. (pp. 436–38)

4 Using Figure 21.5 (p. 436), describe antibody structure. Be sure to point out the light chains, the heavy chains, the disulfide bonds that link the chains, the constant regions, the variable regions, and the antibody binding sites. (p. 436)

5 Describe how the cell-mediated immune response acts to defend the body against antigens such as those found on virus-infected cells. Be sure to include the roles of memory cells, cytotoxic T cells, helper T cells, suppressor T cells, cytokines, and macrophages. (pp. 438–42)

6 Using 21.13 (p. 441), explain the process of clonal selection. (p. 440).

7 Describe the structural similarities and differences among antibodies, T-cell receptors, and MHC molecules. Describe the different functions of MHC I and MHC II molecules. You may find Figures 21.10 and 21.11 (p. 439) useful.

8 Using Figure 21.14 (p. 442), explain the role of macrophages in initiating the immune response. (pp. 440–42).

9 Differentiate between active and passive immunity. For each of the following, indicate whether the resulting immunity would be active or passive: infection with

influenza virus, injection with a vaccine consisting of weakened live viruses, injection with antibodies produced in another animal, injection with inactivated toxins, and ingestion by an infant of mother's milk containing antibodies. (pp. 442–47)

10 Describe the role of the immune system in distinguishing "self" from "nonself." Discuss the role of MHC molecules in the recognition of grafted or transplanted cells. (p. 447)

KEY TERMS

specific immune response (p. 433)
nonspecific immune response (p. 433)
inflammatory response (p. 434)
antigen (p. 435)
humoral immune system (p. 435)
antibody (p. 435)
cell-mediated immune system (p. 436)
lymphocyte (p. 436)
B lymphocyte (p. 436)
T lymphocyte (p. 436)
plasma cell (p. 436)
memory cell (p. 436)
macrophage (p. 437)
natural killer (NK) lymphocyte (p. 438)
T-cell receptor (p. 438)
major histocompatibility complex (MHC) (p. 438)
cytotoxic T cell (p. 439)
helper T cell (p. 438)
cytokines (p. 438)
suppressor T cell (p. 438)
clonal selection (p. 440)
clone (p. 440)
interleukin-1 (p. 441)
interferon (p. 442)
vaccine (p. 446)
active immunity (p. 446)
passive immunity (p. 446)
antiserum (p. 446)
colostrum (p. 447)
autoimmune disease (p. 447)

SUMMARY

The immune response involves two interacting defense systems: nonspecific and specific immunity. The *specific* immune response involves the recognition and destruction of specific foreign substances, whereas the *nonspecific* immune response nonselectively defends the body against foreign invaders.

The nonspecific immune response The nonspecific immune response acts as the first line of defense against infectious agents and most potential pathogens. The epithelial tissue barrier with its normal flora, bodily secretions, phagocytic cells, and the inflammatory response are all part of the nonspecific immune response. The signs of an inflammatory response are *calor* (heat), *rubor* (redness), and *dolor* (pain), all of which are the result of chemicals released from damaged tissue and from invading bacteria. The chemicals increase the blood supply to the infected area, make the capillary walls more permeable to tissue fluid and white blood cells, and attract white blood cells to the area. Many of the white blood cells destroy pathogens by phagocytosis; others produce powerful proteins that help destroy bacteria and detoxify foreign proteins and other potentially dangerous substances.

The specific immune response In the specific immune response, vertebrate animals manufacture highly specific cells and *antibodies* to inactivate or destroy invading *antigens*, large molecules ordinarily foreign to the organism's body. Two interacting systems of specific immunity protect the vertebrate body from foreign antigens: humoral and cell-mediated immunity. *Humoral immunity* defends the body against pathogens living outside human cells whereas *cell-mediated immunity* defends the body against pathogens found inside human cells. Cell-mediated immunity is also involved in regulating the activity of the humoral system. The cells that respond to the antigen are the *lymphocytes*, of which there are two types: the *B cells*, which mature in the bone marrow and mediate the humoral immune response, and the *T cells*, which mature in the thymus and control the cell-mediated immune response.

The humoral immune response Each B cell carries a characteristic membrane-bound protein receptor, called an antibody, which is capable of binding a complementary antigen. Antibodies are Y-shaped proteins. Each molecule consists of four polypeptide chains: two identical "heavy" chains and two identical shorter "light" chains; these chains are linked by disulfide bonds. Most of the antibody molecule is constant in its amino acid sequence and structure; the two binding sites for the antigen are at the ends of the variable portion of the molecule.

When an antigen binds to a complementary antibody on a B cell, that lymphocyte grows and begins dividing. Some of the resulting cells are *memory cells*, which will make possible a faster immunological response during subsequent encounters with the antigen. Others become specialized as *plasma cells*, which secrete circulating antibodies. Each antibody molecule can bind to two antigen molecules simultaneously; hence the antibodies cause agglutination of the antigen-bearing bacteria or viruses. This agglutination aids in the destruction of the antigens by ingestion by *macrophages*, destruction by *natural killer (NK) cells*, or lysis mediated by the enzymes of the complement cascade.

The cell-mediated immune response Each T lymphocyte has many identical membrane-mounted receptors that can bind to antigens. *T-cell receptors* consist of two different polypeptide chains, each with variable and constant regions. The variable region of the two chains forms the antigen-binding site. Unlike antibodies, however, T-cell receptors cannot be secreted, they can bind to only one antigen at a time, and they bind to cell-surface proteins produced by the genes of the *major histocompatibility complex (MHC)*.

There are two general types of MHC products: MHC-II proteins, which are found on the membranes of B cells, cytotoxic T cells, and macrophages; and MHC-I proteins, which are found on all other cells of the body. Cells bearing MHC-II proteins participate in the regulation of the immune system. Those with MHC-I proteins can be recognized and killed by cytotoxic T cells. MHC molecules have a double-chain structure very similar to that of antibodies and T-cell receptors. Cells that have been infected by a virus "present" antigens found in their cytoplasm on the MHC molecules in their cell membrane.

When a T cell with an affinity for a specific antigen encounters an infected cell with that antigen bound to its surface, it is activated and begins to divide. Four types of cells are produced: *memory cells, cytotoxic T cells, helper T cells,* and *suppressor T cells*. Cytotoxic T cells bind to the MHC/antigen complex and kill the infected cell. The helper and suppressor T cells are modulatory T cells and regulate the strength of the immune response. Helper T cells release chemicals called *cytokines,* which activate other cells of the immune response, whereas suppressor T cells inhibit the activity of macrophages and other lymphocytes and shut down the immune response. The balance between activation and inhibition determines the overall level of the immune response.

The clonal selection hypothesis Each lymphocyte, whether of the B or the T type, has just one kind of receptor on its surface. An antigen can bind only to those lymphocytes whose surface receptors are specific for it. According to the *clonal selection hypothesis,* the antigen can bind to only those B or T cells that are capable of responding to it. Antigen binding stimulates cell division and produces a clone of cells that can respond to the antigen.

The role of macrophages When a macrophage engulfs an invader, it processes the antigen, attaches pieces of it to an MHC protein, places the complex on its surface, and presents it to a compatible T cell. In doing so, the macrophage is activated and produces *interleukin-I (IL-I),* which produces symptoms of a fever and turns on T-cell proliferation.

The binding of the presented MHC/antigen complex to the appropriate T cell induces division of that cell to produce a clone. The activated T cells then differentiate to form memory cells and active helper, cytotoxic, and suppressor T cells. The helper T cells have several roles: they secrete cytokines (primarily interleukins) that activate other helper and cytotoxic T cells and macrophages; they produce gamma *interferon,* which inhibits viral reproduction and stimulates phagocytosis by macrophages; and they activate B cells, which then divide and produce memory cells and plasma cells. Cytotoxic T cells and antibodies produced by plasma cells eventually destroy the pathogen.

It is the helper T cells that are infected by the AIDS virus (HIV). The central role of these cells in stimulating the specific immune response accounts for the devastating effect of AIDS on the immune system.

Once the antigen has been eliminated, the suppressor T cells inhibit the activity of macrophages and lymphocytes, and so end the immune response. The memory cells remain in circulation, ready to mount an immediate defense should the antigen be reintroduced.

Active and passive immunity Some diseases can be prevented by immunization with a vaccine or antiserum. A *vaccine* consists of antigenic substances from the pathogen; these produce a long-lasting *active immunity.* An *antiserum* contains antibodies against a specific antigen that have been synthesized by another animal; these produce an immediate, short-term *passive immunity.*

Recognition of "self" Early on in the development of the immune system, the body learns to distinguish between "self" and "nonself." Grafts and transplanted organs are recognized as "nonself" and are rejected. Rejection occurs because the foreign tissues bears many alien MHC proteins. Each person has a unique array of these molecules so that when tissue is transplanted, most of the donor tissue proteins will be unknown to the host's immune system and thus trigger an immune response.

Immunologic disease Occasionally the immunologic mechanisms get out of control and are responsible for disease symptoms. In allergies, the invading microorganisms or other foreign materials cause little if any harm to the host; instead, it is the immune system of the host that actually produces the damage.

Sometimes the ability to distinguish between "self" and "nonself" is impaired. In such cases an *autoimmune disease* results, where the immune system attacks the tissues of the body instead of protecting it. How self-tolerance is overcome in such diseases is not yet understood. Autoimmune diseases include myasthenia gravis, systemic lupus erythematosis, rheumatoid arthritis, multiple sclerosis, Grave's disease, and insulin-dependent diabetes.

CONCEPT MAP

Map 21.1

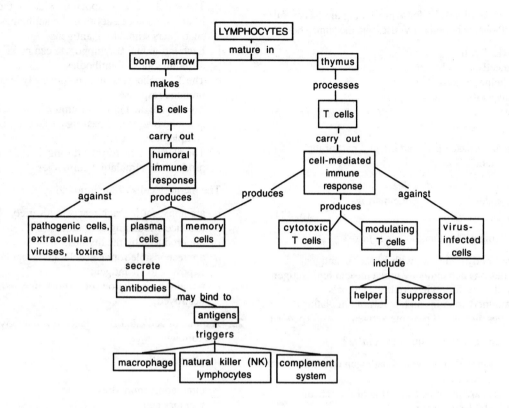

QUESTIONS

Testing recall

For each of the items or characteristics listed below, write T if it is generally associated with T lymphocytes, B if it is generally associated with B lymphocytes, and B, T if it is associated with both.

1. cell-mediated immune response (p. 438) T
2. humoral immune response (p. 436) B
3. plasma cell (p. 436) B
4. circulating antibodies (p. 436) B, T
5. memory cell (p. 436) B, T
6. mature in the thymus (p. 436) T
7. mature in the bone marrow (p. 436) B
8. generally stimulated by bacteria and toxins free in the plasma (p. 435) B
9. particularly effective against virus-infected cells (p. 436) T
10. secrete cytokines (p. 438) T
11. membrane-mounted receptors with one region for antigen and one region for surface marker proteins (pp. 438–39) T
12. antibodies can bind two antigen molecules together (p. 437) B
13. carry MHC-II proteins (p. 438) B, T
14. produce cytotoxic, helper, and suppressor cells (pp. 438–39) T

Testing knowledge and understanding

Choose the one best answer.

15 Which is the first line of defense preventing the successful entry of pathogenic bacteria or viruses into the human body?

 a cytotoxic T cells
 b plasma cells
 c virgin helper T cells
 d the nonspecific immune response
 e memory cells (pp. 433–35)

16 Which organ or tissue is *not* an integral part of the immune system?

 a pancreas
 b lymph nodes
 c thymus
 d spleen
 e bone marrow (p. 434)

17 Which statement about antibodies is *false*?

 a Each antibody combines with a specific antigen.
 b Two sites on each antibody molecule can bind antigen.
 c Antibodies are found in vertebrate animals.
 d Human blood contains high levels of circulating antibodies against all possible antigens. (p. 436)

18 Which statement about antibodies is *false?*

 a Vaccination with a weakened pathogen can stimulate antibody formation.
 b Antibodies are proteins consisting of four chains: 2 light and 2 heavy chains.
 c Each antibody can recognize and bind to many different antigens.
 d Antibodies are highly specific in their recognition of antigens.
 e Circulating antibodies are produced by plasma cells of the B-cell system. (p. 436)

19 An antibody molecule is made up of four polypeptide chains joined by disulfide bonds. Which one of the following statements concerning these polypeptide chains is *correct*?

 a All four chains are of equal length.
 b Only the two light chains have variable regions.
 c Only the two heavy chains have variable regions.
 d Only the heavy chains have constant regions.
 e All four chains have variable regions. (p. 436)

20 Which one of the following statements concerning antibodies is *correct*?

 a Each antibody has two chains.
 b Antibodies are usually proteins, but sometimes they are polysaccharides.
 c Antigens are bound by the variable regions of the antibody molecules.
 d The heavy, but not the light, chains of the antibody molecule have variable and constant regions.
 e The tail region is important in antigen specificity. (p. 436)

21 Which one of the following statements about immunity is *false*?

 a The vertebrate body apparently has all the genetic information necessary to make antibodies before it encounters stimulating antigens.
 b Each individual B lymphocyte can make several different kinds of antibodies.
 c The T lymphocytes function primarily in cell-mediated immune responses.
 d To activate the humoral immune response, an antigen is bound by a B cell bearing surface antibodies complementary to it.
 e The plasma cells formed during an immune response produce the circulating antibodies. (pp. 436–38)

22 The memory cells of the immune system

 a are nonspecific; each reacts to a variety of antigens.
 b produce circulating antibodies.
 c are produced by the thymus.
 d are responsible for an accelerated response on second exposure to an antigen.
 e remain in circulation for a short time and are then destroyed. (pp. 436–37)

23 A clone of activated B lymphocytes may give rise to all the following *except*

 a plasma cells.
 b T cells.
 c circulating antibodies.
 d memory cells. (p. 436)

24 Agglutination can trigger all the following reactions *except*

 a ingestion by phagocytic macrophages.
 b destruction by natural killer cells.
 c formation of suppressor cells.
 d complement system-mediated lysis. (pp. 437–38)

25 Which one of the following is *true* of T lymphocytes?

 a They are responsible for the humoral immune response.
 b They produce circulating antibodies.
 c They are primarily effective against bacteria and free viruses.
 d They secrete chemical mediators called interleukins.
 e They activate the complement system to cause lysis of the cell. (pp. 438–41)

26 A clone of activated T lymphocytes can give rise to all the following *except*

 a cytotoxic T cells.
 b helper T cells.
 c suppressor T cells.
 d memory cells.
 e plasma cells. (pp. 436–38)

27 The receptors of the T lymphocytes differ from the membrane-mounted antibodies on B lymphocytes in that

   a the receptor molecule can bind to only one antigen at a time.
   b each receptor molecule can recognize many different antigens.
   c each receptor molecule is specific for a single antigen.
   d on exposure to an antigen, the receptor molecules are released into circulation to fight the infection.
   (p. 438)

28 The helper T cells of the immune system

   a produce suppressor proteins that inhibit other T cells.
   b produce Interleukin-1 (IL-1) to stimulate other T cells.
   c inactive the cytotoxic T cells.
   d play a central role in activating both the T-cell and B-cell systems.
   e are cytotoxic cells that destroy the invading antigen.
   (p. 438)

29 Which of the following statements best describes the effect on the body's immune system of an antigen when it is first encountered?

   a Antigen molecules cause developing lymphocytes to rearrange their genes.
   b Antigen molecules are used by the lymphocytes as "models" in order to make correctly shaped antibodies.
   c Antigen molecules cause those lymphocytes that carry appropriately shaped antibodies to divide.
   d Antigen molecules catalyze the formation of disulfide bonds between light and heavy chains of the antibody.
   (p. 436)

30 Which one of the following would *not* be considered a *vaccine* that would produce an active immunity?

   a a small amount of active toxin
   b a prescribed amount of inactivated toxin
   c dead cells of the pathogen
   d live, weakened cells of the pathogen
   e antibodies produced in another host
   (p. 446)

31 Which one of the following is *true* concerning the immune system's ability to distinguish "self" and "non-self"?

   a The ability to distinguish between "self" and "nonself" is acquired late in childhood.
   b The ability to distinguish between "self" and "nonself" can be permanently reduced by brief treatment with immunosuppressive drugs.
   c The ability to distinguish between "self" and "nonself" is permanent and never breaks down.
   d The ability to recognize "nonself" is largely determined by the presence of MHC proteins.
   e Two of the above are correct.
   (p. 447)

32 MHC-II may be found on all of the following *except*

   a B cells.
   b cytotoxic T cells.
   c macrophages.
   d body cells for "self" recognition   (p. 438)

33 Transplanted tissue *least* likely to be rejected would e from what age of the organism?

   a embryo   d young adult
   b young child   e mature adult
   c teenager   (pp. 438, 447)

For further thought

1 The Sabin poliomyelitis vaccine consists of three strains of weakened live viruses.

   a Explain the process by which your body develops immunity to the polio viruses after you receive an oral dose of this vaccine.
   b Why are babies usually given two or three separate doses of the vaccine rather than one dose?
   c Children are given a "booster" dose of the vaccine at the age of $1\,^{1}/_{2}$ years and again before they start school. What is the purpose of the booster?

2 Describe the process of transplanted tissue rejection and the techniques used to avoid rejection. Working from your knowledge of biology and your system of values, state your position on the use of embryonic tissue for tissue transplant.

ANSWERS

Testing recall

1	T	5	B, T	9	T	13	B, T
2	B	6	T	10	T	14	T
3	B	7	B	11	T		
4	B	8	B	12	B		

Testing knowledge and understanding

15	d	20	c	25	d	30	e
16	a	21	b	26	e	31	d
17	d	22	d	27	a	32	d
18	c	23	b	28	d	33	a
19	e	24	c	29	c		

Chapter 22

REGULATION OF BODY FLUIDS

A GENERAL GUIDE TO THE READING

This chapter discusses the mechanisms by which animals keep the internaL body fluids constant. You may find it helpful to begin by reviewing the material on osmosis (see Chapter 5, pp. 95–98), since understanding osmosis is essential to understanding much of the material in this chapter (pp. 453–73 for example). As you read Chapter 22 in your text, you will want to concentrate on the following topics.

1 Homeostasis. The concept of homeostasis was introduced in Chapter 20 (see p. 407) and is elaborated on here. It is an important concept for you to master.

2 The differences between plants and animals with respect to the problem of maintaining a constant environment. These differences are presented on pages 453 to 455.

3 The vertebrate liver. The role of the liver in maintaining homeostasis (pp. 455–58) is often unappreciated; you will want to learn its functions. Pay particular attention to the hepatic portal system (Fig. 22.3, p. 456), a most unusual arrangement of blood vessels.

4 Excretion and the balance of salt and water. The discussion (pp. 462–64) is interesting; study the section on osmoregulation in fish carefully. You will find Figure 22.10 (p. 463) a great help in understanding this material.

5 Types of excretory systems. You will want to know the excretory mechanisms of various animals (pp. 464–67). Notice that the insect excretory system (p. 467) is built entirely differently from that of both other invertebrates and vertebrates.

6 The vertebrate kidney. You should concentrate on learning this material thoroughly. Figure 22.18 (p. 470) is most helpful in explaining the functioning of the kidney.

KEY CONCEPTS

1 In order to carry out their life functions, the cells of multicellular animals require a nonfluctuating internal environment similar to the relatively constant seawater in which they evolved. (p. 453)

2 Living organisms have various mechanisms for maintaining a relatively constant internal environment despite changes in the external environment. This constant internal environment is required in order for them to be able to live in a variety of external environments. (p. 453)

3 The extracellular fluid of plants differs greatly from that of animals in that it is not distinct from environmental water and therefore cannot be as well regulated as the tissue fluid and blood of animals. (pp. 454–55)

4 Plant cells can withstand much greater fluctuations in the composition of the fluids bathing them than animal cells can. (pp. 454–55)

5 Multicellular animals have evolved mechanisms for ridding their bodies of metabolic wastes and regulating the salt and water balance of their body fluids. (pp. 464–67)

REGULATION OF BODY FLUIDS • 161

CROSSLINKING CONCEPTS

1. Many unique solutions to the problem of osmotic regulation have evolved since life began. An evolutionary perspective can make the diversity of solutions easier to understand. As organisms moved from saltwater to freshwater environments, natural selection favored a reduction of the osmotic concentration of their body fluids; for this reason modern freshwater animals have osmotic concentrations that are decidedly lower than seawater. Marine bony fish are hypotonic to seawater because they evolved from freshwater bony fish.

2. In this chapter, the liver is described primarily in reference to its role in maintaining the balance of nutrients in the blood; the discussion is thus intimately tied to that of the function of the digestive system (Chapter 17). In Chapter 25, you will see how the liver is influenced by hormones from the endocrine and nervous systems.

3. An understanding of the molecular structure of carbohydrates, fats, and proteins (Chapter 3) and the interlinkages among their biochemical pathways (Figure 8.13) greatly contributes to learning the multiple roles of the liver. An understanding of protein structure will help you follow the deamination reaction that occurs in the liver.

4. In Chapter 20, you learned that the blood was responsible for transporting nitrogenous wastes (among other things). This chapter gives you a deeper understanding of how these nitrogenous wastes are formed and where they are excreted.

5. Your understanding of osmosis, active transport, and ion pumps (Chapter 5) is prerequisite to a comprehension of kidney function.

OBJECTIVES

1. State the principle of homeostasis and explain why the body's health and survival depend on the maintenance of homeostasis. (p. 453)

2. Contrast the extracellular fluids of multicellular plants with those of animals. (pp. 454–55)

3. Explain why plant cells are able to withstand greater fluctuations in the makeup of their extracellular fluids than animal cells and what happens to typical plant and animal cells in a hypotonic environment and in a hypertonic environment. (pp. 454–55)

4. Explain the functional significance of the hepatic portal system. In doing so, describe the circulatory connections between the stomach and intestines and the liver. Trace the flow of blood from the aorta through the intestines and liver and into the vena cava, using Figure 22.3 (p. 456) or Figure 20.5 (p. 457).

5. Discuss the role of your liver in maintaining homeostasis. Include in your answer the liver's function in carbohydrate, protein, and fat metabolism. (pp. 455–59)

6. List the three main nitrogenous waste products. Compare them with respect to toxicity and to the amount of water that must be expelled in order to excrete each one. Indicate where these compounds are synthesized. Specify the main nitrogenous waste product of each of the following animals: freshwater hydra, marine jellyfish, freshwater planarian, freshwater fish, marine fish, mammal, bird, lizard, and insect. Explain the correlation between the nitrogenous waste product and the habitat of an organism. (pp. 458–64)

7. Explain the difference between osmoregulation and excretion and then the difference between excretion and elimination (defecation). (p. 462)

8. Contrast the osmoregulatory problems faced by freshwater bony fish with those of marine bony fish. Discuss the adaptations that enable these animals to cope with their respective problems. (pp. 462–64)

9. For each of the following organisms, specify the type of excretory mechanism they possess, indicate whether the mechanism functions primarily in water regulation or in both water regulation and excretion, and describe the role of the circulatory system in excretion: *Paramecium*, planarian, earthworm, mammal, and insect. (pp. 464–69)

10. Using Figures 22.15 (p. 468), point out the following structures: renal artery, renal vein, kidney, ureter, urethra, and bladder.

11. Draw a diagram of the nephron, labeling the following structures: glomerulus, Bowman's capsule, proximal convoluted tubule, descending limb of the loop of Henle, ascending limb of the loop of Henle, distal convoluted tubule, and collecting duct. (p. 470)

12. Describe in some detail the process of urine formation in humans, discussing:

    a the source of urea and when the urea content of the blood will be highest
    b what happens in the glomeruli, naming the force that powers the filtration process, and indicating what substances remain in the blood and are not filtered into the glomerulus
    c the process by which glucose, amino acids, and some salts are reabsorbed
    d how water is reabsorbed
    e the roles of filtration, reabsorption, and tubular excretion and where and how each of these processes occurs (pp. 469–72)

13. Summarize the role played by the kidney in maintaining homeostasis. (pp. 469–72)

KEY TERMS

homeostasis (p. 453)
extracellular fluid (p. 454)
intracellular fluid (p. 454)
plasmolysis (p. 455)
hepatic portal vein (p. 455)
portal system (p. 455)
deamination (p. 458)
ammonia (p. 458)
urea (p. 458)
uric acid (p. 458)
excretion (p. 462)
elimination (p. 462)
osmoregulation (p. 462)
contractile vacuole (p. 464)
flame-cell system (p. 465)
nephridia (p. 466)
Malpighian tubules (p. 467)
nephron (p. 467)
Bowman's capsule (p. 467)
pelvis (p. 467)
ureter (p. 467)
urinary bladder (p. 467)
urethra (p. 467)
renal artery (p. 467)
glomerulus (p. 467)
renal vein (p. 469)
proximal convoluted tubule (p. 469)
loop of Henle (p. 469)
distal convoluted tubule (p. 469)
collecting tubule (p. 470)
vasopressin (p. 470)
kidney threshold (p. 471)
tubular excretion (p. 472)

SUMMARY

Living cells require a relatively constant environment to carry out their life functions. As complex multicellular animals arose, body fluids developed that could provide a stable environment for the internal cells. A variety of mechanisms evolved for maintaining the *homeostasis* of the body fluids despite changes in the external environment.

The extracellular fluids of plants are not separate from the environmental water. Consequently plants regulate only the composition of their intracellular fluids, whereas animals must regulate the composition of both intracellular and extracellular fluids. Plant cells can withstand much greater fluctuations in their extracellular fluids than animal cells can. Because the rigid cell wall resists expansion, the plant cell can withstand pronounced changes in the osmotic concentration of the surrounding fluids as long as the fluids remain hypotonic to the cell. If the external fluids become decidedly hypertonic to the cell, the cell will lose water and shrink away from the wall in a process called (*plasmolysis*).

The vertebrate liver The liver is important in maintaining a relatively constant environment for the body cells. All the blood from the intestine and stomach is collected in the *portal vein* and conducted to the liver, where it flows into a second set of capillaries before it is collected into the hepatic vein and emptied into the posterior vena cava. Because of this portal system, the products of digestion are brought directly to the liver cells, where the levels of the digestive products can be regulated.

When the incoming blood is high in glucose, the liver removes the excess and stores it as glycogen or fat; when the incoming blood is low in glucose, the liver reconverts glycogen into glucose to maintain the blood-sugar level. The liver is the center for fat metabolism; fatty acids and other lipid materials are processed here and some plasma lipids are synthesized.

The liver also removes many of the amino acids from the blood and temporarily stores small quantities. Excess amino acids are *deaminated* and the remainder of the molecule is converted into carbohydrate or fat. In the deamination process the amino group ($-NH_2$) is converted into *ammonia* (NH_3). The ammonia may be released directly as a waste product or it may be converted into the less toxic compounds *urea* and *uric acid*. These are released into the blood and must be removed from the body.

In addition to its metabolic functions, the liver also detoxifies many injurious chemicals, usually by changing them from nonpopular to polar substances that can be more easily excreted from the body. The liver also manufactures many plasma proteins, stores certain substances, destroys red blood corpuscles, synthesizes bile salts, and excretes bile pigments.

Excretion and osmoregulation Animals need mechanisms to rid their bodies of metabolic wastes (*excretion*) and to regulate their salt and water balance (*osmoregulation*).

The nitrogenous excretory product characteristic of most aquatic animals is ammonia. Ammonia is poisonous but is easily released from the body if the water supply is plentiful. Many marine invertebrates simply release their wastes across their membrane surfaces. Such organisms are isotonic with their environment and thus have no problems with water balance as long as they remain in the sea.

Freshwater animals have an osmotic concentration that is less than that of seawater but greater than that of freshwater. A freshwater organism tends to gain water and lose salts to the surrounding water. To survive, the organism must have an osmoregulatory structure to pump out the excess water and some mechanism to absorb salts. Freshwater bony fishes compensate by drinking very little, actively absorbing salts through specialized gill cells, and excreting copious dilute urine.

Marine bony fishes are hypotonic relative to seawater. They tend to lose water and take in salt. They compensate by drinking constantly, actively excreting salts across their gills, and producing little urine.

The greatest problem for a land animal is desiccation. The water lost by evaporation, elimination, and excretion must be replaced. It is replaced by drinking, eating foods containing water, and oxidating nutrients. Amphibians and mammals produce urea as their nitrogenous waste product; it must be released from the body in solution, and thus drains away needed water. Urea excretion is correlated with bearing live young. Most reptiles, birds, insects, and land snails excrete uric acid. Little water is lost in the excretion of this highly insoluble compound. The uric acid produced by a developing embryo is precipitated and stored in the eggs of these animals; in this form it is not harmful. Uric acid excretion is correlated with egg laying.

Excretory mechanisms in animals Many unicellular and simple multicellular animals have no special excretory structures; the nitrogenous wastes are excreted across membrane surfaces. Some protozoans do have a *contractile vacuole*, which collects fluid and expels it from the cell. Its primary function is the elimination of excess water. Flatworms have a simple, tubular *flame-cell system* to eliminate excess water.

In animals that have evolved closed circulatory systems, the blood vessels have become intimately associated with the excretory organs; this makes possible the direct exchange of materials between the blood and the excretory system. In earthworms, each segment of the body has a pair of excretory organs called *nephridia*, which open to the outside. Tissue fluids move into the open end of this tubular structure and pass through the coiled tubule to the exterior. Blood vessels associated with the tubules also secrete substances into the tubules for excretion.

The excretory organs of insects are called *Malpighian tubules.* They are diverticula of the digestive tract located at the junction between the midgut and hindgut. The tubules are bathed directly by the blood; fluid from the blood moves into the tubules, where the uric acid is precipitated. The concentrated urine moves into the hindgut and rectum, where it is exposed to a powerful water-absorptive action; the urine and feces leave the rectum as dry material.

The vertebrate kidney The vertebrate excretory organ is the kidney, which is composed of *nephrons*. Three processes are important in urine formation: filtration, reabsorption, and tubular secretion. High blood pressure within the *glomerulus* forces small molecules from the blood into the *Bowman's capsule* (filtration). The filtrate passes successively through the *proximal convoluted tubule, loop of Henle, distal convoluted tubule* and *collecting duct.* As the filtrate moves through the tubules, most of the water and substances needed by the body are reabsorbed into a second set of capillaries associated with the tubules. In addition, some chemicals are actively removed from the blood by the tubules and deposited in the filtrate (*tubular secretion*). The osmotic concentration of the urine is regulated by *vasopressin*, a hormone released from the pituitary in response to changes in the amount of water in the body. Vasopressin makes the walls of the collecting ducts more permeable to water. As a result, more water is reabosrbed into the blood, and a more concentrated urine is produced. From the collecting ducts the urine is conducted into the kidney *pelvis*, then down the *ureter* to the urinary bladder for storage, and finally out the *urethra*.

Most substances reabsorbed by the kidney have a *threshold level;* if the concentration of such the substance exceeds the threshold, the excess is not reabsorbed by the tubules but instead appears in the urine. Thus, the kidneys play a vital role in maintaining homeostasis; they help to regulate the blood-sugar level and the concentration of various inorganic ions in the blood.

CONCEPT MAP

Map 22.1

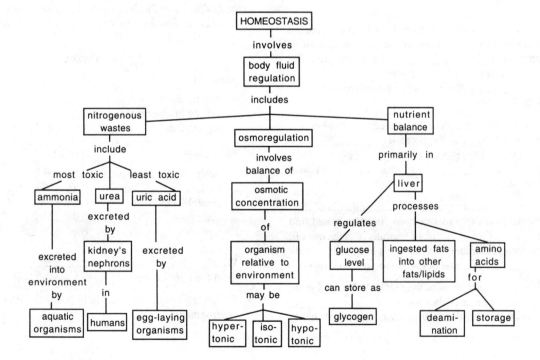

KEY DIAGRAM

In the diagram at right of a human kidney, color in the following, using a different color for each structure. First color the names, then the appropriate structures. (Fig. 22.15, p. 468)

CORTEX

MEDULLA

RENAL PELVIS

URETER

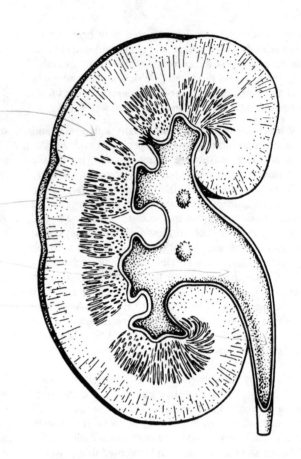

Next, draw in a single nephron in its proper position in the kidney.

QUESTIONS

Testing recall

Complete the following statements with the words greater than, less than, or equal to.

1. If you were to measure the extracellular fluid of a marine alga living in seawater, you would find that its osmotic concentration is ___equal to___ that of seawater. The osmotic concentration of intracellular fluid of the alga would be ___equal to___ that of the extracellular fluid. (pp. 462–64)

2. The osmotic concentration of the extracellular fluid of a marine jellyfish living in the ocean is ___greater___ that of seawater. The osmotic concentration of its intracellular fluid is ___ ___ that of its extracellular fluid. (pp. 462–64)

3. If you were to measure the osmotic concentration of the extracellular fluids of a marine bony fish, you would find it to be ___less___ that of seawater, and the osmotic concentration of the interior of the fish's body cells would be ___ ___ that of the blood. (pp. 462–64)

4. The osmotic concentration of the extracellular fluid in the cortex of a root of a land plant would be _____ that of soil water and the osmotic concentration of the cortex cells would be _____ that of soil water. (pp. 454–55)

For questions 5 to 11 use the following table, which lists the osmotic concentrations of seawater and freshwater and of the body fluids of various animal groups living in these environments. (pp. 462–64)

		Percent
Seawater		3.5
a	marine sharks	3.5
b	marine invertebrates	3.6
c	marine fish	1.5
Freshwater		0.01–0.5
d	freshwater invertebrates	0.04–0.6
e	freshwater bony fish	0.85
f	amphibians	0.85
g	mammals	0.9
h	reptiles	0.9

5. Which organisms will tend to take in excessive water from their environment? e

6. Which organisms are essentially isotonic with their environment? d

7 Which organisms will tend to lose excessive body water to their environment?

8 Which organisms tend to take in excessive salt from their environment?

9 Which organisms would have special mechanisms for actively taking up salts from their environment and releasing them into their body fluids?

10 Which organisms probably use ammonia as their main nitrogenous waste product?

11 Which organisms use mainly urea as their nitrogenous waste product?

12 Label the parts (*a–g*) in the sketch of the nephron. (p. 468)

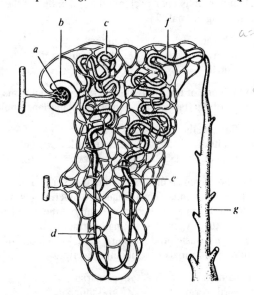

13 Referring to the sketch, indicate by letter in which part of the nephron each process occurs.

  a filtration of the blood (p. 469)
  b passive transport of Na$^+$ and Cl$^-$ (p. 471)
  c reabsorption of most water (p. 470)
  d pumping of Na$^+$ out of the tubule to establish a concentration gradient (p. 471)
  e conducting of urine to pelvis of the kidney (p. 467)

14 Which parts of the nephron are found in the cortex of the kidney? (p. 468)

Testing knowledge and understanding

Choose the one best answer.

15 The term homeostasis is best defined as

  a the maintenance of a constant internal environment.
  b the maintenance of a constant salt concentration in the body.
  c adaptation to a harsh environment.
  d maintenance of a constant body temperature.
      (p. 453)

16 A plant cell placed in a hypotonic environment will

  a lose water and become flaccid.
  b actively transport salts out of the cell.
  c take up water and burst.
  d take up water and become turgid.
  e become impermeable to prevent water loss. (p. 455)

17 An erythrocyte is in an artery leading to the duodenum of a human being. How many capillary beds will it probably pass through before it reaches the right ventricle of the heart?

  a one   *d* four
  b two   *e* five
  c three   (p. 455)

18 In a person who has recently eaten a large meal, which one of the following blood vessels would you expect to contain blood with the highest concentration of sugar?

  a aorta   *d* hepatic portal vein
  b posterior vena cava *e* jugular vein
  c renal artery   (p. 456)

19 Some years ago, Senator John Stennis was shot and severely wounded as he was getting out of his automobile on a Washington street. A later account of the incident published in the *Washington Post* included the following passage:

  The most serious wound, which at the outset many thought would cost him his life, was just at the belt-line on the left side. It affected his pancreas, colon, and portal vein, which supplies blood to the stomach. The vein was almost cut in two.

  There is a biological error in this passage, which is that

  a the portal vein does not supply blood to the stomach.
  b the pancreas is not on the left side of the body.
  c the colon is nowhere near the indicated wound site.
  d a single bullet could not have hit both the pancreas and the colon.
  e the listed wounds would not have been sufficiently serious to endanger his life. (p. 456)

20 In the human being, blood leaving the liver in the hepatic vein ordinarily has a higher concentration of _____ than other blood.

  a bile   *d* erythrocytes
  b urea   *e* leukocytes
  c oxygen   (p. 458)

21 All the following are functions of the human liver *except*

  a removing the excess glucose from the blood and storing it as glycogen.
  b inactivating certain drugs and injurious chemicals.
  c deaminating amino acids and converting the ammonia thus formed into urea.
  d manufacturing many plasma proteins and some plasma lipids such as cholesterol.
  e secreting digestive enzymes into the small intestine.
                                    (pp. 456–59)

22 The vertebrate liver carries out all the following functions *except*

 a regulation of blood-sugar levels.
 b secretion of bile.
 c metabolism of amino acids.
 d production of cholesterol.
 e removal of urea from the blood. (pp. 456–58)

23 Which of the following nitrogen-containing compounds is most toxic to animals, and can be tolerated only in dilute solutions?

 a nitrate ion   d ammonia
 b urea   e uric acid
 c amino acid   (p. 458)

24 The nitrogen present in proteins can be eliminated in the form of ammonia, which is highly toxic and must be removed from the body rapidly. In which of the following habitats would you be most likely to encounter organisms using ammonia as their nitrogenous waste product?

 a underground
 b in water
 c on land
 d in the air
 e in deserts   (p. 458)

25 A biologist experimenting with a protozoan noticed that the contractile vacuole stopped contracting although the other parts of the organism seemed healthy and active. Which of the following experiments was the one most likely to have produced this result?

 a transferring the protozoan from a lighted environment to a dark environment
 b transferring the protozoan from freshwater to seawater
 c transferring the protozoan from seawater to freshwater
 d changing the pH of the medium from 7.0 to 6.5
 e cooling the medium from 20° to 15°C (p. 463)

26 A trout in a freshwater stream

 a loses water and salts by diffusion, and compensates by drinking large quantities of water and actively absorbing salts via the gills.
 b takes in much water and loses salts by diffusion, and compensates by excreting copious dilute urine and actively absorbing salts via the gills.
 c loses water and takes in large quantities of salts by diffusion, and compensates by actively absorbing water via the gills and by excreting salty urine.
 d takes in much water and salt by diffusion, and compensates by excreting copious urine and actively excreting salts via the gills.
 e takes in much water and loses salt by diffusion, and compensates by nearly constantly drinking and by actively excreting both water and salts via the gills. (p. 463)

27 A freshwater fish is placed in a marine environment. What would you expect to happen to this fish eventually?

 a It will dehydrate.
 b It will take up water.
 c It will neither gain nor lose water.
 d It will be in osmotic balance with the seawater.
 e It will actively excrete salt. (p. 463)

28 A marine bony fish living in the ocean must

 a excrete copious amounts of dilute urine.
 b actively transport salts into the body.
 c drink all the time.
 d produce a hypotonic urine.
 e actively transport water out through the gills. (p. 463)

29 Dehydration is *not* a problem for which one of the following animals?

 a saltwater fish   d human
 b freshwater fish   e desert iguana
 c camel   (p. 463)

30 Which one of the following excretory mechanisms do many of the flatworms, such as the planarian, employ?

 a flame cells   d bladder
 b nephridia   e contractile vacuole
 c loop of Henle   (pp. 464–66)

31 You are given an unknown animal to study. It is bilaterally symmetrical and has a complete digestive system. It has a circulatory system, but the blood carries very little oxygen. The animal's principal nitrogenous excretory waste product is uric acid. Which one of the following animals best fits this description?

 a salmon   d planarian
 b earthworm   e grasshopper
 c sparrow   (p. 467)

32 The excretory organs of insects are called

 a Malpighian tubules. d Bowman's capsules.
 b nephridia.   e glomeruli.
 c flame cells.   (p. 467)

33 All the following are functions of the mammalian kidney *except*

 a removal of excess sodium ions from the blood.
 b reabsorption of salts and sugar from blood filtrate.
 c filtration of the blood.
 d synthesis of urea.
 e the concentration of urine. (pp. 467–72)

34 Hydrostatic pressure powers the process of

 a filtration across the glomerulus.
 b reabsorption of water and dissolved substances at the venule end of the capillary.
 c sodium gradient maintenance in the kidney tubule.
 d salt and glucose reabsorption in the kidney tubules.
 e Na^+ recycling in the loop of Henle. (p. 469)

35 The liquid collected by Bowman's capsules in a human kidney may best be characterized as

 a concentrated urine.
 b dilute blood.
 c a solution of urea.
 d blood minus cells, corpuscles, platelets, and plasma proteins. (p. 469)

36 The second of the two capillary beds in the kidney

 a reabsorbs nutrients.
 b is called the glomerulus.
 c filters the blood.
 d goes to the liver.
 e carries less-concentrated blood than does the first bed. (p. 470)

37 A person with kidney failure who is undergoing blood dialysis would probably be restricted to a diet low in

 a protein.   d calories.
 b carbohydrate.   e bulk.
 c fat.   (p. 458)

38 The kidneys of vertebrates function in all the following ways *except* in the

 a excretion of metabolic wastes.
 b regulation of water concentration.
 c regulation of ion concentration.
 d elimination of undigested wastes.
 e elimination of materials in the blood that are in oversupply. (pp. 467–71)

39 Large molecules such as uric acid and penicillin require ATP energy to be excreted by the kidney. These substances enter the nephron by the process of

 a filtration.   d diffusion.
 b osmosis.   e tublar secretion.
 c reabsorption.   (pp. 471–72)

40 The production of urine by the cells of the mammalian kidney requires energy, which is used mainly to

 a actively reabsorb solutes from the urine.
 b actively reabsorb water from the loop of Henle.
 c filter the blood into Bowman's capsule.
 d actively "pump" water into the urine.
 e actively "pump" wastes into the urine. (p. 471)

41 On the basis of your knowledge of the anatomy and physiology of the kidney, in which of the following blood vessels would you expect to find the highest concentration of plasma proteins?

 a the renal artery
 b the arteriole entering Bowman's capsule
 c the vessel leaving Bowman's capsule
 d the venule leaving a nephron
 e the renal vein (p. 469)

42 Which of the following processes that occur in the kidney directly requires active transport?

 a movement of Na^+ into the descending limb of the loop of Henle
 b movement of Na^+ out of the ascending limb of the loop of Henle
 c filtration into Bowman's capsule
 d osmosis of water in the collecting duct
 e movement of urea into Bowman's capsule (pp. 470–71)

For further thought

1 Lake Baikal in Siberia is the only known habitat of freshwater seals; all other seal species inhabit marine environments. What changes are most likely to have occurred in the kidney functions of these freshwater seals? (pp. 469–71)

2 A patient was admitted to a hospital with a preliminary diagnosis of viral hepatitis (inflammation of the liver). The following tests were performed:

 a Fasting blood sugar. The blood-sugar level was measured 6 hours after an injection of glucose. Result: blood sugar was 58 mg/100 mL (normal: 70–120 mg/100 mL).
 b Serum albumin test. The level of albumin in the blood was determined. Result: serum albumin was lower than normal.
 c Prothrombin time. The time it takes for blood to clot was measured. Result: clotting time was higher than normal.
 d Bromsulfalein test. Dye was injected into the blood and the amount of dye retained was measured. Result: 50 percent of the dye was retained in 30 minutes (normal: 10 percent or less). (This dye is not excreted by the kidneys.)

 On the basis of what you know about liver function, explain the above results. What symptoms do you think the patient might present? (pp. 456–59)

3 A molecule of water is absorbed into a capillary in the small intestine of a human being. Trace the route this molecule will follow as it goes by the shortest possible route to the urinary opening. (Name all the blood vessels and parts of the kidney, etc.) (pp. 467–70)

4 Would the urine composition of a vegetarian differ from that of a heavy meat eater? (p. 458)

5 Approximately 180 liters of fluid are filtered through the kidney every 24 hours, but only about 1.5 liters of urine are produced. What is the average rate of filtration in mL/minute (1 liter = 1,000 mL)? What will happen to the rate of filtration if the blood pressure is increased? What is the rate of urine production in mL/minute? If the two kidneys together have two million glomeruli, how many milliliters of fluid are filtered by each glomerulus in 24 hours?

 a 9.0 mL   d 8.9 mL
 b 0.9 mL   e 0.89 mL
 c 0.09 mL   f 0.089 mL
                            (p. 469)

6 What do kidneys and nephridia have in common? Why must insects have an excretory system that operates on an entirely different principle? (pp. 466–67)

7 Many states have an organ donor system linked to obtaining a driver's license. On the basis of your biological understanding from this chapter, explain the significance of liver and kidney transplants. On the basis of your own value structure, decide if you would want to have your organs donated in the event of a fatal automobile accident.

ANSWERS

Testing recall

1 equal to, equal to
2 equal to, equal to
3 less than, equal to
4 equal to, greater than
5 (d) freshwater invertebrates, (e) freshwater bony fishes, (f) amphibians
6 (a) marine sharks, (b) marine invertebrates
7 (c) marine bony fishes, (g) mammals, (h) reptiles
8 (a) marine sharks, (c) marine bony fishes
9 (d) freshwater invertebrates, (e) freshwater bony fishes, (f) amphibians
10 (a to e) sharks, fishes, invertebrates
11 (f) amphibians, (g) mammals
12 *a* glomerulus
   b Bowman's capsule
   c proximal convoluted tubules
   d loop of Henle
   e loop of Henle
   f distal convoluted tubules
   g collecting duct
13 a *a*
   b *d*
   c *c*
   d *e*
   e *g*
14 *a, b, c, f*

Testing knowledge and understanding

15	*a*	22	*e*	29	*b*	36	*a*
16	*d*	23	*d*	30	*a*	37	*a*
17	*b*	24	*b*	31	*e*	38	*d*
18	*d*	25	*b*	32	*a*	39	*e*
19	*a*	26	*b*	33	*d*	40	*a*
20	*b*	27	*a*	34	*a*	41	*c*
21	*e*	28	*c*	35	*d*	42	*b*

Chapter 23

CHEMICAL CONTROL IN PLANTS

A GENERAL GUIDE TO THE READING

This chapter is the first of six concerned with the response of organisms to changes in the external environment. Responding to environmental changes requires a complex flow of information: reception of stimuli, communication, and response. This chapter and Chapter 24 examine development and chemical control in plants. The remaining four chapters focus on animals. You will learn that hormonal control in plants differs from hormonal control in animals in a variety of ways; the fundamental one, which you should try to keep in mind, is that plant hormones are involved in nearly all aspects of growth and development, whereas animal hormones are primarily involved in maintaining homeostasis. As you read Chapter 23 in your text, you will want to concentrate on the following topics.

1. The differences between plant and animal hormones. The major differences are summarized on page 476. They will be referred to again in Chapter 25.

2. Auxins. Considerable emphasis is placed on auxins; the other plant hormones are not discussed in much detail. Auxins have been emphasized for two reasons: first, because they were the earliest investigated and are the best known of the groups of plant hormones, and second, because they can serve as a model for hormonal activity in plants; by studying the activity of one group of hormones you can learn the fundamental principles of hormonal control in plants. (pp. 476–83).

3. Chemical control. A summary of chemical control in plants is on page 487; you will find it extremely helpful.

KEY CONCEPTS

1. Chemical control mechanisms play an important role in the coordination of the myriad functions of living organisms. Hormonal control is common to plants and animals. (p. 475)

2. Plant hormones are not highly specific in their action but rather participate in nearly all aspects of growth and development. (p. 476)

3. A plant's response to a given hormone depends on the tissue, the concentration of the hormone, and the concentration of other plant hormones that may be present. (p. 476)

4. Cell division is stimulated by cytokinins and auxins. (p. 485)

5. Control of cell enlargement involves auxins and gibberellins, which promote elongation. (pp. 484–85)

6. The various plant hormones, by their mutual interactions and differential effects on various parts of the plant body, help integrate and coordinate the development of form and function. (p. 487)

7. Ethylene contributes to many changes characterizing the aging process in a plant or parts of a plant. (p. 486)

CROSSLINKING CONCEPTS

1 It is interesting to note that plant gibberellins are derived from the same biosynthetic pathway as vertebrate steroid hormones (Chapters 25 and 26) and probably function in the same way: by entering a cell and turning particular genes on or off.

2 The type of hormonal cooperation exhibited by auxins (in cell elongation) and cytokinins (in cell division) is similar to the hormonal interaction you will see in vertebrates (Chapter 25).

OBJECTIVES

1 List at least three ways in which plant hormones differ from animal hormones. (p. 476)

2 Discuss the role of auxin in phototropism, describing the effect of light on the distribution of auxin and how auxin distribution affects phototropism. (pp. 478–79)

3 Describe the significance of the experiments on phototropism by the Darwins (Fig. 23.6, p. 479) and by Went (Fig. 23.7, p. 479).

4 Discuss the role of auxin in gravitropism in stems and roots, explaining the possible mechanism by which plants sense gravity, the effect of gravity on auxin distribution, and how auxin distribution affects gravitropism. (pp. 480–81)

5 Discuss the role of auxin in apical dominance, using Figure 23.14 (p. 482) to explain this phenomenon. Indicate why pinching off the growing tip of some plants causes the plants to become more bushy. (p. 482)

6 Name a major effect that each of the following has on plant growth and development: gibberellins, cytokinins, abscisic acid, and ethylene. (pp. 484–86)

7 Contrast the effect of auxin and gibberellins on cell elongation in the plant stem. (pp. 476–77, 484–85)

8 Explain how auxin and cytokinins influence cell division and growth. (pp. 476–77, 485)

9 Describe the process of leaf abscission and how auxin and ethylene influence this process. (pp. 482–83)

KEY TERMS

hormone (p. 475)
auxin (p. 476)
acid-growth hypothesis (p. 477)
phototropism (p. 478)
tropism (p. 478)
coleoptile (p. 478)
indoleacetic acid (p. 480)
gravitropism (p. 480)
amyloplasts (p. 481)
apical dominance (p. 482)
abscission layer (p. 482)
gibberellin (p. 484)
gibberellic acid (p. 484)
cytokinin (p. 485)
tissue culture (p. 485)
abscisic acid (p. 485)
ethylene (p. 486)

SUMMARY

The flow of information Three steps are involved in the flow of information in organisms: reception of the stimulus, communication of the stimulus information to the site of response, and the response itself. Plants transmit information by means of chemicals; control chemicals called hormones move from one part of the plant to the target cells, where responses are elicited. Plant hormones are produced most abundantly in the actively growing parts of the plant body: the apical meristems, young growing leaves, and developing seeds. The known plant hormones are primarily involved in regulating growth and development.

Auxins Auxins are a class of plant hormones that produce a variety of effects, the most important of which is control of cell elongation. Experiments show that auxin is produced by the apical meristem, and moves downward, and promotes cell elongation in the stem. According to the *acid growth hypothesis,* auxins induce the transport of hydrogen ions in the cell from the cytoplasm to the cell wall. The acid pH activates enzymes in the wall that break the crosslinkages between the cellulse fibrils, and the wall becomes more extensible.

Tropisms are turning responses of plants caused when one side of a plant stem (or root) grows faster than the other, and so causes it to bend. Experiments have shown that auxins are involved in the tropic responses of plants to light (*phototropism*) and gravity (*gravitropism*). Light causes migration of the auxin away from the lighted side. Consequently the illuminated side of the plant grows more slowly than the shaded side and the plant turns toward the light. Gravity also causes an unequal distribution of auxin within the stem. The plant shoot turns away from the pull of gravity, and so shows negative gravitropism. Roots turn toward the pull of gravity. The mechanism by which gravity is detected and the process by which the detection of gravity is translated into changes in hormonal gradients are still not completely understood.

Auxin produced in the terminal bud moves downward in the shoot and inhibits the development of the lateral buds (*apical dominance*). Auxin also acts to prevent the abscission (dropping) of flowers, fruits, and leaves by inhibiting the formation of the *abscission layer*. In early spring it stimulates renewed cell division in the cambium. It also initiates the formation of lateral roots and promotes the development of roots from cuttings. Auxinlike chemicals such as 2,4-D are used as broad-leaved weed killers.

Gibberellins Plant hormones of a different class, the *gibberellins,* bring about dramatic stimulation of rapid stem elongation in dwarf plants. Because gibberellins can move freely throughout the plant, most of their effects on the pattern of growth are different from those of auxin. Gibberellins are also active in inducing the germination of seeds, stimulating the production of a starch-hydrolyzing enzyme in germinating seeds, inducing flowering in some plants, and stimulating fruit development.

Cytokinins Together with auxins, the hormones of yet another class, the *cytokinins,* promote cell division. In the normal growing plant, cytokinins and auxins may act cooperatively in some situations and antagonistically in others. Among their other functions, cytokinins stimulate the development of chloroplasts, release lateral buds from apical dominance, and delay the aging of leaves.

Abscisic acid The hormone *abscisic acid* promotes the transport of the products of photosynthesis to the growing embryo in seeds and induces the synthesis of proteins for storage in seeds. It also can cause the stomata to close when there is a shortage of water.

Ethylene The hormone *ethylene* is a gas that induces the ripening of fruit in mature plants. It also contributes to leaf abscission and various other changes involved in the aging process.

CONCEPT MAP

Map 23.1

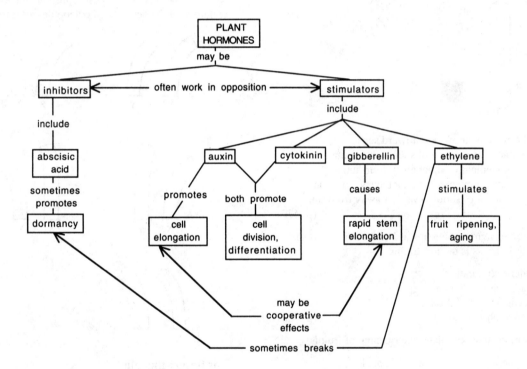

QUESTIONS

Testing recall

Match each function below with the associated plant hormone. Answers may be used once, more than once, or not at all.

a abscisic acid
b auxin
c cytokinin
d ethylene
e gibberellin

1 induces ripening of fruit and aging of the plant (p. 486) d

2 stimulates cell division (p. 485) b, c

3 can cause rapid elongation when applied to intact stems of dwarf plants (p. 484) e

4 inhibits growth of lateral buds (p. 482) b

5 inhibits seed germination (p. 485) a, e

6 causes closing of stomata (p. 485) a

7 induces bending of coleoptiles (p. 479) b

8 participates in phototropism and gravitropism in stems (pp. 478–81) b

Testing knowledge and understanding

9 Plant hormones, in contrast to most animal hormones,

    a always stimulate and never inhibit various processes.
    b are required in large amounts.
    c are not produced in specialized glands.
    d do not coordinate activities of the organism.
    e are pigmented green. (p. 476)

10 In the experiment shown in the following drawing, a transparent cap is placed on a grass shoot from which the apical meristem has been removed; also, an opaque cylinder is placed around the base. If the sun is positioned as shown, what result would you expect?

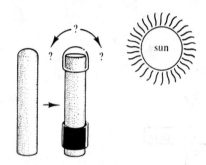

    a slight movement of shoot toward sun
    b pronounced movement of shoot toward sun
    c slight movement of shoot away from sun
    d pronounced movement of shoot away from sun
    e no movement of shoot toward or away from sun
    (p. 479)

11 A horizontal stem turning upward under conditions of constant darkness is demonstrating

    a ethylene stimulation.
    b negative gravitropism.
    c abscisic acid inhibition.
    d negative phototropism. (p. 480)

12 The hormone that stimulates the ripening of fruit is

    a ethylene.   *d* abscisic acid.
    b auxin.   *e* gibberellin.
    c cytokinin. (p. 486)

13 Phototropism

    a cannot occur unless tissues are capable of differential growth.
    b occurs because the concentration of auxin (indoleacetic acid) is higher on the side of the apex that is illuminated, and thus inhibits cell growth.
    c occurs because the tolerance limits of the shoot apex to auxin concentrations are very low.
    d results in the elongation of internodes due to an increase in cell division and elongation.
    e does none of the above. (pp. 478–79)

14 An agar block, divided into X and Y halves by an impermeable partition, is inserted underneath the growing tip of an oat coleoptile. The coleoptile is then exposed for a day

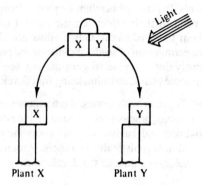

to light from the right, as shown in the drawing above. The agar block is next removed, and the X and Y halves are placed asymmetrically atop two decapitated coleoptiles and left for two days in the dark. Which one of the following diagrams, *a* to *e*, best represents the expected condition of the decapitated coleoptiles after two days?

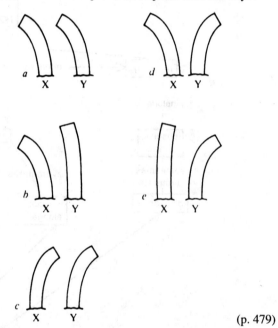

(p. 479)

For further thought

1 Explain the role of the hormones auxin, gibberellin, abscisic acid, and cytokinin in the germination of the seed and the growth of the young plant. (pp. 476–77, 484–86)

ANSWERS

Testing recall

1 d   3 e   5 a   7 b
2 c   4 b   6 a   8 b

Testing knowledge and understanding

9 c   11 b   13 a
10 e   12 a   14 b

Chapter 24

HORMONAL CONTROL OF REPRODUCTION AND DEVELOPMENT IN FLOWERING PLANTS

A GENERAL GUIDE TO THE READING

This chapter continues the examination of development and chemical control in plants begun in Chapter 23. It focuses on the control of flowering, the development of the seed, and the growth and differentiation of the plant. As you read Chapter 24, you will want to concentrate on the following topics.

1 Photoperiodism. This concept is introduced on pages 489 to 490. Notice that the critical element in the photoperiodism of flowering is actually the length of the night, not the length of the day.

2 The seed and its germination. You will want to learn the embryonic stages of development in a plant and the structure of the seed. Also focus on the process of germination, including the hormonal changes that accompany it. (pp. 495–99)

3 Growth and differentiation of the plant body. Pay close attention to this section and note how the hormones discussed in Chapter 23 interact to control plant development. (pp. 498–503)

KEY CONCEPTS

1 In many plants, flowering is affected by photoperiodism: the response to the duration and timing of light and darkness. (pp. 489–90)

2 The reproductive adaptations evolved by the flowering plants have contributed to their success in exploiting the terrestrial environment. (pp. 490–92)

3 The hormones cytokinin, auxin, gibberellin, and abscisic acid help to control the developmental processes of plants. (p. 497)

4 In plants, a few cells remain forever embryonic; thus new organs can be formed and growth will continue throughout the plant's life. (p. 497)

5 As development proceeds, individual plant cells become more and more committed to one particular course of differentiation. (pp. 502–503)

CROSSLINKING CONCEPTS

1 Chapter 9 introduced you to the concept of the plant life cycle and alternation of generations. This chapter further explores how plants develop between their haploid and diploid phases.

2 In this chapter, you are introduced to the exciting idea of biological clocks. This concept will take on added significance when you study animal behavior (Chapter 35).

3 The mechanisms by which gene expression is controlled, described in Chapter 12, are the basis for the process of cellular differentiation described in this chapter and in Chapter 27.

OBJECTIVES

1 Differentiate among short-day plants, long-day plants, and day-neutral plants, specifying which will flower when nights are shorter than a certain critical length and which will flower when nights are longer. Describe the effect of a flash of light in the middle of the night on the flowering of a short-day plant. (pp. 489–90)

2 Discuss the relationship among the hypothetical hormone florigen, phytochrome, and photoperiodism in the control of flowering. (pp. 490–92)

3 Using Figure 24.4 (p. 498), identify the parts of a flower and describe the process of reproduction in angiosperms. (Figs. 24.5, p. 494; 24.7, p. 495; and 24.8, p. 496, may be helpful.)

4 Using Figure 24.12 (p. 498), describe the early embryology of an angiosperm plant. In doing so, explain the function of the cotyledons, epicotyl, hypocotyl, radicle, and apical meristems; then specify what each of the following tissues gives rise to in the mature plant: the protoderm, the provascular tissue, and the ground tissue. (pp. 497–98)

5 Explain how changing levels of auxin, cytokinin, gibberellin, and abscisic acid interact to control the development and germination of a seed. (pp. 497–99)

6 Using a Figure 24.13 (p. 498), identify the different parts of a seed: the cotyledon, endosperm, radicle, seed coat, epicotyl, and hypocotyl. Give two differences between the seed of a dicot and that of a monocot. Then, using Figures 24.14 (p. 498), 24.15 (p. 500), and 24.16 (p. 500), describe patterns of seed germination. (pp. 499–501)

7 Using Figures 24.17 (p. 501) and 24.19 (p. 502), point out the areas of cell division, cell enlargement, and cell maturation. (pp. 499–502)

8 Describe the process by which leaves are produced along the young growing stem. Using Figure 24.20 (p. 502), point out a terminal bud, axillary bud, node, leaf scar, internode, bundle scar, and bud-scale scar. Specify what the axillary bud gives rise to. (pp. 501–502)

9 Using Figure 24.21 (p. 503), explain how differentiation proceeds in a young root. Be sure to mention the role of the cambium. (pp. 502–505)

KEY TERMS

short-day plant (p. 490)
long-day plant (p. 490)
day-neutral plant (p. 490)
photoperiodism (p. 490)
florigen (p. 490)
phytochrome (p. 492)
sepals (p. 492)
petals (p. 492)
corolla (p. 493)
stamens (p. 493)
filament (p. 493)
anther (p. 493)
pistil (p. 493)
ovary (p. 493)
style (p. 493)
stigma (p. 493)
carpel (p. 493)
spore (p. 493)
gametophyte (p. 493)
sporophyte (p. 493)
ovule (p. 493)
megaspore (p. 493)
microspore (p. 494)
pollination (p. 494)
double fertilization (p. 495)
embryo (p. 495)
endosperm (p. 495)
seed coat (p. 495)
seed (p. 496)
fruit (p. 497)
suspensor (p. 497)
protoderm (p. 497)
provascular tissue (p. 497)
ground tissue (p. 497)
cotyledon (p. 497)
hypocotyl (p. 497)
apical meristem of the shoot (p. 497)
apical meristem of the root (p. 497)
epicotyl (p. 498)
radicle (p. 498)
germination (p. 498)
perennial plant (p. 499)
annual plant (p. 499)
node (p. 501)
internode (p. 501)
bud (p. 502)
axillary bud (p. 502)
differentiation (p. 503)
totipotency (p. 504)

SUMMARY

Control of flowering The flowering response in many plants is initiated by *photoperiodism,* the length of light and dark periods. Most plants can be classified as (1) short-day plants, which flower only when the night is *longer* than a critical value; (2) long-day plants, which flower only when the night is *shorter* than a critical value; and (3) day-neutral plants, which flower independently of day and night length. The critical night length is a *maximum* value for a long-day plant and a *minimum* value for flowering in a short-day plant.

It is believed that an inducing photoperiod causes the leaves to produce the hormone *florigen,* which moves to the buds and stimulates flowering. A noninducing photoperiod causes the leaves of many plants to inhibit florigen in some way. All attempts to isolate florigen have failed and many botanists suggest that flowering is controlled by a balance of different hormones—that there is no florigen.

Light is detected by a receptor pigment in the plasma membrane called *phytochrome,* which exists in two forms, one that absorbs red light (P_r) and one that absorbs far-red light (P_{fr}). Red and far-red light interconverts the phytochrome between its two forms. P_{fr} is the less stable form in the dark; it reverts to P_r over time, and some of it is enzymatically destroyed.

When phytochrome is exposed to both red and far-red light simultaneously (as in sunlight), the red light dominates and most of the pigment is converted into P_{fr}. During the night the P_{fr} supply dwindles as a result of reversion and destruction. Thus the plant has a way to detect whether it is day or night.

The mechanism that enables the plant to measure the length of the dark period is apparently tied to an "internal clock." The phytochrome mechanism determines whether it is day or night, whereas the internal clock measures the length of the dark period. According to the florigen hypothesis, once these mechanisms have indicated to the plant that the photoperiod is appropriate, florigen is produced and flowering is initiated.

Phytochrome participates in many other light-induced functions such as the germination of seeds, cell elongation, expansion of new leaves, breaking of dormancy, the formation of plastids.

Sexual reproduction in flowering plants Angiosperms are the dominant land plants. Their reproductive structures are flowers. In a flower the outer leaflike *sepals* protect the inner floral parts, the *petals* which collectively form the *corolla* often attract animal pollinators, the *stamens* (*filament* and *anther*) are the male reproductive organs, and the central *pistil* is the female reproductive organ. The pistil is made of one or more *carpels,* each of which consists of a *stigma, style,* and *ovary*. Within the ovary are one or more sporangia called *ovules*. Meiosis occurs in aeach ovule, and produces one functional *megaspore,* which divides to produce (in many species) a haploid, seven-celled, eight-nucleate, female *gametophyte,* the embryo sac.

Each anther has four pollen sacs in which meiosis occurs and produces haploid microspores. These develop into thick-walled, two-nucleate pollen grains—the male gametophytes. *Pollination* is the transfer of pollen from the anther to the stigma of a carpel. A pollen grain germinates when it lands on the stigma; a pollen tube grows through the style, enters the ovary, and discharges two sperm nuclei into the female gametophyte. *Double fertilization* occurs: one sperm nucleus fertilizes the egg and the other combines with the two polar nuclei to form a triploid nucleus, which will give rise to the *endosperm.*

After fertilization, the ovule matures into a seed, which consists of a *seed coat,* stored food (endosperm), and *embryo.* The seeds are enclosed in *fruits* that develop from the ovaries and associated structures. The fruit protects the seeds from drying out and often also facilitates their dispersal.

Development of an angiosperm plant During seed development, the embryo at first grows rapidly and then enters a period of arrested development. In many plants there is a period of dormancy before growth resumes. Auxin, cytokinin, gibberellin, and abscisic acid all play important roles in the developmental process.

The first cell division of the zygote gives rise to two equal cells, a terminal cell and a basal cell. The terminal cell will give rise to the embryo proper. Three types of tissues begin to differentiate: a surface layer of *protoderm,* which will form epidermal tissue; an inner core of *provascular tissue,* which will form the cambium and vascular tissues; and a middle layer of *ground tissue,* which will form the cortex and pith. Next, two embryo leaves, the *cotyledons,* arise. The part of the embryonic axis below the cotyledons is the *hypocotyl.* Small clumps of tissue at each end of the embryonic axis remain undifferentiated and become the *apical meristems of the shoot and root.* The embryo, together with the the endosperm, becomes enclosed in a seed coat; the resulting structure is the seed. Unlike the embryo of an animal, the plant embryo does not have all the organs of the adult organism even in rudimentary form.

Eventually development of the embryo halts, and the seed begins to lose water. The later stages of seed development are marked by a decline in the concentration of auxin and gibberellin and a rise in the level of abscisic acid, which is believed to stimulate the maturation of the embryo while inhibiting germination of the seed.

Germination of the seed begins with a massive uptake of water by the seed. As a result, the embryo releases gibberellin, which in turn stimulates the production of enzymes that hydrolyze the food stored in the endosperm. When the seed germinates, the hypocotyl emerges and turns downward; the *radicle* at its lower end forms the root. The *epicotyl* then begins to elongate; it forms most of the shoot. Growth of the shoot or root involves the production of new cells by the apical meristem, and then the elongation and differentiation of these cells. *Perennial plants* (those that live from year to year) grow throughout their lives, whereas *annual plants* stop growing at maturity and die at the end of the growing season.

In stems, certain areas of the apical meristem give rise to *nodes,* swellings where leaf primordia arise. The length of stem between two successive nodes is an *internode.* Most additions in length to the stem result from the elongation of the cells in the young internodes. At the tip of the stem is a *bud,* which consists of the apical meristem and unelongated internodes enclosed within leaf primordia. When the bud opens, the internodes elongate and mitosis within the leaf primordia produces the leaves. Small areas of meristematic tissue (the *axillary buds*) arise between the leaf and internode. The cells produced by the apical meristems will differentiate to form the various primary tissues of the plant. Increase in the circumference of a root or stem depends on the formation of secondary tissues derived from the lateral meristems.

Developing cells are influenced by hormones and by the physical environment (e.g., light, temperature, gravity). As development proceeds, the individual cells become more and more committed to one particular course of *differentiation.* Experiments have shown that in many plant cells, differentiation can be reversed and *totipotency* (full developmental potential) resumed. The various growth patterns are coordinated and integrated by plant hormones.

QUESTIONS

Testing recall

1 Which of the following pairs of terms are closely related? (pp. 495–98)

    a hypocotyl—stem
    b endosperm—stored food
    c cotyledons—embryonic leaves
    d seed dormancy—increased abscisic acid concentration
    e epicotyl—primary root
    f meristematic cells—differentiated cells
    g ground tissue—cortex
    h terminal cell—plant embryo
    i apical meristem—secondary tissues
    j shoot growth—cell division and cell elongation
    k node—leaf primordia
    l vascular cambium—elongation of the stem
    m seed—embryo, endosperm, seed coat

Testing knowledge and understanding

2 A response to changes in the length of daylight is known as

    a photoperiodism.   *d* phototaxis.
    b phototropism.   *e* photolysis.
    c photosynthesis.   (p. 490)

3 The pigment responsible for detecting the presence or absence of light is

    a auxin.   *d* gibberellin.
    b phytochrome.   *e* florigen.
    c chlorophyll.   (p. 490)

4 Suppose you took two plans, a long-day plant (LDP) with a critical night length of 12 hours and a short-day plant (SDP) also with a critical night length of 12 hours, and exposed the two to several "days," each consisting of a 14-hour dark period and a 10-hour light period. You would predict that

    a the LDP would flower and the SDP would not.
    b the SDP would flower and the LDP would not.
    c neither plant would flower.
    d both plants would flower.   (pp. 489–90)

5 A certain short-day plant flowers only when the night is at least 14 hours long. Under which one of the following light regimes will this plant flower? (pp. 489–90)

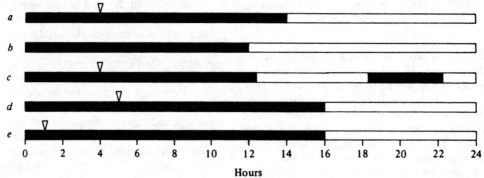

Key
White bars: daylight
Black bars: night
White triangles: intense flashes of light
Black triangles: intense flashes of far-red light

6 A certain short-day plant has a critical night length of 14 hours. Under which one of the following light regimes will this plant flower? (pp. 489–90)

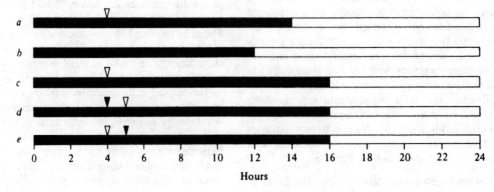

7 In angiosperms, the gametophytes are produced within the

   a stamens and pistils.
   b sepals and petals.
   c sepals and stamens.
   d anthers and petals.
   e sepals and pistils. (p. 495)

8 Which one of the following is a correct sequence of processes that takes place when a flowering plant reproduces?

   a meiosis—fertilization—ovulation—germination
   b fertilization—meiosis—nuclear fusion—formation of endosperm
   c meiosis—pollination—nuclear fusion—formation of embryo and endosperm
   d growth of pollen tube—pollination—germination—fertilization
   e meiosis—mitosis—nuclear fusion—pollination (pp. 494–95)

9 A fruit is a mature

   a embryo.
   b seed.
   c ovary.
   d stem.
   e root. (p. 497)

10 In tobacco plants, the diploid number is 48. How many chromosomes do tobacco endosperm cells have?

   a 12
   b 24
   c 48
   d 72
   e 96 (p. 495)

Match each function below with the associated part of the flower. Answers may be used once, more than once, or not at all.

   a anther
   b ovary
   c ovule
   d sepal
   e stigma

11 contains the female gametophyte (p. 493)
12 produces the male gametophyte (p. 494)
13 matures to form the fruit (p. 497)
14 site on which pollen is deposited by pollinator (p. 495)
15 forms the seed (p. 493)

Questions 16 to 18 refer to the following diagram of a plant embryo.

16 From what part will the cotyledons be produced? (p. 497)
17 From what part will the apical meristem arise? (p. 497)
18 Which part will give rise to xylem and phloem? (p. 497)

Choose the one best answer

19 Which one of the following areas is closest to the tip of a plant root.

   a axillary bud
   b zone of cell elongation
   c apical meristem
   d zone of cell differentiation
   e zone of cell maturation (pp. 497–98)

20 The primary function of the cotyledons is to

   a give rise to the epicotyl.
   b define the first node of the embryo.
   c absorb nutrients from the endosperm.
   d protect the hypocotyl as it grows out of the seed coat.
   e do none of the above (p. 497)

21 Provascular tissue

   a is found only in the root.
   b gives rise to highly modified epidermal cells that include leaf hairs and root hairs.
   c differentiates into the primary xylem and phloem.
   d is found between the protoderm and the ground tissue.
   e is or does none of the above (p. 497)

22 Which of the following seed structures is *not* part of the embryo?

   a cotyledons
   b endosperm
   c shoot apical meristem
   d radicle
   e hypocotyl (p. 495)

23 Which of the following statements about plant development is *incorrect*?

   a All the organs of the mature plant are present in the embryo.
   b Cell differentiation occurring during development can often be experimentally reversed.
   c Some plant cells continue to divide as the plant grows.
   e The root–shoot axis is established in the embryo before seed germination. (p. 498)

24 Which one of the following statements about a bean seed and its germination is *false*?

   a In the dormant state the seed consumes oxygen, but at a very low rate.
   b As germination begins, much water is absorbed and oxygen consumption rises.
   c The part of the embryonic axis called the epicotyl grows down and gives rise to the plant root.
   d The two cotyledons contain stored food and provide the principal source of nourishment on which the early growth of the embryo depends.
   e As the young seedling develops, its increase in length is due to mitotic activity in the apical meristems of the root tips and terminal buds, followed by cell elongation in regions immediately behind the meristems.
   (pp. 497–98)

Questions 25 to 28 refer to the following diagram of a root.

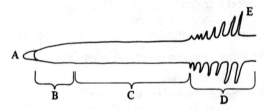

25 Which is the region of elongation? (p. 501)

26 Which is the region of cell division? (p. 501)

27 Which is the region where xylem and phloem are differentiating? (p. 501)

28 Which is the region that produces the root-cap cells? (p. 501)

Use the diagram at the top of the next column to answer questions 29 to 31.

29 Which tissue carries sugar from the shoot to the root? (p. 501)

30 Which tissue carries water from the root to the shoot? (p. 501)

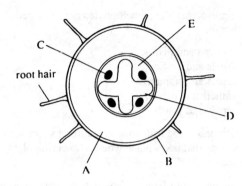

32 From which zone is this root cross section taken?

   a zone of cell division
   b root cap
   c zone of cell maturation
   d zone of elongation
   e apical meristem (p. 503)

For further thought

1 The poinsettia is a short-day plant with a critical night length of 12 hours. How could you get the plant to flower in June? (pp. 489–91)

2 It is terribly important for scientific researchers to be able to "look sideways," and not just blindly straight ahead. Use an example from photoperiodism research to illustrate this generality. (pp. 489–91)

ANSWERS

Testing recall

1 *a, b, c, d, g, h, j, k, m*

Testing knowledge and understanding

2	*a*	10	*d*	18	*C*	26	*B*
3	*b*	11	*c*	19	*c*	27	*D*
4	*b*	12	*a*	20	*c*	28	*B*
5	*e*	13	*b*	21	*c*	29	*C*
6	*e*	14	*e*	22	*b*	30	*D*
7	*a*	15	*c*	23	*a*	31	*c*
8	*c*	16	*B*	24	*c*		
9	*c*	17	*A*	25	*C*		

Chapter 25

CHEMICAL CONTROL IN ANIMALS

A GENERAL GUIDE TO THE READING

This chapter describes the major hormonal control mechanisms in animals. It begins with a discussion of the differences between plant and animal hormones and goes on to consider the major mechanisms of hormonal action. The remainder of the chapter is devoted to a consideration of hormones in vertebrates; especially humans. As you read Chapter 25 you will want to pay particular attention to the following material.

1. The major differences between plant and animal hormones. Table 25.1 (p. 508) summarizes these differences and reviews and amplifies the corresponding discussion at the beginning of Chapter 23.

2. The mechanism of the action of hormones. The section on this topic (pp. 508–11) is indispensable and interesting. Pay particular attention to the material on second messengers.

3. The location of the major human endocrine organs. Figure 25.1 (p. 508) will help you to learn this material.

4. The functions of important mammalian hormones. Table 25.2 (p. 509) is a wonderful summary of these functions; you will want to refer to it frequently.

5. The actions of insulin are described on pages 512 to 514. You will find the various effects of this hormone easier to learn if you keep in mind that the overall effect of insulin in the bloodstream is to lower its sugar content.

6. Double glands. Notice that both the adrenals (pp. 515–18) and the pituitary (pp. 522–25) glands are double: each is composed of two different glands that are anatomically and functionally distinct.

7. The role of the anterior pituitary and the hypothalamus in homeostatic control. This topic is very important; you will want to study it (pp. 523–25) carefully with special attention to Figure 25.19 (p. 523).

8. Local chemical mediators. Notice that hormones are not the only control chemicals. The chapter concludes with a discussion of the effect of local chemical mediators such as prostaglandins and endorphins.

KEY CONCEPTS

1. Chemical control mechanisms play an important role in the coordination of the myriad functions of living organisms. Hormonal control is common to plants and animals. (p. 507)

2. Animal hormones, notably those found in mammals, not only guide growth and development, but also play a large role in regulating metabolism and in maintaining general homeostasis. (p. 507)

3. Numerous hydrophilic hormones act on target cells indirectly by interacting with specific receptors on the cell membrane and thereby opening an ion channel or activating cytoplasmic enzyme systems in the cell. Other hormones called hydrophobic hormones, act more directly, by moving into the target cell and interacting with its genetic material. (pp. 508–12)

4. The secretion of many animal hormones is regulated by a negative-feedback control mechanism; low concentrations of a particular hormone result in increased secretion of

that hormone, and when the concentration of the hormone reaches a certain level in the blood, its secretion is reduced. Feedback loops are an extremely important mechanism for maintaining homeostasis. (pp. 514–15)

5 The endocrine tissues that produce animal hormones are often closely associated with the nervous system. (p. 525)

6 Local chemical mediators are secreted by cells that are not part of specialized chemical-control organs: in most cases these are so rapidly destroyed that they affect only cells in their immediate neighborhood. (pp. 526–27)

CROSSLINKING CONCEPTS

1 A thorough understanding of proteins and steroids is helpful when studying the structure and function of hormones.

2 To understand how levels of blood sugar are maintained, you should recall your study of digestion of carbohydrates (Chapter 17), the role of the liver in carbohydrate and fat metabolism, the portal circulatory connection between the intestine and the liver, fat synthesis and storage in adipose cells (Chapter 22), and the interlocking biochemical pathways of carbohydrates, fats, and proteins (Fig. 8.13).

3 In this chapter you learn how the parathyroid hormone and calcitonin regulate the concentration of calcium ions in the blood. The significance of this balance will become more apparent when you study nerve impulse transmission (Chapter 28) and muscle contraction (Chapter 30).

4 You have previously studied portal systems in the kidney and the liver (Chapter 22). In this chapter you find an example of another portal system; this one links the hypothalamus and the anterior pituitary.

5 Earlier you studied gene-control mechanisms. In this chapter, you will learn how hormones exercise chemical control both directly iby controlling gene transcription and indirectly through second-messenger systems.

OBJECTIVES

1 Define the terms hormone and endocrine. Explain how most of the hormones in higher animals are transported from the site of synthesis to the target tissue. (p. 507)

2 List the major endocrine organs in the human body and indicate their location. (p. 508)

3 Describe the second-messenger model of hormonal control. In doing so, specify the type of hormones that use the two-messenger system, indicate the intracellular effects of the extracellular binding of the hormone, and show why only certain cells are influenced by a given hormone. (pp. 508–10)

4 Explain how steroid hormones enter their target cell, and discuss their mode of action. (pp. 511–12)

5 Give the location of the endocrine gland that secretes insulin; using a drawing such as Figure 25.6 (p. 512), point out and name the cells where insulin is produced, describe the function of the α islet cells, and indicate what other cells are present in the pancreas and what their function is. (pp. 512–13)

6 List four actions of insulin that reduce the concentration of glucose in the blood. Contrast its action with that of a second hormone from the pancreas, glucagon. Name one other hormone that acts as an insulin antagonist. (pp. 513–14)

7 Give the location of the adrenal glands, and in a cross section like that in Figure 25.9 (p. 515), point out the cortex and the medulla. Indicate which portion secretes adrenalin and which secretes steroid hormones. (pp. 515–18)

8 Describe the effect of adrenalin on the blood pressure, heartbeat, blood-sugar level, oxygen consumption by cells, blood supply to the skeletal and heart muscle, and blood supply to the skin and the digestive tract. (pp. 515–17)

9 Name the three categories of steroid hormones produced by the adrenal cortex, and give a major function of each. (p. 518)

10 Discuss the role of the thyroid hormones (T_3 and T_4, together known as TH) in the human body. In doing so, specify the location of the thyroid gland, the effect of TH on oxidative metabolism of the cells of the body, the symptoms of excessive TH secretion (hyperthyroidism), and the symptoms of insufficient TH secretion (hypothyroidism). (pp. 519–20)

11 Discuss the roles of calcitonin and parathyroid hormone (PTH) in the control of calcium and phosphate levels in the blood. In doing so, specify where each hormone is synthesized; describe the effect of PTH on the blood-calcium level, the effect of calcitonin on the blood-calcium level, and the effect of PTH on the blood-phosphate level. Name the three target organs of PTH. (pp. 520–21)

12 Describe the relationship between the hypothalamus and the posterior pituitary. Indicate where the hormones oxytocin and vasopressin are produced, from what organ they are released, and how they get from one place to another. Give one function of vasopressin and oxytocin. (p. 522)

13 For each of the following hormones produced by the anterior pituitary, describe a major function and specify the target organ: prolactin (PRL), growth hormone (GH), thyrotropic hormone (TSH), adrenocorticotropic hormone (ACTH), and gonadotropic hormone (FSH, LH). (pp. 523–25)

14 Explain the structural and functional relationships between the hypothalamus and the anterior pituitary. In doing so, be sure to describe the portal system between the hypothalamus and the anterior pituitary; specify the way releasing hormones produced in the hypothalamus affect the activity of the anterior pituitary. Using Figure 25.21 (p. 526), describe the basic features of a negative feedback loop involving the thyroid gland. (pp. 525–26)

15 Given the location of the pineal gland, contrast its function in lower vertebrates with its function in mammals. (p. 526)

16 Explain how local chemical mediators differ from hormones. For prostaglandins and endorphins, indicate where each is produced and its target cells, and give one function for each. (pp. 526–28)

KEY TERMS

endocrine tissue (p. 507)
exocrine glands (p. 507)
cyclic AMP (p. 508)
adenylate cyclase (p. 508)
second-messenger model (p. 508)
G-protein (p. 510)
islet cell (p. 512)
insulin (p. 512)
glucagon (p. 512)
negative feedback (p. 514)
somatostatin (p. 515)
adrenal medulla (p. 515)
adrenal cortex (p. 515)
adrenalin (p. 515)
noradrenalin (p. 515)
glucocorticoid (p. 518)
mineralocorticoid (p. 518)
goiter (p. 518)
thyroxin, or T_4 (p. 519)
triiodothyronine, or T_3 (p. 519)
thyroid hormone, or TH (p. 519)
calcitonin (p. 520)
parathyroid hormone, or PTH (p. 520)
hypothalamus (p. 522)
posterior pituitary (p. 522)
oxytocin (p. 522)
vasopressin (p. 522)
anterior pituitary (p. 522)
prolactin, or PRL (p. 523)
growth hormone, or hGH (p. 524)
growth factors (p. 524)
tropic hormone (p. 524)
thyrotropic hormone, or TSH (p. 524)
adrenocorticotropic hormone, or ACTH (p. 524)
gonadotropic hormone (p. 524)
releasing hormone (p. 525)
pineal (p. 526)
melatonin (p. 526)
local chemical mediator (p. 527)
histamine (p. 527)
prostaglandin (p. 527)
endorphin (p. 528)

SUMMARY

The tissues that produce and release hormones in animals are termed *endocrine tissues*. The hormones are secreted directly into the blood, transported through the circulatory system, and exert highly specific effects on target tissue.

Mechanisms of hormonal action Animal hormones can be classified as water-soluble (hydrophilic) or water-insoluble (hydrophobic). Most hormones are water-soluble, and are derived from amino acids or proteins. They interact with receptor molecules on the target cell membrane surface. The hydrophobic hormones, being lipid-soluble, readily diffuse through the cell membrane and bind to specific receptors inside the cell.

The hydrophilic hormones cannot pass through the cell membrane. According to the *two-messenger model*, the hormone acts as an extracellular first messenger; it binds to a specific receptor site on the outer membrane surface of the target cell. The binding results in the intracellular activation of the enzyme *adenylate cyclase,* which catalyzes the production of a second messenger, cAMP, inside the cell. The presence or absence of receptors on the membrane determines whether a cell will respond to a given hormone.

Some hydrophilic hormones do not use the cAMP system. For example, insulin induces a rise in cGMP. Ions can also act as second messengers inside the cell. The binding of certain hormones to specific receptors on the cell membrane may open an ion (e.g., calcium ion) channel. The ions then rush in and bind to and activate one or more specific enzymes in the cell.

The hydrophobic hormones—the thyroid and steroid hormones—can easily pass through the cell membrane. Steroid hormones (S) bind to a receptor molecule (R) and within the nucleus the complex (S-R) interacts directly with specific genes to influence transcription.

Most hydrophilic hormones circulate through the blood for only a few minutes before being degraded; furthermore the enzyme systems that these hormones influence are activated and deactivated very rapidly. In contrast, hydrophobic hormones such as steroids remain in the blood for many hours. Because they regulate gene activity, their effects are slower but longer-lasting than those of hydrophilic hormones.

The pancreas The *islet cells* of the pancreas secrete *insulin,* which reduces the blood-glucose concentration by stimulating glucose absorption in liver, muscle, and adipose cells, by promoting glucose oxidation and glycogen synthesis in liver and muscle cells, and by inhibiting glycogen hydrolysis. It also promotes the synthesis of fat and proteins while inhibiting their breakdown. a deficiency in insulin production or insensitivity of the tissues to insulin results in diabetes, a disease characterized by an abnormally high blood-sugar level.

The islet cells of the pancreas secrete two other hormones, *Glucagon*, which is *antagonistic* to insulin; cause an increase in the blood-glucose concentration. The interaction between these hormones and the blood-glucose concentration is an example of *negative feedback*. A third pancreatic hormone, *somatostatin*, inhibits glucagon and insulin secretion and prolongs the time food remains in the digestive tract.

The adrenal glands The two *adrenal glands*, located above the kidneys, consist of an inner *medulla* and outer *cortex*, which remain functionally distinct. The adrenal medulla secrets two hormones, *adrenalin* and *noradrenalin*, whose effects are similar. Both help to prepare the body for emergencies by stimulating reactions that increase the supply of glucose and oxygen to the skeletal and heart muscles ("fight-or-flight" response).

The adrenal cortex produces many different steroid hormones. The cortical hormones may be grouped into three functional categories: (1) those regulating carbohydrate and protein metabolism, the *glucocorticoids*; (2) those regulating salt and water balance, the *mineralocorticoids*; and (3) those that function as sex hormones.

The thyroid gland The *thyroid gland* is located just below the larynx. It has a great affinity for iodine, which is used to synthesize two thyroid hormones (TH), *thyroxin* and *triiodothyronine*. These hormones stimulate the oxidative metabolism of most tissues in the body and have a role in regulating many aspects of development. Hyperthyroidism—excessive TH secretion—produces an increase in the metabolic rate with high body temperature, high blood pressure, profuse perspiration, irritability, and weight loss. Hypothyroidism—decreased TH secretion—leads to the opposite symptoms. It can be caused by dietary iodine insufficiency or by malfunction of the thyroid itself. Continued iodine deficiency causes an enlargement of the gland, known as *goiter*. Untreated hypothyroidism in newborn children is called cretinism; such children show retarded physical, sexual, and mental development.

The thyroid also secretes the hormone *calcitonin*, which prevents the excessive rise of calcium ions in the blood.

The parathyroids The *parathyroids* are four small pealike organs located on the surface of the thyroid. The *parathyroid hormone* (PTH), antagonistic to calcitonin, regulates the calcium-phosphate balance between the blood and other tissues; it acts primarily on the kidneys, the intestine, and the bones to lower blood-calcium levels.

The pituitary and the hypothalamus As with the adrenal gland, the pituitary gland is made of two sections that are functionally distinct. The *posterior pituitary* is connected to a part of the brain, the *hypothalamus*, by a stalk. It stores and releases two hormones, *oxytocin* and *vasopressin*, which are produced in the hypothalamus and flow along nerves in the stalk to the posterior pituitary. The hormones are released on nervous stimulation from the hypothalamus. Oxytocin stimulates the contraction of uterine muscles. Vasopressin causes constriction of the arterioles with a consequent rise in blood pressure and stimulates the kidney tubules to reabsorb more water.

The *anterior pituitary* produces some hormones with far-reaching effects. The hormone *prolactin* stimulates milk production in the mammary glands and also participates in reproduction, osmoregulation, growth, and metabolism of carbohydrates and fats. *Growth hormone* (GH) promotes normal growth. A serious deficiency in a child results in stunted growth; an oversupply results in a giant. Growth hormone is a powerful inducer of protein synthesis and acts as an insulin antagonist.

The anterior pituitary also secretes a number of hormones that help control other endocrine organs (it is often called the master gland). *Thyrotropic hormone* stimulates the thyroid, *adrenocorticotropic hormone* (ACTH) stimulates the adrenal cortex, and the two *gonadotropic hormones* (FSH and LH) act on the gonads. The interaction between the anterior pituitary and these glands is another example of negative feedback. For example, when the thyroxin level in the blood is low, the anterior pituitary releases the thyrotropic hormone, which stimulates the thyroid to increase production of TH. The increased TH level then inhibits the secretion of more thyrotropic hormone by the pituitary.

The activity of the anterior pituitary is, in turn, regulated by the hypothalamus, which produces special peptide *releasing hormones*. These hormones are carried by a special blood portal system to the anterior pituitary, where they stimulate its secretory activity. The hypothalamus is the point at which information from the nervous system influences the endocrine system and is also one of the major sites of negative feedback from the endocrine system.

The pineal gland The *pineal*, a lobe in the forebrain, secretes a hormone called *melatonin*. In some lower vertebrates melatonin lightens the skin and is involved in the control of circadian rhythms. In mammals, the amount of light influences the pineal's secretion of melatonin in an inverse relationship: the more light, the less melatonin. Melatonin, in turn, influences the secretion of gonadotropic hormones.

Local chemical mediators Prostaglandins and endorphins are secreted by cells that are not part of specialized chemical-control organs. In most cases these are so rapidly destroyed that they affect only cells in their immediate neighborhood. Normally they do not enter the blood in significant amounts.

Prostaglandins are synthesized from phospholipids in the cell membrane and exert a wide variety of effects on their target sites. *Endorphins* are chemical mediators that bind to opiate receptors on nerve cells in the part of the brain concerned with pain perception an d mood.

CHEMICAL CONTROL IN ANIMALS • 183

CONCEPT MAP

Map 25.1

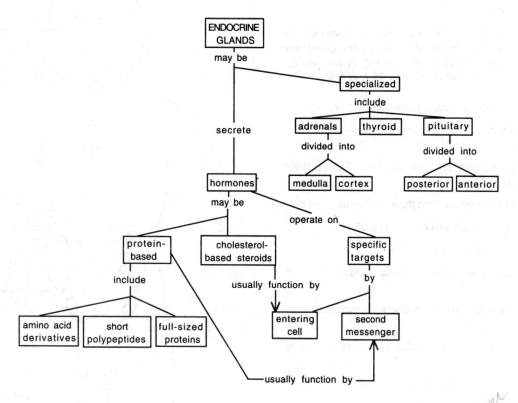

KEY DIAGRAM

The following coloring exercise is designed to familiarize you with the anatomical relationships between the hypothalamus and anterior and posterior pituitary. First color the names and then the appropriate structures, using a different color for each structure. (Fig. 25.19, p. 523)

HYPOTHALAMUS
ANTERIOR PITUITARY
POSTERIOR PITUITARY
NERVE CELLS
CAPILLARIES

Neurosecretory cells producing:

RELEASING HORMONES
OXYTOCIN, VASOPRESSIN

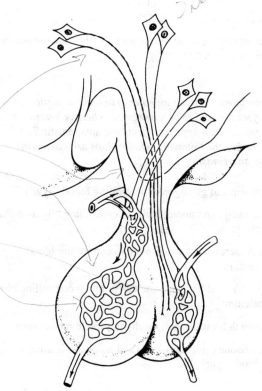

184 • CHAPTER 25

QUESTIONS

Testing recall

Match each function below with the associated endocrine gland or glands shown in the figure. Several glands may match the same statement. Answers may be used once, more than once, or not at all. (Fig. 25.1, p. 508)

1 secretes a variety of steroid hormones (p. 517) *l*
2 secretes a hormone that prevents an excessive rise in blood calcium (p. 520) *e*
3 secretes the hormones that stimulate the "fight-or-flight" reaction (p. 517) *i*
4 regulated by hormones from the anterior pituitary (p. 524) *k, l, e, i*
5 produces vasopressin and oxytocin (p. 522) *a*
6 produces growth hormone (p. 524) *c*
7 secretes a hormone that lowers blood-glucose concentration (p. 513) *a, h, i*
8 secretes a hormone that raises blood-glucose concentration (p. 514) *c, h, i*
9 secretes a hormone that stimulates oxidative metabolism (p. 520) *e*
10 secretes a hormone that regulates calcium-phosphate balance (p. 520) *d, e*
11 secretes melatonin (p. 526) *b*
12 produces releasing hormones that stimulate the anterior pituitary (p. 525) *a*

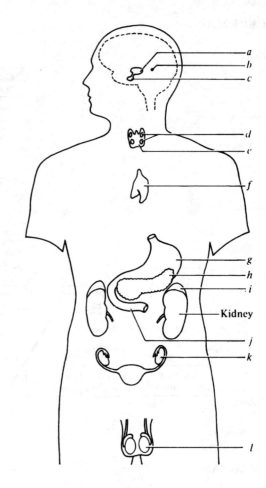

Hormones that act together to achieve a particular effect are said to be synergistic, *whereas hormones that produce opposite effects are* antagonistic. *Tell whether the following relationships are synergistic or antagonistic.*

13 insulin and glucagon in controlling blood sugar *A*
14 adrenalin and noradrenalin in controlling "fight-or-flight" response *S*
15 aldosterone and vasopressin in controlling blood pressure *S*
16 parathyroid hormone and calcitonin in controlling blood calcium *A*
17 growth hormone and insulin in protein metabolism *S*
18 cortisone (glucocorticoid) and adrenalin in controlling blood sugar *S*

Testing knowledge and understanding

Choose the one best answer.

19 All the following are characteristic of hormones *except* that

 a they act within the gland that synthesizes them.
 b they are present in small quantities.
 c they are carried in the bloodstream.
 d they are sometimes synthesized by nervous tissue.
 e they are secreted by glands without ducts.
 (p. 507)

20 According to the proposed second-messenger model for the mode of action of animal hormones,

 a the hormone enters the target cell by combining with a carrier molecule and moves to the nucleus, where it reacts with genetic material.
 b the hormone enters the target cell, where it causes production of prostaglandins, which then interact with the genetic material in the nucleus.

c the hormone interacts with a specific receptor site on the outer surface of the cell so as to influence the concentration of cyclic AMP within the cell.
d the first hormone is produced by the pituitary and it in turn causes the secretion of a second hormone from another endocrine gland. (pp. 508–11)

21 Which one of the following hormones does *not* work by the second-messenger system?

a TSH   d glucagon
b aldosterone   e ACTH
c adrenalin   (p. 511)

22 Which one of the following hormones requires the involvement of cAMP in order to bring about a change in the target cells?

a mineralocorticoids   d thyroxin
b sex hormones   e adrenalin
c all plant hormones   (p. 510)

23 A polypeptide hormone affects muscle but not liver cells. The most likely reason is that

a liver cells lack a receptor for that hormone.
b liver cells do not have adenylate cyclase in their plasma membrane.
c liver cells do not make cAMP.
d liver cells do not contain enzymes. (p. 510)

24 Which one of the following statements concerning steroid hormones is *true*?

a They bind to a receptor that is located on the plasma membrane.
b They form a complex with a receptor within the target cell and bind to DNA in the cell nucleus.
c They activate the enzyme adenylate cyclase.
d They travel within the body in the lymphatic rather than the circulatory system. (p. 511)

25 Which one of the following may function as a second messenger in a chemical control system?

a calcium ions
b adenylate cyclase
c ATP
d adrenalin
e thyrotropic hormone (TSH)   (p. 511)

26 By the technique of autoradiography it is possible to determine the location of hormones in a target tissue. Where do steroid hormones act to alter the physiology of target cells?

a plasma membrane   d cytosol
b mitochondria   e endoplasmic reticulum
c nucleus
d cytosol
e endoplasmic reticulum   (p. 511)

27 The pancreas

a has a duct leading to the intestine and therefore is not an endocrine gland.
b is a major gland in mammals in which digestive enzymes are made.
c produces a hormone that lowers the blood-sugar level.
d is located next to the pineal gland.
e is or does two of the above   (p. 512)

28 Which one of the following endocrine organs has no known function except hormone secretion?

a pancreas   c gonads (sex organs)
b adrenals   d hypothalamus   (pp. 515–18)

29 In a normal person, insulin is secreted by the pancreas immediately after ingestion of carbohydrates. Which one of the following is *not* an effect of insulin?

a It promotes the transport of glucose from the blood into muscle cells.
b It stimulates glycogen synthesis from glucose by the liver.
c It promotes the synthesis of fats from glucose in adipose cells.
d It stimulates the convoluted tubules of the kidney to reabsorb more sugar.
e It acts to lower blood-glucose levels. (p. 513)

30 In persons who have abnormally high levels of sugar in their urine, the production of insulin by the pancreas is usually

a abnormally high.   b abnormally low.   (p. 513)

31 Which one of the following organs does not play an important role in the regulation of blood-glucose levels?

a parathyroid   d anterior pituitary
b pancreas   e hypothalamus
c adrenal cortex   (pp. 520–22)

32 Which one of the following is *not* an effect of insulin?

a Stimulates absorption of glucose from the blood by muscle and adipose cells.
b Promotes oxidation of glucose in liver and muscle cells.
c Inhibits breakdown of stored glycogen in liver and muscle cells.
d Promotes synthesis of glucose from fats in adipose cells.
e Alleviates symptoms of diabetes mellitus.   (p. 513)

33 Glucocorticoids are produced by the

a thyroid.   d adrenal medulla.
b pituitary.   e pancreas.
c adrenal cortex.   (p. 518)

34 An iodine deficiency in the human diet is most likely to lead to

   a an elevated blood-calcium level.
   b an elevated blood-glucose level.
   c excessive inflammation reactions.
   d a lowered metabolic rate.
   e an increase in insulin secretion. (pp. 518–19)

35 An increase of thyrotropic hormone in the blood of a mammal causes

   a an increased blood supply to the thyroid.
   b increased secretion of thyroxin.
   c regression of the thyroid gland.
   d increased nervous stimulation of the thyroid.
   e reduced secretion of thyroxin. (pp. 524–25)

36 A physician examines an obese patient with low blood pressure. The patient complains of feeling very lethargic and mentally dull much of the time. Of the following endocrinological malfunctions, the patient's condition is most likely to be

   a diabetes.
   b hypoparathyroidism.
   c hyperparathyroidism.
   d hypothyroidism.
   e hyperthyroidism. (p. 520)

37 When a person takes thyroid tablets to increase his blood level of thyroid hormone,

   a the secretion of TRH by the hypothalamus is inhibited.
   b the secretion of thyrotropic hormone increases.
   c the thyroid gland enlarges.
   d the anterior pituitary is stimulated to produce more hormones.
   e the thyroid gland produces more TH. (p. 525)

38 A patient has an endocrine malfunction that results in excessive irritability of the muscles and nerves, which respond even to minor stimuli with tremors, cramps, and convulsions. Blood tests show an abnormally high concentration of phosphate and an abnormally low concentration of calcium. This condition is probably

   a hyperthyroidism.
   b hypothyroidism.
   c hyperparathyroidism.
   d hypoparathyroidism.
   e hyperadrenocorticalism. (pp. 521–22)

39 Which one of the following hormones regulates the rate of cellular respiration?

   a insulin
   b thyroid hormone (TH)
   c parathyroid hormone
   d adrenocortical hormones
   e vasopressin (p. 520)

40 Parathyroid hormone

   a is under the control of the adrenal gland.
   b is a steroid.
   c serves to increase the concentration of calcium in the blood.
   d mobilizes the release of calcium and phosphate from bone.
   e is or does two of the above (p. 521)

41 Which one of the following hormones is mismatched with the stated function?

   a melatonin—inhibits gonadotropin secretion
   b parathyroid hormone—regulates calcium-phosphate balance
   c ACTH—stimulates the adrenal cortex
   d oxytocin—stimulates water reabsorption by the kidneys
   e prolactin—stimulates milk production by the mammary glands (p. 509)

42 The primary connection between the nervous system and the endocrine system is

   a adrenalin.   *d* the hypothalamus.
   b the pancreas.   e cyclic AMP.
   c the brain. (p. 525)

43 All the following hormones are secreted by the anterior pituitary *except*

   a follicle-stimulating hormone (FSH).
   b thyrotropic hormone (TSH).
   c cortisone.
   d luteinizing hormone (LH).
   e growth hormone (GH). (p. 524)

44 Vasopressin is synthesized in the hypothalamus. Where is it stored?

   a pineal gland   d parathyroids
   b anterior pituitary e adrenal cortex
   c posterior pituitary (p. 522)

45 The injection of a small quantity of posterior pituitary extract into the blood is likely to cause

   a an increase in the volume of urine.
   b a decrease in the volume of urine. (p. 522)

46 A person with a hypofunctioning anterior pituitary would probably show all the following symptoms *except*

   a decreased metabolic rate.
   b decreased activity of the adrenal cortex.
   c sexual immaturity.
   d increased urine output.
   e decreased growth-hormone secretion. (p. 524)

47 All the following are chemical mediators *except*

   a histamine.   c endorphins.
   b prostaglandins.   *d* cortisone. (pp. 527–28)

For further thought

1. List the hormones that help control (*a*) carbohydrate metabolism and (*b*) protein metabolism.

2. Discuss the relationship between the hypothalamus and the anterior and posterior pituitary. Describe the role of the anterior pituitary and hypothalamus in coordinating the activity of the thyroid gland.

3. At the age of 9 years a girl was taken to a doctor who suspected that she had some sort of endocrine deficiency. At that time she was only 36 inches tall, though correctly proportioned. By 17, she was only as tall as a 9-year-old. There was no sexual development during her adolescent years.

   a What type of endocrine deficiency might this child have?
   b What hormones is she not producing?
   c What other symptoms might she have?
   d What hormonal therapy might be prescribed?

4. On the basis of your knowledge of the circulatory and endocrine systems, give some benefits and problems associated with cholesterol.

5. It is not terribly unusual for important finds to go unacclaimed. Describe one such find that was overlooked in the research on the thyroid gland.

6. Do you suffer from "winter depression"? Based on your reading from this chapter, what do you think you could do about it?

ANSWERS

Testing recall

1. *i*
2. *e*
3. *i*
4. *e, i, k, l*
5. *a*
6. *c*
7. *h*
8. *c, h, i*
9. *e*
10. *d, e*
11. *b*
12. *a*
13. antagonistic
14. synergistic
15. synergistic
16. antagonistic
17. synergistic
18. synergistic

Testing knowledge and understanding

19 *a*	27 *e*	34 *d*	41 *d*
20 *c*	28 *b*	35 *b*	42 *d*
21 *b*	29 *d*	36 *d*	43 *c*
22 *e*	30 *b*	37 *a*	44 *c*
23 *a*	31 *a*	38 *d*	45 *b*
24 *b*	32 *d*	39 *b*	46 *d*
25 *a*	33 *c*	40 *e*	47 *d*
26 *c*			

Chapter 26

HORMONAL CONTROL OF VERTEBRATE REPRODUCTION

A GENERAL GUIDE TO THE READING

The human reproductive system is as fascinating as it is intricate. It will be interesting for you to compare the reproductive methods of different vertebrates; these methods for propagation of species are related to physical structure, which is, in turn, related to an organism's environment. As you read Chapter 26 in your text, concentrate on the following material.

1. The introductory section, "The Process of Sexual Reproduction." This section (pp. 531–34) presents many important terms and concepts that you will need to know.

2. The eggs of land vertebrates. Figure 26.6 (p. 534), shows a so-called amniotic, or land egg; an important evolutionary advance in that it provides a fluid-filled chamber in which the embryo can develop even when the egg is laid in a dry place. The evolution of the amniotic egg enabled land vertebrates (first reptiles and later birds and mammals) to be truly terrestrial, since reproduction no longer required the organisms to lay eggs in moist places.

3. The human male and female reproductive tracts. These are thoroughly explained in this chapter. Figures 26.7 and 26.8 (p. 535) will help you learn the anatomy of the male reproductive tract, and Figure 26.14 (p. 541) will help you learn the anatomy of the female reproductive tract.

4. The role of hormones in mammalian reproduction. The material on hormonal control of sexual development and function in the male is relatively straightforward and is summarized nicely by Figure 26.13 (p. 538). The section "Hormonal Control of the Female Reproductive Cycle" (pp. 545–49) is more complicated. You will want to study the latter section carefully and refer frequently to, Figures 26.20 and 26.21 (p. 546).

KEY CONCEPTS

1. Sexual reproduction in higher animals always involves the union of two gametes; an egg and a sperm. Once united in fertilization, these form the first cell of the new individual. (pp. 531)

2. The hormones of both the ovaries and testes are controlled by a negative feedback loop involving the hypothalamus and anterior pituitary gland. (pp. 537–38, 545)

CROSSLINKING CONCEPTS

1. Your study of evolution (Chapter 31) will provide a framework for viewing adaptations of organisms as they moved to terrestrial environments. The amniotic egg and internal fertilization were particularly important evolutionary adaptations for life on land.

2. In Chapter 25, you learned how negative feedback mechanisms regulate substances in the blood. In this chapter, the negative feedback loop connecting the hypothalamus, anterior pituitary, and sex organs is particularly important.

OBJECTIVES

1. Differentiate between self-fertilization and cross-fertilization, and specify which is more common in animals. (p. 531)

2. Explain the differences between external fertilization and internal fertilization, and show how the number of gametes produced by a particular organism, the organism's need for water, and the type of environment the organism lives in are related to this difference. Indicate whether each of the following organisms uses external or internal fertilization: fish, bird, snake, frog, butterfly, earthworm, mammal, and lizard. (pp. 532–33)

3. Using a diagram of an amniotic egg, Figure 26.6 (p. 534), identify the amnion, chorion, allantois, and yolk sac, and give the function of each.

4. Using Figures 26.8 (p. 535) and 26.14 (p. 541), trace the path of a sperm from the testis of a male to an egg in the oviduct of a female. In doing so, identify the vagina, ovary, cervix, uterus, oviduct, bladder, rectum, urethra, penis, seminal vesicle, vas deferens, and Cowper's gland. Give two functions of the seminal fluid produced by the seminal vesicles, prostate, and Cowper's glands. (p. 535).

5. Explain the role of the hypothalamus, GnRH, LH, and FSH in the human male at the time of puberty. State the effects of testosterone. Describe the negative feedback loops controlling testosterone secretion and spermatogenesis. Using structural formulas as in Figure 26.11 (p. 537), point out one difference between the male sex hormone testosterone and the female sex hormone progesterone. (pp. 537–38)

6. Using Figures 26.20 and 26.21 (p. 546), describe the sequence of events in the menstrual cycle of a human female. Be sure to specify the successive levels of the hormones involved, and the effect that each hormone has on the follicle, the uterine lining, or the production of other hormones. (pp. 545–47)

7. For each of the birth-control methods listed below, indicate the way in which it prevents conception, its relative effectiveness, and whether it is normally reversible.

    diaphragm with spermicidal jelly or cream
    rhythm method
    vasectomy
    tubal ligation
    intrauterine device (IUD)
    the pill   (pp. 537, 543, 548)

8. Outline the events that occur between the time an egg is fertilized in the oviduct and the birth of the baby. In doing so, use the following terms: corpus luteum, oxytocin, placenta, oviduct, progesterone, luteinizing hormone (LH), human chorionic gonadotropin (hCG), uterine lining, follicle-stimulating hormone (FSH), lactation, estrogen, and implantation. (pp. 549–54)

9. Using Figures 26.6 (p. 534) and 26.23 (p. 552), contrast the embryonic membranes in a mammal such as a human with those of a reptile or a bird. Specify which of the following are present in both: amnion, chorion, shell, allantois, placenta, and yolk sac. (pp. 533–34, 552)

KEY TERMS

hermaphroditic (p. 531)
external fertilization (p. 532)
internal fertilization (p. 532)
amniotic egg (p. 533)
amnion (p. 533)
allantois (p. 533)
yolk sac (p. 533)
chorion (p. 533)
testes (p. 534)
scrotal sac (p. 534)
inguinal canal (p. 534)
seminiferous tubule (p. 535)
interstitial cells of Leydig (p. 535)
epididymis (p. 535)
vas deferens (p. 535)
urethra (p. 535)
penis (p. 535)
semen (p. 535)
seminal vesicle (p. 535)
prostate (p. 535)
Cowper's gland (p. 535)
testosterone (p. 537)
GnRh (p. 537)
FSH (p. 537)
LH (p. 537)
inhibin (p. 538)
ovary (p. 541)
oocyte (p. 541)
follicle (p. 541)
ovulation (p. 542)
oviduct (p. 542)
uterus (p. 543)
vagina (p. 543)
cervix (p. 543)
hymen (p. 543)
vulva (p. 543)
clitoris (p. 545)
estrogen (p. 545)
progesterone (p. 545)
menstrual cycle (p. 545)
estrous cycle (p. 545)
menstruation (flow phase) (p. 545)
follicular phase (p. 547)
LH surge (p. 547)
corpus luteum (p. 547)
luteal phase (p. 547)
menopause (p. 548)
zygote (p. 549)
umbilical cord (p. 552)
placenta (p. 552)
human chorionic gonadotropin (hGC) (p. 552)

SUMMARY

Fertilization and embryonic development Sexual reproduction in higher animals involves bringing together two *gametes*, an egg, and a sperm, which then unite in the process of fertilization to form the first cell of the new individual.

Most aquatic organisms use *external fertilization*. The gametes are shed directly into the water, and the sperm must swim to the egg. Generally, these animals release large numbers of gametes and often go through elaborate behavioral sequences to ensure that both sexes shed gametes simultaneously.

Most land animals use *internal fertilization*, in which the egg cells remain in the female reproductive tract until they have been fertilized by sperm inserted by the male. The sperm swim through the fluid in the female reproductive tract. Once fertilized, the egg in its fluid-filled chamber is enclosed by membranes. In reptiles, birds, and mammals the membranes are the *amnion*, which surrounds a fluid-filled chamber housing the embryo; the *allantois*, which functions in waste storage and gas exchange; the *yolk sac*, which stores food; and the *chorion*, which encloses the embryo and its membranes. The embryo and its membranes are then either surrounded by a protective shell and released, or held within the female's body until embryonic development is complete.

Male reproductive system The male gonads are the *testes*, which lie in the *scrotal sac*. Each testis has two functional components: the *seminiferous tubules*, in which the sperm are produced, and the *interstitial cells of Leydig*, which secrete male sex hormone. Mature sperm move into the much-coiled *epididymis*, where they are stored and activated. The sperm then move into the *vas deferens*, which conducts them to an area near the prostate gland. During orgasm the sperm move into the urethra, which passes through the *penis* and empties to the outside. Seminal fluid from the *seminal vesicles*, the *prostate*, and the *Cowper's glands* is added to the sperm to form the *semen*. Semen provides a fluid medium for transport, lubrication, and protection of the sperm from acids in the female genital tract. Its sugar provides energy for the active sperm.

During embryonic development, the testes secrete small amounts of the male sex hormone, *testosterone*, which is crucial to the differentiation of male structures. At puberty, the hypothalamus sends to the anterior pituitary releasing hormone (*GnRH*), which stimulates the anterior pituitary to release *FSH* and *LH*. The LH induces the interstitial cells of Leydig to produce more testosterone; this plus the FSH induces the maturation of the seminiferous tubules and causes sperm production to begin.

Testosterone stimulates the development of the secondary sexual characteristics. Testosterone secretion is governed by a negative feedback loop involving the hypothalamus and anterior pituitary. The hormone *inhibin* inhibits FSH production and thereby limits spermatogenesis.

Female reproductive system The female gonads are the *ovaries*, located in the lower abdominal cavity. The ovaries produce the egg cells (*oocytes*) and secrete sex hormones. Each oocyte is enclosed within a small *follicle*. During maturation, the follicle fills with fluid. When ovulation occurs, the outer wall ruptures and the oocyte and fluid are expelled into the abdominal cavity. Cilia lining the adjacent *oviduct* create a current drawing the egg into it. If sperm are present, fertilization occurs in the oviduct.

Each oviduct empties into the muscular *uterus*. If the egg is fertilized it becomes implanted in the uterine wall, where the embryo develops. At its lower end the uterus connects with the tubular *vagina*, which leads to the outside. The vagina is the receptacle for the penis during copulation.

Puberty in the female begins when the hypothalamus sends more GnRH to the anterior pituitary, which stimulates it to release FSH and LH. These hormones cause maturation of the ovaries, which begin secreting the female sex hormones, *estrogen* and *progesterone*. Estrogen stimulates maturation of the reproductive structures and development of the secondary sexual characteristics. The changing hormonal balance triggers the onset of the menstrual cycles.

Rhythmic variations in the secretions of gonadotropic hormones in most mammals lead to *estrous cycles*—rhythmic variations in the reproductive tract and sex urge. The reproductive cycle differs somewhat in humans: the female is receptive to the male throughout her cycle, and the thickened uterine lining is not completely reabsorbed (as in most other mammals) if no fertilization occurs; instead part of the lining is sloughed off during *menstruation*.

At the beginning of the menstrual cycle, the uterine lining is thin and there are no ripe follicles. The hypothalamus responds to the low level of sex hormones by releasing GnRH, which stimulates the pituitary to increase FSH secretion. FSH stimulates the maturation of the follicles, which begin to secrete estrogen. Estrogen stimulates the uterine lining to thicken: this *follicular phase* lasts about 9 days. The high level of estrogen apparently stimulates a surge of LH from the pituitary, which triggers ovulation. The LH converts the follicle into the *corpus luteum*, which continues to secrete estrogen and begins to secrete *progesterone*. Progesterone prepares the uterus to receive the embryo by activating its many glands and by inducing other chemical changes. The high level of sex hormones also suppresses the growth of new follicles. If no fertilization occurs, the corpus luteum atrophies about 11 days after ovulation and progesterone secretion falls. When this happens, the thickened uterine lining can no longer be maintained, and menstruation occurs and lasts about 4 to 6 days. The hypothalamus again responds to the low level of sex hormones and another cycle begins.

Pregnancy The egg cell must be fertilized within 24 hours after ovulation. The fertilized egg, or zygote, moves down the oviduct and becomes implanted in the uterine wall 8 to 10 days after fertilization. The embryonic membranes then develop and the *placenta* is formed. The exchange of materials between the blood of the mother and that of the embryo takes place by diffusion through the placenta. The placenta soon begins to secrete *human chorionic gonadotropin* (hCG), which maintains the corpus luteum and its secretion of progesterone, and thus sustains the pregnancy. Later the placenta secretes estrogen and progesterone directly.

In late pregnancy, estrogen secretion by the placenta increases. Both oxytocin (secreted by the posterior pituitary) and prostaglandins (secreted by the uterus) are necessary to stimulate uterine contractions in the birth process.

The development of the breasts is regulated primarily by estrogen and progesterone. Initiation and maintenance of lactation by mature mammary glands after birth seems to be controlled primarily by prolactin and glucocorticoids. Oxytocin from the posterior pituitary causes the milk to be ejected into the ducts of the nipple.

CONCEPT MAP

Map 26.1

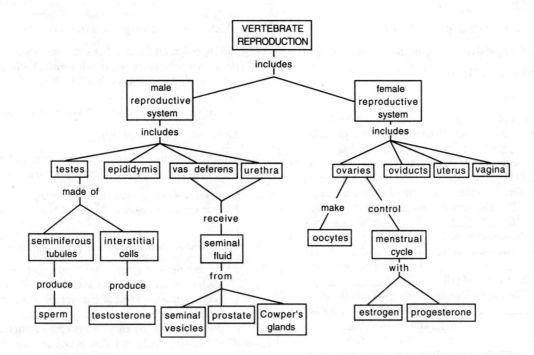

QUESTIONS

Testing recall

Questions 1 to 8 refer to the following diagram of the human female reproductive system.

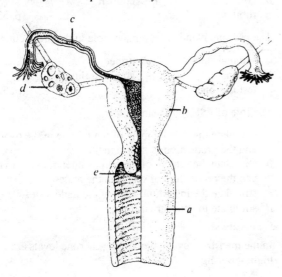

1. Where is the egg produced? (p. 541) d
2. Where does fertilization occur? (p. 549) c
3. Where would implantation of a fertilized egg take place? (p. 549) b
4. Where is a tubal ligation performed? (p. 543) c
5. Where does a diaphragm act to prevent conception? (p. 543) e
6. Where are estrogen and progesterone produced? (p. 545) d
7. Where does an IUD act to prevent conception? (p. 543) b
8. What part receives the male penis during copulation? (p. 543) a

Match each function below with the associated part or parts of the human male reproductive system shown in the figure.

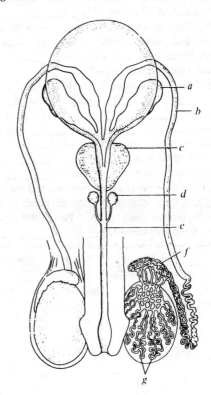

9 produces sperm (p. 535) _g_

10 conducts the sperm through the penis to the outside of the body (p. 535) _e_

11 produces seminal fluid (p. 535) _a, c, d_

12 connects the epididymis with the urethra (p. 535) _b_

13 stores sperm (p. 535) _f_

Fill in the blanks.

The females of most species of mammals show rhythmic variations in the reproductive tract and sex urges called the (14) _estrous_ cycle (p. 545).

At the beginning of the menstrual cycle in humans the hormone (15) _FSH_ from the anterior pituitary stimulates the maturation of the follicle in the ovary (p. 547). The growing follicle begins to secrete the hormone (16) _estr._, which causes the lining of the uterus to thicken (p. 547). High levels of this hormone stimulate an abrupt production of high levels of (17) _LH_ from the pituitary; this triggers the process of (18) _ovulat._ (p. 547). The empty follicle is converted into the (19) _corp. lut_, which secretes the hormones (20) _progest._ and _estr._ (p. 547).

These hormones cause the uterine lining to become thicker and more glandular. High levels of the hormone (21) _progest_ suppress the release of (22) _GnRH_ from the hypothalamus, and thereby limits (23) _FSH_ and _LH_ secretion by the pituitary (p. 547).

If no fertilization occurs the corpus luteum begins to atrophy and its secretion of (24) _proges._ falls (p. 547). Now the thickened lining of the uterus can no longer be maintained; part of it is reabsorbed and part of it is sloughed off during (25) _menst._ (p. 547). The low levels of the sex hormones free the hypothalamus and immature follicles from inhibition, and the pituitary begins to secrete more (26) _FSH_, beginning a new cycle (p. 547).

The follicular phase of the cycle lasts about (27) _9–10_ days, the luteal phase about (28) _13–15_ days, and the flow phase about (29) _4–6_ days (p. 546).

If fertilization occurs, the embryo becomes implanted in the wall of the uterus about (30) _6–10_ days after fertilization (p. 549). The (31) _placenta_ forms from the embryonic membranes and the uterine lining; here the exchange of materials between mother and embryo takes place (p. 552). Part of this organ begins to secrete the hormone (32) _hCG_; this preserves the corpus luteum, which continues to secrete (33) _progest._ (p. 552). Later this organ secretes the hormones (34) _estr._ and _progs._, which sustain the pregnancy (p. 552).

Near the time of birth the secretion of the hormone (35) _oxytocin_ increases (p. 553). The hormone (36) _oxy._, secreted by the posterior pituitary, causes powerful uterine contractions, which aid in the birth process (p. 553).

37 List, in order, all the structures through which a sperm passes from the place where it is formed to the place where fertilization occurs. (pp. 534–43)

Testing knowledge and understanding

Choose the one best answer.

38 In comparison with organisms that use external fertilization, organisms that use internal fertilization usually

   a produce fewer gametes.   *c* live on land.
   b produce fewer zygotes.   *d* do all the above
                                                          (p. 532)

39 The hormones testosterone and progesterone act by

   a combining with receptor molecules and moving to the nucleus, where they react with the genetic material.
   b combining with a receptor molecule on the membrane surface, which activates adenylate cyclase.
   c directly activating enzyme systems within the cell.
   d stimulating the hypothalamus and anterior pituitary to release hormones.
   e inducing a rise in cAMP within the cell.
                                   (p. 537; Chapter 25, p. 511)

40 Which one of the following animals produces eggs that lack an amnion?

   a human   *d* black snake
   b cow   *e* codfish
   c robin   (p. 533)

41 The prostate gland plays a major role in the production of

   a seminal fluid.   *d* urine.
   b sperm.   *e* amniotic fluid.
   c testosterone.   (p. 535)

42 The role of FSH in humans is to

   a stimulate the development of the follicle in the ovary and the production of sperm in the testes.
   b stimulate the growth of the corpus luteum in the ovary and the production of sperm in the male.
   c stimulate the interstitial cells of the male testes.
   d stimulate the menstrual flow.
   e stimulate uterine contractions at birth.   (p. 547)

43 In the menstrual cycle, blood progesterone levels are highest during

   a the follicular (growth) phase.
   b the time of ovulation.
   c the luteal (secretory) phase.
   d the first part of the flow phase.
   e the last part of the flow phase.   (p. 547)

44 Which of the following is the correct sequence of organs involved in producing progesterone?

   a hypothalamus—anterior pituitary—ovary
   b hypothalamus—posterior pituitary—ovary
   c anterior pituitary—adrenal cortex—uterus
   d posterior pituitary—adrenal medulla—ovary
   e hypothalamus—ovary—uterus   (p. 547)

45 Which of the following hormones stimulates the maturation of reproductive structures in *both* male and female humans?

   a estrogen
   b follicle stimulating hormone (FSH)
   c progesterone
   d testosterone
   e oxytocin   (p. 545)

46 Which of the following statements is *false* regarding progesterone in the human female menstrual cycle?

   a It promotes the development of follicles.
   b It maintains the thickened uterine lining that forms in preparation for embryo implantation.
   c It is secreted by the corpus luteum.
   d It inhibits release of hormones by the hypothalamus.
   e It is a component of birth-control pills. (p. 547)

47 Which of these contraceptives does *not* prevent fertilization?

   a condom
   b diaphragm
   c intrauterine device (IUD)
   d pill (oral contraceptive)
   e spermicidal foram   (p. 543)

48 Which one of the following statements is *false*?

   a Most birth-control pills are mixtures of estrogen and progesterone (or close analogs) designed to inhibit the FSH and LH production.
   b Vasectomy does not alter a man's endocrine system, and hence should not reduce either sexual drive or sexual competency.
   c Tubal ligation does not alter a woman's endocrine system, and hence should not reduce either sexual drive or sexual responsiveness.
   d The diaphragm is a birth-control device that, when properly fitted, covers the cervix and prevents sperm from entering the uterus.
   e In the rhythm method of birth control, sexual abstinence on days 14 to 16 in each menstrual cycle reliably prevents pregnancy.   (p. 548)

49 The birth-control pills most commonly used in the United States today are mixtures of estrogenlike and progesteronelike compounds. They probably prevent pregnancy mainly by

   a stimulating the release of FSH from the anterior pituitary.
   b preventing the glands of the uterine lining from becoming secretory.
   c decreasing the mobility of the sperm by inhibiting uterine contractions.
   d blocking secretion of FSH and LH from the anterior pituitary.
   e causing premature release of the egg from the follicle. (p. 548)

50 Which one of the following describes a function the placenta does *not* perform?

   a removes waste materials from the fetus
   b supplies oxygen to the fetus
   c secretes hormones
   d provides the fetus with nutrients
   e replaces used fetal erythrocytes with new ones from the mother   (p. 552)

51 If gonadotropins (hCG, LH) from human placenta were injected into a mature male, one might expect to see

   a an increase in estrogen and progesterone.
   b an increase in testosterone.
   c a decrease in the activity of the testes.
   d a 28-day cycling of male hormones.
   e increased sperm production by the testes.   (p. 552)

52 Fertilization of the human egg occurs

   a externally.   d in the uterus.
   b in the vagina.   e in the oviduct.
   c in the cervix.   (p. 549)

ANSWERS

Testing recall

1 d   5 e   8 a   11 a, c, d
2 c   6 d   9 g   12 b
3 b   7 b   10 e   13 f
4 c

14 estrous
15 FSH
16 estrogen
17 LH
18 ovulation
19 corpus luteum
20 estrogen, progesterone
21 progesterone
22 GnRH
23 FSH, LH
24 progesterone
25 menstruation
26 FSH
27 9 to 10
28 13 to 15

29 4 to 6
30 6 to 10
31 placenta
32 chorionic gonadotropin (hCG)
33 progesterone
34 progesterone, estrogen
35 estrogen
36 oxytocin
37 seminiferous tubules, ducts in epididymis, vas deferens, urethra, vagina, uterus, oviduct

Testing knowledge and understanding

38	d	42	a	46	a	50	e
39	a	43	c	47	c	51	b
40	e	44	a	48	e	52	e
41	a	45	b	49	d		

Chapter 27

DEVELOPMENT OF MULTICELLULAR ANIMALS

A GENERAL GUIDE TO THE READING

This chapter examines the principal events in the development of multicellular animals. Because of their autotrophic, nonmobile nature, plants develop along a considerably different pathway than animals. The latter half of this chapter discusses the patterns of cell differentiation. We will see that in both plants and animals, chemical signals modify cell's patterns of gene expression. These chemicals underly the cell differentiation and morphogenesis that determine an organism's path of development. As you read Chapter 27 in your text, you will want to concentrate on the following topics.

1. Fertilization. You may be surprised to learn from the section on this topic (pp. 558–59) that development begins with penetration of the sperm rather than with fertilization itself.

2. Early cleavage and morphogenetic stages. Many new terms are presented in the discussion of this topic (pp. 559–67). You will find it helpful to take the time to learn them before reading further, as these terms are used frequently. Careful attention to Figures 27.3 to 27.9 (pp. 560–64) will also prove helpful, particularly as you study the material on gastrulation and neurulation, these are the crucial processes that establish the shape and form of the embryo.

3. Development of the human embryo. The fascinating account on pages 567 to 570, illustrated by Figures 27.5 (p. 562) and 27.14 (p. 568), details some of the principal events in human embryology.

4. Larval development and metamorphosis. The text (pp. 570–73) describes a pattern of development characteristic of many animals. You will want to understand the difference between gradual and complete metamorphosis.

5. Polarity of eggs, zygotes, and blastomeres. The distribution of yolk and chemicals in the fertilized egg helps determine important axes of the embryo. (pp. 574–75)

6. Induction in embryogenesis. The examples in Figures 27.24 to 27.26 (pp. 576–77) will help you understand induction.

7. Cell migration. Figures 27.30 and 27.31 (pp. 579–80) present graphically how cell-adhesion molecules (CAMs) bind, and how the molecules' differential affinities allow the cells to move forward until their CAMs find the best matches.

8. Pattern formation. After the embryo has undergone gastrulation and neurulation, the embryo is subdivided into a series of domains, each of which develops independently. Focus on the role of morphogens and homeotic genes. (pp. 580–81)

9. The organization of neural development. The development of the nervous system provides an excellent example of many aspects of development. By learning how neurons migrate to their proper place (p. 584), you will gain some understanding of how development as a whole proceeds.

KEY CONCEPTS

1 Development in sexually reproducing organisms begins with the penetration of the sperm into the ovum and ends with the organism's death. (pp. 558–59)

2 In animals, the developmental processes of cell division, cell growth, cell differentiation, and morphogenetic movements convert the fertilized egg into the mature organism. (pp. 558–70)

3 The postembryonic organism is not a static entity. It continues to change, and hence to develop, until death brings the developmental process to an end. (pp. 572–73)

4 All the cells of a single organism arise by repeated division of the fertilized egg and are genetically identical. Which genes are active, and hence which potentialities are expressed, is determined in part by the nonuniform distribution of cytoplasmic substances in dividing cells. (pp. 574–75)

5 Developing cells are subject to many influences external to their own cytoplasm, including neighboring cells, hormones, and environmental factors that act on the complete organism. (pp. 575–78)

6 Mesodermal tissue seems to play a dominant role throughout the embryo, by migrating to new locations and inducing adjacent endoderm and ectoderm to differentiate. (pp. 575–82)

7 Cell-adhesion molecules (CAMs) enable cells of the same class to recognize and bind to each other. Cells migrating from one place to another recognize their pathways by means of particular ratios of CAMs. (pp. 579–80)

8 Pattern formation is an important aspect of development. Morphogens play an important role in establishing domains in the developing embryo. Homeotic genes code for control substances that bind to DNA and direct development within each domain. (pp. 580–81)

9 In some organisms, particularly those that grow throughout their life cycle, differentiation appears to be reversible. (pp. 582–83)

10 The highly integrated functional network characterizing the nervous system is formed by the organized migration of neurons to their proper location and the subsequent growth of axons to specific target cells. (p. 584)

CROSSLINKING CONCEPTS

1 Your previous knowledge of animal reproduction (Chapter 26) has established a context for your examination of animal development in this chapter.

2 This chapter focuses on animal development. Plant development was addressed in Chapter 24.

3 During morphogenesis, cells often move about and change shape, probably through interactions between actin and myosin filaments that were first introduced in Chapter 6.

4 Comparisons of embryological stages of different organisms further our understanding of the evolutionary relationships among species.

5 The chemicals, such as homeotic gene products, that play an important role in determining development interact intimately with mechanisms that control gene expression (Chapter 12).

6 Cell-adhesion molecules (CAMs) are closely related to specialized proteins of the immune system (Chapter 21). This developmental recognition system may have provided the basis for the evolution of the molecules of the mammalian immune system.

7 The importance of hormones in controlling vertebrate sexual development was described in Chapter 26. This chapter provides additional information on the role of sex hormones as inducers of the embryonic development of the gonads and genitalia.

8 The organization of neural development is introduced. A thorough investigation of the nervous system follows in Chapter 28.

OBJECTIVES

1 Describe the events triggered by the penetration of an animal egg by the sperm, and the process of fertilization. Explain what happens to prevent more than one sperm from fertilizing the egg. (pp. 558–59)

2 Using Figure 27.3, describe the principal events that occur during the cleavage stages of animal embryos. Indicate whether the embryos increase in size during cleavage, and explain how the amount of yolk in the egg affects the cleavage pattern. (pp. 560–62)

3 Using Figures 27.7 (p. 563), 27.8 (p. 564), and 27.9 (p. 564), describe gastrulation in amphioxus. Indicate how the process differs in amphioxus, in frogs, and in birds. (pp. 562–64)

4 Discuss the process of neurulation in amphioxus, and name the structures that the neural tube gives rise to (Fig. 27.12, p. 566, may be helpful). (pp. 566–67)

5 Using Figures 27.3 (p. 560), 27.7 (p. 563), 27.8 (p. 564), and 27.12 (p. 566), point out the blastula, gastrula, ectoderm, endoderm, mesoderm, archenteron, blastopore, and neural folds. (pp. 560–66)

6 Name the three primary cell layers, and indicate which primary cell layer gives rise to each of the following adult structures or tissues:

    fingernails
    brain
    notochord
    lungs
    anus
    connective tissue
    blood
    hair
    lining of digestive tract
    nerve cord
    muscle
    skin (epithelial portion)
    bone
    liver
                            (pp. 565–67)

7 Describe the major events of human embryological development, indicating when the embryo is most susceptible to factors that can cause malformations, when it has some chance of survival outside the uterus, and what developmental changes in the circulatory system occur at the time of birth. (pp. 567–69)

8 Describe the role of growth in the postembryonic development of an animal. (pp. 569–70)

9 Define metamorphosis, and explain the adaptive significance of the larval stage. Distinguish between complete and gradual metamorphosis in insects. (pp. 570–72)

10 Describe the process of aging, indicating which kinds of cells tend to age and which do not. List three factors that appear to contribute to aging. (pp. 572–73)

11 Using Figures 27.22 and 27.23 (pp. 574–75), explain how the polarity of an egg cell and the location of its plane of cleavage influence development. (pp. 574–75)

12 Explain what is meant by induction, using as an example the role of the dorsal lip of the blastopore in a salamander (as shown in Fig. 27.24, p. 576) or the role of optic vesicles in the induction of lenses in a frog or mammalian eye (see Figs. 27.25, p. 577, and 27.26, p. 577). (pp. 575–77)

13 Using Figures 27.27 (p. 578) and 27.28 (p. 578), discuss the role of hormones in the development of the gonads and genitalia in humans. (pp. 578–79)

14 Describe the role of cell-adhesion molecules in cell migration. (pp. 579–80)

15 Cite evidence to indicate that cells have positional information to ensure proper pattern formation. (p. 579)

16 Fully describe the process of pattern formation, and the role of morphogens in establishing domains. Describe the role of homeotic genes in differentiation within domains. (pp. 580–81)

17 Distinguish between determination and differentiation and cite evidence that differentiation is sometimes reversible in animals. (pp. 581–82)

18 Describe the process by which neurons become organized into an integrated, functional network, including the factors involved in guiding migrating cells and in forming connections to target cells. Also describe the role of cell death in development. (p. 584)

KEY TERMS

yolk (p. 558)
capacitation (p. 558)
acrosome (p. 558)
pronucleus (p. 559)
fertilization membrane (p. 559)
in vitro (p. 559)
zygote (p. 559)
cleavage (p. 559)
morula (p. 559)
blastocoel (p. 559)

blastula (p. 560)
animal hemisphere (p. 560)
vegetal hemisphere (p. 560)
blastodisc (p. 560)
larval stage (p. 561)
blastocyst (p. 561)
inner cell mass (p. 561)
morphogenesis (p. 562)
gastrula (p. 562)
gastrulation (p. 562)
blastopore (p. 563)
archenteron (p. 563)
ectoderm (p. 564)
endoderm (p. 564)
mesoderm (p. 564)
notochord (p. 566)
neurula (p. 566)
neurulation (p. 566)
neural plate (p. 566)
neural folds (p. 566)
fetus (p. 568)
metamorphosis (p. 570)
pupa (p. 570)
imaginal discs (p. 570)
complete metamorphosis (p. 570)
gradual metamorphosis (p. 572)
differentiation (p. 574)
pattern formation (p. 574)
gray crescent (p. 574)
indeterminate (p. 575)
totipotent (p. 575)
determinate (p. 575)
dorsal lip of the blastopore (p. 576)
chordamesoderm (p. 576)
organizer (p. 577)
embryonic induction (p. 577)
cell-adhesion molecules (CAMs) (p. 579)
determined (p. 580)
homeotic gene (p. 581)
homeobox (p. 581)
determination (p. 581)
differentiated (p. 582)
dedifferentiation (p. 582)

SUMMARY

Fertilization The process of development in a sexually reproducing multicellular animal begins with the maturrration of the egg. Penetration of the activates the developmental process. The membrane of the sperm fuses with that of the egg, and the sperm nucleus moves into the egg. This fusion of membranes initiates changes in the egg membrane that lead to the formation of the *fertilization membrane*, which prevents penetration by other sperm. *Fertilization* occurs when the two gamete nuclei fuse.

Embryonic development The zygote then undergoes a rapid series of mitotic divisions called *cleavage*. The cytoplasm of the one large cell is partitioned into many new smaller cells. As cleavage continues, the cells become arranged in a hollow ball called a *blastula*. Two regions are often evident: the *animal hemisphere*, made up of small cells, and the *vegetal hemisphere*, made up of larger yolky cells.

Next begins a series of complex movements that are important in establishing the shape and pattern of the developing organism (*morphogenesis*). The blastula is converted into an embryo called the *gastrula*. Cleavage and gastrulation are greatly influenced by the amount of yolk in the egg. Gastrulation first produces an embryo with two layers, an outer *ectoderm* and an inner *endoderm*. A third layer, the *mesoderm*, forms between them. The ectoderm gives rise to the outermost layers of the body, the nervous system, and the sense organs, the endoderm to the lining of the digestive tract and associated structures, and the mesoderm to the supportive tissue—muscles and connective tissues.

The morphogenetic movements of gastrulation and neurulation give form and shape to the embryo and bring masses of cells into proper position for their later differentiation into the principal tissues of the adult body. The developmental processes of cell division, cell growth, cell differentiation, and morphogenetic movement convert the gastrula into a young animal ready for birth.

Postembryonic development The predominant factor in postembryonic development in most animals is growth in size. Growth does not occur at the same rate and at the same time in all parts of the body. Many aquatic animals and certain groups of terrestrial insects go through *larval* stages that bear little resemblance to the adult. The complex series of developmental stages that convert an immature animal into an adult is called *metamorphosis*. Some insects undergo *complete metamorphosis*, which begins with a wormlike larval stage quite unlike the adult. After a period of growth the larva enters the *pupal stage*, during which it is extensively reorganized to form the adult. Other insects undergo *gradual metamorphosis*; the young, which already resemble the adult, go through a series of molts that make them more and more like the adult.

Aging is another aspect of development. It is a complex series of developmental changes that lead to the deterioration of the mature organism and ultimately to its death. The aging process seems to be correlated with the degree of cellular specialization. Cells that remain relatively unspecialized and continue to divide do not age as rapidly as cells that have lost the capacity to divide. Aging of the whole multicellular organism results from the deterioration and death of irreplaceable cells and tissues.

Polarity of eggs, zygotes, and blastomeres Although the genetic content of all embryonic cells is identical, their cytoplasm is not. Therefore, the daughter cells produced by cleavage of the egg cell may not share equally in all cytoplasmic materials, depending on the orientation of the first plane of cleavage. If the cleavage partitions the cytoplasmic constituents equally, the cleavage is *indeterminate*. The new cells are *totipotent*; that is, they have the full developmental potential of the zygote. If, however, the first cleavage partitions critical cytoplasmic constituents unequally, the cleavage is *determinate*; then the daughter cells, if they were to be separated, would not both have the potential to develop into a normal embryo. Most mammalian embryos remain totipotent for the first three cleavages, whereas most molluscs and segmented worms divide asymmetrically from the first cleavage.

Certain differentially distributed cytoplasmic substances must play a prominent role during early embryological development. The nonuniform distribution of yolk, various proteins, and mRNA along the animal-vegetal axis appears to establish the dorsal-ventral axis and leads to the expression of different traits in different parts of the embryo. The anterior-posterior axis is often defined by the point of sperm penetration of the egg; in frogs the *gray crescent* appears on the egg opposite this point.

Induction in embryogenesis Developing cells are influenced by neighboring cells, by diffusible chemicals they secrete, by hormones, and by the physical environment (e.g., light, temperature, pH, gravity). In frogs, the cells of the gray crescent, which later form the *dorsal lip* of the blastopore, play a crucial role in gastrulation. These cells then move inward to form the *chorda-mesoderm*, a structure that gives rise to the notochord and other mesodermal structures. Signals from these cells induce the formation of the neural plate, which is important in establishing the longitudinal axis of the embryo and in inducing formation of other structures. Because of its particularly strong inductive function, the dorsal lip is an example of an *organizer* area of the embryo. At the organ level, the differentiation of epidermal tissue into lens tissue depends on some inductive stimulus from the underlying optic vesicle. Most developmental interactions are probably mediated by chemicals, including hormones.

Migrating cells can recognize the pathways to follow by their differential attraction to *cell-adhesion molecules (CAMs)* that are embedded in cell membranes. Each type of CAM bonds only to others of the same class. Since different cells have different types and proportions of CAMs, they bind to each other with different strengths. When a pseudopod extends from a cell, its CAMs attach to neighboring cells. When the cell withdraws the pseudopod, the cell is dragged toward the cell to which its CAMs have attached most firmly. CAMs are closely related to the MHC molecules, T-cell receptors, and antibodies of the immune system.

Pattern formation After neurulation, the embryo is subdivided into a series of domains, which develop limbs and organs largely independently.

In vertebrates, once the spinal cord has fully formed, the most dorsal region of mesoderm differentiates into blocks of tissue called *somites*. Each somite is *determined*. It is committed to give rise to a specific vertebra and the bones, muscles, and skin associated with the vertebra. The spatial information necessary to give cells a sense of position and to guide them as development proceeds often depends on measuring the concentration of chemicals called *morphogens*, which are secreted from specific points in the embryo and diffuse away.

Experiments with mutant strains of the fruit fly *Drosophila melanogaster* led to the discovery of *homeotic genes*, whose products are generally control substances that bind to DNA and orchestrate the operation of many other genes, and thus cause each segment of the fly to express its unique character. Mutations of such genes cause regions of cells to misinterpret their position and produce an inappropriate structure.

Cell determination and differentiation A cell becomes determined when its range of potential specializations is restricted. *Differentiation*, which usually takes place after determination, involves changes in the cell's morphology and chemistry. As cells become more differentiated before determination fully fixes their fate and can thus dedifferentiate. In mammals and birds, generally only embryonic cells can dedifferentiate. Some cells in amphibians, fish, and reptiles, however, retain this ability even in adulthood, and so allow the organisms to regenerate complicated structures such as limbs and tails. In plants the changes undergone by cell nuclei during differentiation appear to be entirely reversible. In animals, the situation is less clear, but at least in come cases differentiation may be reversible.

The organization of neural development Nerve cells are born, migrate to their proper places, send axons to specific target locations, and form a highly integrated functional network. A newly formed presumptive neuron will move from where it was formed to where it is supposed to be. Such movement is well oriented and appears to involve diffusing chemicals, CAMs, and tactile clues. In most animal nervous systems many more cells are produced than are used; those not needed die.

CONCEPT MAPS

Map 27.1

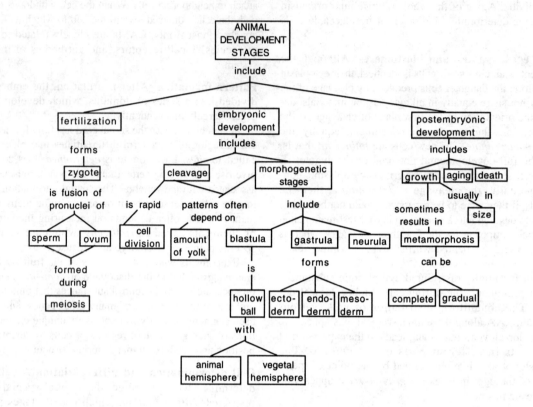

Map 27.2

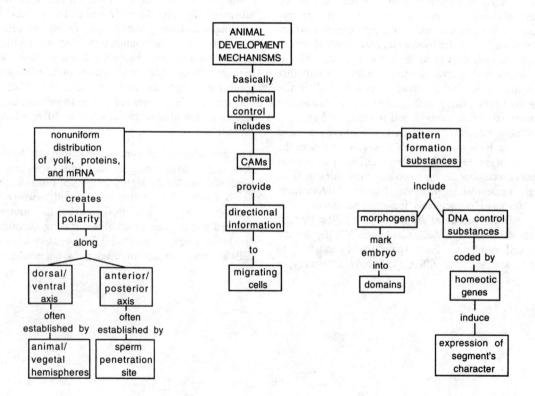

DEVELOPMENT OF MULTICELLULAR ANIMALS • 201

KEY DIAGRAM

The following coloring exercise is designed to help you learn the early embryology of amphioxus, frog, and chick. Selected diagrams from the blastula through the neurula stage are shown below. Color the names and the appropriate structures in the order given, using a different color for each. (Hint: In most embryology books ectoderm is colored blue, endoderm yellow, and mesoderm red. You may want to follow this convention when you do your coloring. We suggest you color in each required structure, if present, on all the diagrams so that you will see the relationships among them. For example, the first item to be colored in is ectoderm, so color in all the ectoderm in each of the diagrams, then go on to endoderm, etc. You will find that some of the structures will be found on only one or two of the diagrams and others will be on almost every one. (Figs. 27.7, p. 563; 27.9, p. 564; 27.9, p. 564; and 27.12, p. 566)

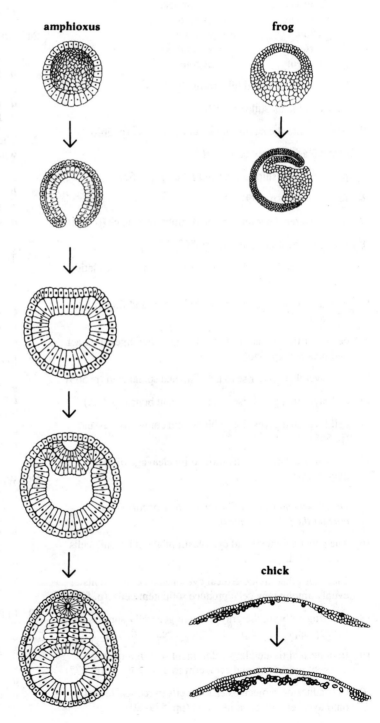

Ask yourself

What primary cell layers are the notochord and spinal cord derived from?

PRIMARY CELL LAYERS
Derivatives:

ECTODERM

ENDODERM

MESODERM

CAVITIES (color lightly):

BLASTOCOEL

ARCHENTERON

COELOM

NERVOUS AND SUPPORTIVE STRUCTURES:

SPINAL CORD

NOTOCHORD

QUESTIONS

Testing recall

Below are listed some important terms in animal embryology. For questions 1 to 15, match each statement with the proper term.

a	animal pole	h	endoderm
b	archenteron	i	mesoderm
c	blastocoel	j	morphogenesis
d	blastopore	k	morula
e	blastula	l	neurula
f	gastrula	m	vegetal pole
g	ectoderm	n	zygote

1. stage in which neural tube forms (p. 566)
2. cavity of the blastula (p. 559)
3. opening into the cavity of the digestive tract (p. 563)
4. embryo as a hollow ball (p. 561)
5. process establishing shape and pattern (p. 562)
6. fertilized egg (p. 559)
7. a two-layered, later three-layered, embryonic stage (p. 562)
8. cavity of the digestive tract (p. 563)
9. part of the embryo where the invagination of gastrulation usually occurs (p. 563)
10. cell layer that gives rise to the hair, skin, and fingernails (p. 565)
11. cell layer that gives rise to the lining of the digestive tract and pancreas (p. 565)
12. cell layer that gives rise to the brain and spinal cord (p. 565)
13. cell layer that gives rise to muscles and bones (p. 565)
14. cell layer that gives rise to blood and connective tissue (p. 566)
15. grapelike cluster of cells formed by cleavage of the zygote (p. 559)

Decide whether the following are true or false and correct the false statements.

16. The genetic content and cytoplasm of all embryonic cells are identical. (p. 574)
17. Cleavages that divide critical cytoplasmic constituents evenly among new cells produce totipotent cells. (p. 575)
18. In a frog embryo, the gray crescent generally appears on the egg opposite the point of sperm penetration. (p. 576)
19. Endodermal tissue plays a dominant role in inducing adjacent mesoderm and ectoderm to differentiate. (p. 577)
20. Cell-adhesion molecules enable cells to recognize and bind to others of the same class. (pp. 579–80)
21. Once the embryo is divided into domains, each domain develops largely independently of the others. (p. 581)
22. Morphogens are produced by homeotic genes. (pp. 580–81)
23. Nervous-system cells that are not needed atrophy and die. (p. 584)

Testing knowledge and understanding

Choose the one best answer.

24. An egg cell differs from a sperm cell in that the egg cell
    a has cytoplasm.
    b has mitochondria.
    c contains a haploid set of genes.
    d is a product of meiosis.
    e has much greater energy reserves. (p. 558)

25. The fertilization process can be regarded as completed when
    a a sperm comes into contact with an egg.
    b a fertilization membrane is formed.
    c the sperm penetrates the cell membrane.
    d the sperm enters the egg cytoplasm.
    e the sperm nucleus fuses with the egg nucleus. (p. 558)

26. The process of development begins when
    a the sperm nucleus fuses with the egg nucleus.
    b the sperm penetrates the egg.
    c egg oogenesis is completed.
    d cleavage begins.
    e the cell enters the M stage. (p. 558)

27. In early cleavage, with each successive division the
    a number of chromosomes per cell is reduced by half.
    b number of chromosomes is doubled.
    c volume of cytoplasm per cell is reduced.
    d cytoplasmic contents are equally distributed.
    e extensive growth in size of the embryo begins. (p. 559)

28. Which one of the following represents a correct sequence in the developmental process?
    a zygote → gastrula → neurula → blastula
    b fertilization → cleavage → blastula → gastrula
    c fertilization → gastrula → blastula → neurula
    d zygote → neurula → blastocoel → gastrula
    e fertilization → zygote → blastula → neurula → gastrula (pp. 559–62)

29. In the early-cleavage embryo of the chick,
    a the blastocoel forms by invagination.
    b all cells have an equal amount of yolk.
    c the heart has already begun to beat.
    d there is no cleavage through the yolk.
    e cell divisions are restricted to the egg white. (p. 560)

30 In most embryos, the archenteron

  a usually disappears during gastrulation.
  b is filled with mesoderm.
  c is formed by invagination.
  d is present in the blastula.
  e eventually forms the neural tube. (p. 563)

31 Which one of the following statements is *false*?

  a In a vertebrate embryo, the blastula is formed by mitotic cell divisions without significant cell growth.
  b The embryonic archenteron becomes the cavity of the digestive tract in the adult.
  c Cleavage is restricted to the cytoplasmic disc of the ovum in birds.
  d The details of the gastrulation process vary considerably from species to species, depending on the amount of yolk in the ovum.
  e The embryonic blastocoel opens to the outside via the blastopore. (pp. 559–64)

32 Which one of the following is *false* concerning the process of morphogenesis?

  a It refers to the production of form and pattern.
  b It involves cell movements.
  c It plays a major role in the development process.
  d It is synonymous with differentiation.
  e It is exhibited in the formation of the neural tube. (p. 562)

Questions 33 to 38 refer to the following diagram of an amphioxus embryo during gastrulation.

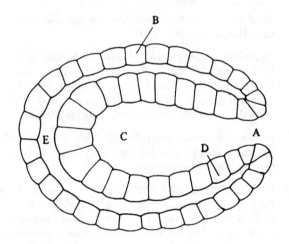

33 The area marked A

  a is called the blastopore.
  b forms during neurulation.
  c opens into the blastocoel.
  d forms during cleavage.
  e is or does two of the above (p. 563)

34 From which cell layer (B or D) will the nervous system develop? (p. 566)

35 From which cell layer will the lining of the digestive tract develop? (p. 563)

36 Which letter marks the blastocoel? (p. 563)

37 Which letter marks the archenteron? (p. 563)

38 All the following are shown in the diagram *except*

  a ectoderm.   d archenteron.
  b endoderm.   e two of the above.
  c mesoderm.   (p. 563)

39 Which one of the following is an *incorrect* match?

  a neural tube—neurula
  b three germ layers—gastrula
  c archenteron—gastrula
  d blastopore—blastula
  e hollow ball—blastula   (p. 563)

40 Which one of the following sets of structures is derived primarily from mesoderm?

  a eye, brain, spinal cord
  b epidermis, hair, fingernails
  c blood, skeleton, muscle
  d lining of gut, lining of lungs, liver
  e epidermis, nervous system, skeleton (pp. 565–66)

41 Many insects undergo a complete metamorphosis during development. If the following four stages were placed in order, which one would be third?

  a pupa   c larva
  b egg   d adult   (p. 570)

42 Groups of cells present in the larva that give rise to the adult are called

  a blastomeres.
  b imaginal discs.
  c blastodiscs.
  d morphogenetic cells.
  e protoderm.   (p. 570)

43 The dorsal–ventral axis of a fertilized egg is established by

  a the animal–vegetal polarity.
  b the point at which the sperm penetrated the egg.
  c morphogens.
  d homeotic genes.   (p. 574)

44 In silkworms, silk gland cells are devoted to the synthesis of large quantities of the protein known as silk fibroin. These same cells do not make blood-specific proteins. One would expect that silk gland cells have

  a only silk fibroin genes.
  b the genes for both blood protein and silk fibroin.
  c silk fibroin genes and some other genes, but not blood protein genes.
  d blood protein genes but not silk fibroin genes.
  e fewer genes than the zygote.   (p. 574)

45 The gray crescent region is very important in the early development of a frog embryo. If the first cleavage division divides this region in two and these two cells are separated,

   a both cells will die.
   b one cell will produce a normal tadpole and the other will produce a mass of unspecialized cells.
   c both cells will produce an abnormal tadpole.
   d both cells will produce a normal tadpole.
   e both cells will produce a mass of mesoderm tissue.
   (p. 576)

46 Which one of the following is *least* likely to be a significant factor in determining the differentiation pattern of embryonic cells?

   a cytoplasmic distribution during early cleavage stages
   b distance from the surface of the embryo
   c inducer substances secreted by neighboring cells
   d gene recombination during the cleavage stages
   (pp. 574–78)

47 Which of the following statements about induction is *false*?

   a Mesodermal tissue plays a dominant role in inducing adjacent endoderm and ectoderm to differentiate.
   b Cells from the dorsal lip of the blastopore induce formation of the neural plate.
   c The neural plate induces formation of the limbs.
   d Differentiation of epidermal tissue into lens tissue depends on a stimulus from the underlying optic vesicle.
   (p. 566)

48 The dorsal lip of the frog blastopore gives rise to mesodermal tissue, which normally underlies the neural tube. What would happen if the dorsal lip were removed prior to invagination?

   a A normal frog would develop.
   b A frog without lips would develop.
   c No neural tube would develop.
   d The gut tract underlying the neural tube would not develop.
   e Two neural tubes would develop.
   (p. 576)

49 During development of the eye, the optic vesicles form as lateral outpockets from the brain. Early during this process the cells of the optic vesicle do not appear to be specialized, but when they are cut from the brain and transplanted to another site, they continue to develop as eyes. At this early stage the cells of the optic vesicles are

   a determined and fully differentiated.
   b fully differentiated but not determined.
   c determined but not fully differentiated.
   d neither determined nor fully differentiated.
   (pp. 581–82)

50 Cells move themselves in the developing organism using

   a inducers.   c CAMs.
   b hormones.   d morphogens.
   (p. 579)

51 Homeotic genes

   a produce permissive inducers.
   b produce homeoboxess.
   c metabolize morphogens.
   d produce DNA control substances.
   (p. 581)

52 The nuclei of highly differentiated human cells

   a have lost certain genes.
   b have specialized functions as a result of selective mutation.
   c transcribe messenger RNA from all the genes.
   d are in a state of specific gene repression.
   e can easily undergo a process of dedifferentiation.
   (p. 582)

53 Before nerve cells join together in a network, many must move from their place of origin to their final location. All the following are probably involved in this movement *except*

   a gradients of diffusing chemicals.
   b selective death of cells.
   c CAMs.
   d tactile clues.
   (p. 584)

54 A case in which differentiation precedes determination is found in the

   a optic vesicle inducing lens formation.
   b yolk distribution in frog eggs.
   c a salamander regenerating a lost leg.
   d imaginal disc forming in insects.
   e neurulation in guinea pigs.
   (p. 583)

For further thought

1 What would happen if more than one sperm fertilized an egg? How is this prevented?

2 Aging has a profound effect on cerebral function in human beings. From midlife on, there is a continuing loss of neurons from the brain and parts of the spinal cord. Suggest factors that may be involved in this aging process. (Neurons normally do not divide after very early childhood.)

3 Explain the various factors in embryonic development that cause cells to have different characteristics despite their identical genetic content.

4 Normally, the lens of a frog's eye begins to differentiate from the outer layer (ectoderm) of the embryo when part of the brain touches the ectoderm. If the brain is transplanted to the tail of a frog embryo, a lens will form where the brain touches the tail ectoderm. If a piece of plastic is placed between the growing brain and the ectoderm, the lens will not develop. What do these experiments tell you about differentiation of the lens?

ANSWERS

Testing recall

1	*l*	5	*j*	9	*m*	13	*i*
2	*c*	6	*n*	10	*g*	14	*i*
3	*d*	7	*f*	11	*h*	15	*k*
4	*e*	8	*b*	12	*g*		

16 false—the genetic content is identical, but the cytoplasm is not
17 true
18 true
19 false—mesodermal tissue plays the inductive role
20 true
21 true
22 false—homeotic genes produce substances that help each segment express its unique character
23 true

Testing knowledge and understanding

24	*e*	31	*e*	37	C	43	*a*	49	*c*
25	*e*	32	*d*	38	*c*	44	*b*	50	*c*
26	*b*	33	*a*	39	*d*	45	*d*	51	*d*
27	*c*	34	B	40	*c*	46	*d*	52	*d*
28	*b*	35	D	41	*a*	47	*c*	53	*b*
29	*d*	36	E	42	*b*	48	*c*	54	*c*
30	*c*								

Chapter 28

NERVOUS CONTROL

A GENERAL GUIDE TO THE READING

In Chapter 25 we learned that there are two principal types of control mechanisms in animals—chemical and nervous. This chapter and the one that follows are concerned with the second of these mechanisms, nervous control. As you read Chapter 28 in your text, you will want to concentrate on the following topics.

1. The contrast between hormonal control and nervous control. The first two paragraphs of the chapter presents this contrast and explain why hormonal control is inadequate for active multicellular animals.

2. The evolution of nervous systems. You will want to concentrate on the evolutionary trends (p. 593) and on the differences between the central nervous system of annelids and arthropods and that of vertebrates (pp. 593–94).

3. The structure of the neuron. To understand the discussions of reflex arcs, nerve-impulse transmission, and synaptic transmission later in the chapter, you need to know the basic structure of neurons and the names of their various parts (pp. 588–89).

4. Nerve-impulse transmission. This topic is indispensable and interesting. Figures 28.11 and 28.13 (p. 597) are particularly helpful. Pay special attention to the material on voltage-gated channels (see Fig. 28.14, p. 598).

5. Restoration of the initial ionic balance in the neuron. Study Figures 28.14 (p. 598) and 28.15 (p. 601) with care as these will help you understand the roles of diffusion, electrostatic attraction, and the sodium-potassium pump. (p. 599)

6. Transmission across synapses. It is important that you understand the process of synaptic transmission and the processes of summation and integration, since these are what is involved in processing information in the nervous system. (pp. 599–604)

7. The reflex arc. You need to understand the components of the reflex arc. Figures 28.23 to 28.25 (pp. 608–10) are helpful.

8. The autonomic nervous system. You should read the section on the autonomic nervous system (pp. 610–11) carefully and study Figure 28.26 (p. 611); these describe motor programs, which are responsible for much of an organism's behavior.

KEY CONCEPTS

1. Responses to stimuli from the environment generally involve four components: detection of the stimulus, conduction of a signal, processing of the signal, and response. (p. 588)

2. The nervous system is composed of cells that are specialized to detect changes in the environment and to conduct information to the effectors, which produce a rapid response. (p. 588)

3. The brains of animals consist almost entirely of interneurons arranged in a complex, highly specialized network. An interneuron typically collects excitatory or inhibitory input from many cells and passes on their information to its target cells. (pp. 589–90)

4. Inhibition is essential to information processing since the nervous system usually operates using an antagonist strategy; both excitatory and inhibitory signals are sent, and the ratio between them determines the target cell's response. (p. 605)

5. Increasing the number of conducting cells in the nervous pathway makes possible a greater number of alternative routes for nerve stimuli and increases the flexibility of the response to a stimulus. (pp. 591–92)

6. Animals with more advanced nervous systems have evolved more complex neural pathways, a greater degree of centralization, and better-developed sense organs than animals with simple nervous systems. (p. 593)

7. Nerve impulses are conducted along the neuron by electrochemical changes; they are transmitted across synapses by chemicals called neurotransmitters. (pp. 594–96)

8. The propagation of an action potential is a membrane phenomenon; it caused initially by an electrostatic gradient across the membrane, followed by a coordinated series of ion-specific changes in the membrane's permeability. (pp. 596–97)

9. The membrane proteins responsible for creating the action potential are voltage-gated channels; they open and close in response to changes in the electrostatic gradient across the membrane. When an impulse passes and the membrane is depolarized, diffusion and electrostatic attraction restore the electrochemical balance between Na^+ outside and K^+ inside almost instantaneously. In addition, the sodium-potassium pump actively extrudes Na^+ ions and takes up K^+ ions. (pp. 598–99)

10. The development of sodium-potassium pumps, combined with the evolution of ion-specific voltage-gated channels, has been the basis for the evolution of neural transmission. (p. 599)

11. Synapses are points of resistance in neural circuits; the receiving cell integrates all the excitatory and inhibitory signals it receives and either fires or remains silent. (pp. 605–606)

12. Reflex arcs are simple neural pathways linking receptors and effectors. They also interconnect with other neural pathways; interneurons synapse with pathways leading to other parts of the spinal cord and to the brain, and the brain can send impulses that modify the reflexes. (pp. 608–609)

13. The central nervous system of vertebrates is a coordinating system for two kinds of pathways—somatic and autonomic. Somatic pathways control the voluntary activities, whereas autonomic pathways innervate involuntary activities. (pp. 608–12)

14. The basis of the more complex behavior patterns in most animals is motor programs. Motor programs are self-contained neural circuits that coordinate muscle movement; they are automatically fine-tuned by sensory feedback and under direct control of the brain. (pp. 613–14)

CROSSLINKING CONCEPTS

1. The flow of information from the environment to response was first introduced when you studied chemical control (Chapter 25). This chapter on nervous control furthers your understanding of how animals respond to stimuli.

2. The concept of antagonistic control was introduced in your study of chemical control, where you saw that hormones often work in opposition to each other (Chapter 25). In this chapter, you learn that there are often contradictory impulses being sent to the central nervous system. Interneurons are able to inhibit excitatory impulses to modify the strength of the response. The sympathetic and parasympathetic nervous systems work in opposition to excite or slow down the autonomic nervous system responses.

3. The sodium-potassium pump, first introduced in Chapter 5, is crucial for maintaining nerve-cell function. The importance of sodium and potassium ions has been previously noted in the discussions of nutrition, kidney function, and mineralocorticoid function. These two very important ions will come up again when you study muscle function.

4. Your comprehension of transmission of a nervous impulse across a synapse rests on your understanding of channels, exocytosis (Chapter 5), and hormone receptors (Chapter 25).

5. There are strong linkages between the endocrine and nervous control systems. In the chapter on chemical control (Chapter 25), you learned how the hypothalamus of the brain synthesizes releasing hormones and hormones that are stored in the posterior pituitary gland. In this chapter, you learn how a variety of neurotransmitters activate the adenylate

cyclase system of nerve cells, a process similar to how hormones operate. In addition, you learn that the sympathetic nervous system produces the same general effects—the fight or flight reactions—as the adrenal medulla. The reason for the similarity is that the transmitters of the sympathetic system and the hormones of the adrenal medulla are identical (adrenalin and noradrenalin).

6 The concept of a motor program is introduced in this chapter. Its significance will be enhanced when you study animal behavior (Chapter 35).

7 Analogy: An action potential is like falling dominoes.

OBJECTIVES

1 Describe the structure of a typical neuron and, using Figure 28.1 (p. 581), point out the axon, dendrite, cell body, and myelin sheath in the different types of neurons. Indicate the path of information flow and point out a synapse and neuromuscular junction. (p. 588)

2 Using Figures 28.2 and 28.3 (p. 590), explain the role that the neuroglia and Schwann cells play in the nervous system and how the myelin sheath is formed. Indicate the function of the myelin sheath. (pp. 588–89)

3 Using Figures 28.6 (p. 592), 28.7 (p. 592), 28.8 (p. 593), and 28.9 (p. 594), compare the nervous systems of cnidarians (such as hydra), flatworms, annelids, and arthropods. In doing so, indicate for each type of organism whether it is radially symmetrical or bilaterally symmetrical, what degree of centralization its nervous system has reached as shown by the absence or presence of major longitudinal nerve cords and by their location—dorsal or ventral—what degree of cephalization it has reached, and whether ganglia are present. (pp. 592–94)

4 Contrast the nervous system of annelids and arthropods with that of vertebrates with respect to the location, form, and structure of the longitudinal nerve cord or cords, the degree of centralization, and the extent of dominance of the brain. (pp. 593–94)

5 Explain how a nerve impulse is conducted along the neuron, using the terms stimulus, threshold, action potential, and all-or-none response. (pp. 594–96)

6 Discuss the basis for the polarization of the nerve cell membrane, considering the relative amounts of sodium, potassium, and negatively charged ions inside and outside the neuron, and state whether the outside of the neuron is charged positively or negatively with respect to the inside (see Fig. 28.11, p. 596). (pp. 596–97)

7 Explain in some detail how an impulse is transmitted along a neuron fiber; specify which ions move and in what order when the fiber is stimulated, and explain what is meant by voltage-gated channels. Using a diagram such as Figure 28.13 (p. 597), explain how the nerve impulse is propagated along the neuron. (pp. 596–98)

8 Explain how diffusion, electrostatic attraction, and the sodium-potassium pump act to reestablish the original ionic balance and keep the neuron functioning. Figures 28.14 (p. 598) and 28.15 (p. 601) may be helpful. (p. 599)

9 Using Figure 28.18 (p. 602), identify the synaptic terminal, the presynaptic membrane, postsynaptic membrane, synaptic cleft, and synaptic terminal. Describe the events occurring at a synapse when an action potential arrives, and explain how the impulse is transmitted across the synapse and what must happen for an action potential to be induced in the postsynaptic neuron. Name three transmitter substances. (pp. 599–604)

10 Compare events at excitatory and inhibitory synapses. Explain what is meant by summation and integration, and describe how neurological drugs can affect nerve-impulse transmission. (pp. 605–606)

11 Contrast impulse transmission across the synapses with impulse transmission across the neuromuscular junction. (pp. 599–607)

12 Using Figure 28.24 (p. 609), trace the flow of information through a reflex arc, beginning with a sensory receptor cell and ending with an effector cell. In doing so, be sure to identify the sensory neuron and indicate where the cell body for this neuron is located, specify the part of the nervous system in which information is relayed to interneurons, describe the role of interneurons in the reflex arc, and explain how information gets to the brain. Identify a motor neuron and indicate where the cell body for this neuron is located. (pp. 608–609)

13 Compare the somatic and autonomic pathways, specifying for each whether it is under voluntary or involuntary control, whether the muscle innervated is skeletal or smooth, and how many motor neurons are in the pathway; then indicate which system (somatic or autonomic) would control each of the following: heartbeat, peristalsis of the digestive tract, walking, sweating, lifting a heavy object in your arms, and the response to stepping on a tack. (pp. 608–12)

14 Using Figure 28.27 (p. 612), compare the four different arrangements for chemical control and explain why the nervous and endocrine systems are so closely related.

15 Compare the sympathetic and parasympathetic nervous systems in structure and function. Give one reason why the sympathetic nervous system and the hormones of the adrenal medulla have similar effects on the body. (pp. 610–12)

16 Explain what motor programs are, how they are fine-tuned and controlled, and what their importance in the control of an organism's behavior is. (pp. 613–14)

KEY TERMS

irritability (p. 587)
stimulus (p. 588)
effector (p. 588)
neuron (p. 588)
cell body (p. 588)
dendrite (p. 588)
axon (p. 588)
synapse (p. 588)
neuroglia (p. 588)
myelin sheath (p. 588)
Schwann cell (p. 589)
node (p. 589)
sensory, or afferent, neuron (p. 589)
motor, or efferent, neuron (p. 589)
interneuron (p. 589)
inhibition (p. 590)
reflex (p. 591)
adaptation (p. 592)
receptor (p. 592)
conductor (p. 592)
nerve net (p. 592)
central nervous system, or CNS (p. 593)
brain (p. 593)
cephalization (p. 593)
ganglion (p. 594)
nerve impulse (p. 594)
threshold value (p. 595)
all-or-none response (p. 595)
potassium leak channels (p. 596)
depolarize (p. 596)
repolarize (p. 596)
action potential (p. 597)
electrostatic gradient (p. 598)
gated channel (p. 598)
voltage-gated channel (p. 598)
refractory period (p. 599)
sodium-potassium exchange pump (p. 599)
Na^+K^+—ATPase (p. 599)
synaptic terminal, or synaptic bouton (p. 601)
electrical synapse (p. 601)
chemical synapse (p. 601)
synaptic cleft (p. 601)
transmitter chemical (p. 601)
synaptic vesicle (p. 601)
postsynaptic membrane (p. 603)
acetylcholine (p. 603)
excitatory synapse (p. 603)
inhibitory synapse (p. 603)
acetylcholinesterase (p. 603)
adrenalin (p. 603)
noradrenalin (p. 603)
serotonin (p. 603)
dopamine (p. 603)
glycine (p. 603)
glutamate (p. 603)
gamma-aminobutyric acid, or GABA (p. 603)
excitatory postsynaptic potential, or EPSP (p. 604)
inhibitory postsynaptic potential, or IPSP (p. 604)
adenylate cyclase (p. 604)
summation (p. 605)
integration (p. 605)
neuromuscular junction (p. 607)
peripheral nervous system (p. 608)
nerve (p. 608)
somatic pathway (p. 608)
reflex arc (p. 608)
dorsal-root ganglion (p. 608)
mixed nerve (p. 609)
autonomic pathway (p. 610)
sympathetic system (p. 610)
parasympathetic system (p. 610)
motor program (p. 613)

SUMMARY

Because hormonal control is too slow for the rapid integration of sensory information and coordination of movement, electric communication has evolved to fill this need. Irritability—the capacity to respond to stimuli—involves reception of the stimulus, conduction of a signal, processing of the signal, and response by an effector. All multicellular animals (except sponges) have evolved some form of nervous system. Most nervous pathways have at least three separate cells: receptor, conductor, and effector cells. More complex pathways have additional conductor cells; these increase the flexibility of response.

Neurons The typical nerve cell, or *neuron*, consists of the *cell body*, which contains the nucleus, and one or more long *nerve fibers* which extend from the cell body. The *dendrites* receive impulses and conduct them to the cell body, and the *axons* conduct impulses away from the cell body. Within the central nervous system (CNS) the neurons are associated with vast numbers of *neuroglia* cells. Some glia provide the neurons with nutrients. In at least some areas of the central nervous system glia provide, during development, a framework along which neurons migrate and axons grow to reach their targets. One class of glia gives rise to the *myelin sheath*, a heavily lipid insulator against "cross talk" between adjacent axons. Outside the central nervous system, many vertebrate axons are enveloped by *Schwann cells*, which may also form a myelin sheath. Myelin sheaths speed up the rate of conduction.

Evolution of nervous systems All multicellular animals (except sponges) have evolved some form of nervous system. Most nervous pathways have at least three separate cells: receptor, conductor, and effector cells. More complex pathways have additional conductor cells; these increase the flexibility of response. *Sensory neurons* lead from receptor cells, *motor neurons* lead to effector cells, and *interneurons* lie between the sensory and motor neurons. Junctions between neurons are called *synapses*.

A *nerve* consists of a number of neuron fibers bound together. Radially symmetrical animals, such as cnidarians, have a simple nerve net with little central control. We can see six major trends in the evolution of the nervous system of bilaterally symmetrical animals:

1. increased centralization by the formation of longitudinal nerve cords
2. one-way conduction
3. increased complexity of pathways
4. formation of distinct functional areas and structures
5. increased cephalization
6. increased number and complexity of sense organs

These evolutionary trends are most developed in the vertebrates and in the annelids and arthropods. The main difference between the central nervous system (brain and spinal cord) of vertebrates and those of annelids and arthropods is that vertebrates have a single dorsal, hollow nerve cord, whereas annelids and arthropods have two ventral, solid nerve cords. The vertebrate brain is also more highly developed and exerts more dominance over the entire nervous system. The vertebrate spinal cord is not so obviously ordered into a series of ganglia and connecting tracts.

Action potential A nerve impulse is a wave of electrical activity moving along a nerve fiber. The potential stimulus must be above a critical intensity, or *threshold*, to initiate an impulse. If the axon fires, it will fire maximally or not at all; this is called an *all-or-none* response.

The inside of a resting nerve fiber is negatively charged relative the outside because the ratio of negative to positive ions is higher inside the cell than outside. The inside has a high concentration of potassium ions (K^+) and negative organic ions; the outside has a high concentration of sodium ions (Na^+). *Potassium leak channels*, which allow K^+ ions to diffuse out of the cell, are an important factor in maintaining the electrical gradient across the membrane.

When a fiber is stimulated and the membrane depolarizes to the threshold value or more, specific gates open and expose Na^+ channels, and Na^+ ions cross the membrane into the cell and make the inside positively charged relative to the outside. Once enough Na^+ ions have flowed through to depolarize the membrane completely, the Na^+ channel gates close and the K^+ gates open and expose K^+ channels; K^+ ions then rush out of the cell, and restore the original charge. This cycle of electrical charges is known as the *action potential*. The action potential at the point of stimulation alters the permeability at adjacent points and initiates the same cycles of changes there. Diffusion, potassium leak channels and the *sodium-potassium exchange pump* restore the original ion and charge distribution.

Synapse transmission The axon of one neuron usually synapses with the dendrites or cell bodies of other neurons. Each tiny branch of an axon usually terminates in a *synaptic terminal*. A few synapses are electrical; in these a special connection between the membrane of the synaptic terminal and the membrane of the adjoining cell permits direct electric transmission from the first neuron to the second. Most synapses, however, are chemical. When an impulse traveling along the axon reaches the synaptic terminal, special voltage-gated calcium channels open, and Ca^{++} ions diffuse into the terminal and cause the *synaptic vesicles* to discharge their stored *transmitter chemical* into the synaptic cleft. The transmitter molecules diffuse across the cleft and alter the polarization of the postsynaptic membrane of the next neuron. Synaptic transmission is slower than impulse conduction along the neuron. It is the chemical synapses that make transmission along the neural pathways one-way.

The transmitter chemical in the *peripheral nervous system* is most often *acetylcholine*. Transmitter chemicals inside the central nervous system include acetylcholine, *noradrenalin, serotonin, dopamine*, and *GABA*. Synaptic transmitter substances act by binding to receptor proteins in the membrane, and thus open the gates of a channel and allow specific ions to cross. This ion movement results in the alteration of the postsynaptic neuron's membrane potential. Transmitter chemicals can be excitatory or inhibitory. An *excitatory* transmitter opens Na^+ gates, and thus slightly reduces the polarization of the postsynaptic membrane and creates an excitatory postsynaptic potential (*EPSP*). If the EPSP reaches threshold, it triggers an impulse. An *inhibitory* transmitter increases the polarization of the postsynaptic membrane, a condition called an inhibitory postsynaptic potential (*IPSP*). This makes the neuron harder to fire.

Synapses are points of resistance in neural circuits; enough excitatory transmitter must be released within a short time to build up sufficient EPSP to initiate an impulse. The cell integrates all the excitatory and inhibitory signals it receives and either fires or remains silent.

Once the transmitter has been released, it must be promptly removed from the synaptic cleft. Some transmitters are lost by diffusion, others are taken back up for reuse by the presynaptic neuron, and still others are destroyed by specific enzymes.

Synapses are responsible for information processing in the nervous system. Their operation depends on a delicate balance between transmitter substance, deactivating enzyme, and membrane sensitivity. Synaptic malfunctions have been implicated in several mental disorders. Neurological drugs can alter synaptic function in a variety of ways.

The gap between the axon and the muscle it innervates is called the *neuromuscular junction*. Transmission across this gap is also via transmitter chemicals. Acetylcholine is the transmitter at neuromuscular junctions of vertebrate skeletal muscle.

Nervous pathways/systems A *reflex arc* is a simple neural pathway linking a receptor and effector. Most somatic reflex arcs begin with a *sensory neuron*, whose cell body is always outside the spinal cord in a *dorsal-root ganglion* and whose axons always enter the spinal cord dorsally, where the sensory neuron synapses with *interneurons*. These in turn synapse with *motor neurons* in the cord, which leave the spinal cord ventrally and conduct impulses along their axons to the effectors (usually skeletal muscles), which respond to the stimulus. Some nerves have only sensory fibers and are therefore *sensory nerves*, others are purely *motor nerves*, and still others have both types of fibers and are *mixed nerves*.

Reflex arcs always interconnect with other neural pathways; the interneurons connect with pathways leading to the brain, and the brain can send impulses that modify the reflexes.

The central nervous system of vertebrates is a coordinating system for two kinds of pathways—somatic and autonomic. Somatic pathways control the voluntary activities, whereas autonomic pathways innervate involuntary activities. The *autonomic nervous system* (ANS) consists of nervous pathways that conduct impulses from the central nervous system to various internal organs. The autonomic nervous system regulates the body's involuntary activities. Autonomic pathways usually have two motor neurons. The first neuron exits from the central nervous system and synapses with a second that innervates the target organ. The autonomic nervous system is separated into two parts, *sympathetic* and *parasympathetic* systems. Most internal organs are innervated by both, with the two systems usually functioning in opposition to each other. The sympathetic system prepares an animal for emergency action, whereas the parasympathetic system restores order or passivity after the crisis is over.

There is a close relationship between the nervous and endocrine systems. The sympathetic system and the adrenal medulla have similar effects because they both release adrenalin and noradrenalin. Synaptic transmission also depends on the release of chemicals.

Motor program The control of complex rhythmic or sequential behaviors in vertebrates and invertebrates is based on *motor programs*. Motor programs are self-contained neural circuits that coordinate muscle movement. The program is automatically fine-tuned by sensory feedback and is under the direct control of the brain. Much of an organism's behavior is accomplished by interacting groups of neural circuits.

CONCEPT MAP

Map 28.1

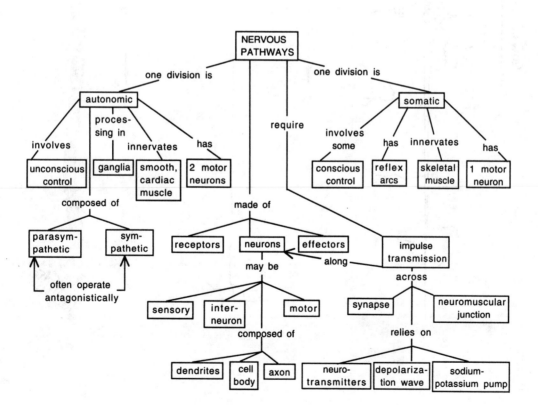

KEY DIAGRAM

The following coloring exercise is designed to help you learn the anatomy of vertebrate neurons. Three types of neurons are shown below. Color the names and the appropriate structures, using a different color for each. (Fig. 28.1, p. 589)

DENDRITE   CELL BODY   MYELIN SHEATH
AXON   NUCLEUS   MUSCLE

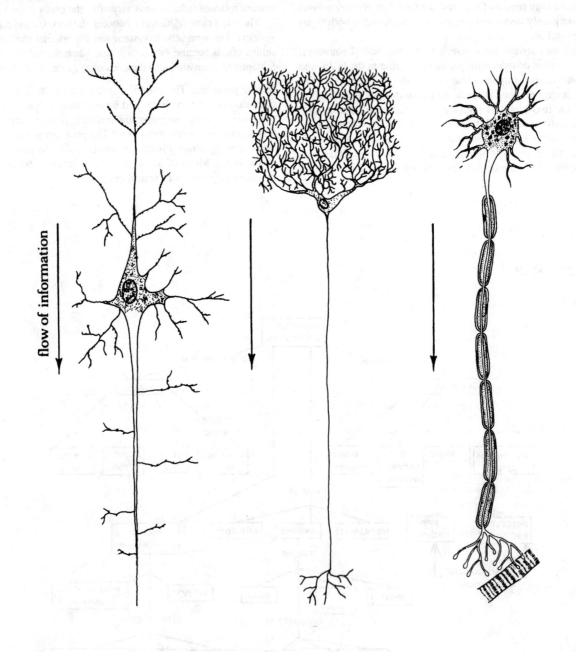

QUESTIONS

Testing recall

Select the correct term or terms to complete each statement.

1. The components of irritability are (sensory reception, conduction, response). (p. 588)

2. Increasing the number of conductor cells in a nervous circuit (increases, decreases) the flexibility of response. (p. 591)

3. The central nervous system of vertebrates consists of the (brain, spinal cord, autonomic nerves, spinal nerves). (p. 594)

4. Radially symmetrical animals such as cnidarians have a nervous system that is (highly centralized, a diffuse nerve net, organized into longitudinal nerve cords). (p. 592)

5. The nervous system of the most-advanced flatworms is characterized by (a nerve net, two ventral nerve cords, a dorsal nerve cord, a highly developed brain, simple sense organs at the anterior end). (p. 593)

6. The nerve cord(s) of an arthropod is (are) (single, double), (solid, hollow), and (dorsal, ventral). (p. 594)

7. The brain of an arthropod exerts (limited, extensive) dominance over the entire nervous system. (p. 594)

8. In the vertebrate neuron, the (axons, dendrites) conduct impulses toward the cell body and the (axons, dendrites) conduct impulses away from the cell body. Each neuron usually has (one, many) dendrite(s) and (one, many) axon(s). (Sensory neurons, Motor neurons, Interneurons) conduct impulses to the central nervous system. (Sensory neurons, Motor neurons, Interneurons) conduct impulses within the central nervous system. (pp. 588–90)

9. Within the central nervous system special satellite cells called (Schwann cells, neuroglia) are associated with the neurons; frequently these wrap around and around the neuron and form the (myelin sheath, axon, nerve). (pp. 588–89)

10. The nerves of the autonomic nervous system innervate the (skeletal muscles, blood vessels, digestive tract, respiratory system, reproductive system). (p. 610)

11. The autonomic nerve pathways usually have (one, two, three, many) motor neuron(s). The parts of the autonomic nervous system are the (somatic, sympathetic, parasympathetic) divisions. (pp. 610–11)

12. The resting nerve fiber is polarized, with the inside having a (positive, negative) charge compared to the outside. When a stimulus is above threshold, the membrane first becomes permeable to (Na^+, K^+, organic) ions, which rush (into, out of) the cell; the inside of the cell is now charged (positively, negatively) with respect to the outside. Next the membrane becomes permeable to (Na^+, K^+, organic) ions and these rush (into, out of) the cell, restoring the original polarization. (pp. 596–97)

13. The original ionic balance is restored by (the sodium-potassium pump, active transport, electrostatic attraction, voltage-gated pumps, diffusion). (p. 599)

14. Myelinated fibers conduct impulses (slower, faster) than unmyelinated fibers. (pp. 597–98)

15. Most synapses are (electrical, chemical). When an impulse traveling down the axon reaches the terminal, it makes the membrane of the terminal permeable to (Na^+, K^+, Ca^{++}), which diffuses into the terminal and promotes the release of (Na^+, K^+, transmitter chemicals) into the synaptic cleft. Synaptic transmission is much (slower, faster) than impulse conduction along the axon. (pp. 599–604)

16. The transmitter substance at neuron–neuron synapses outside the central nervous system is generally (noradrenalin, serotonin, acetylcholine). (p. 603)

17. An excitatory transmitter substance slightly (reduces, increases) the polarization of the postsynaptic membrane; this makes it (easier, more difficult) to trigger an impulse. If the transmitter had been inhibitory, the membrane would have become (depolarized, more polarized), a condition called an (IPSP, EPSP). (pp. 601–603)

18. If both excitatory and inhibitory transmitters converge on a cell at the same time, their effects are combined in the process called (addition, integration). (p. 605)

19. The gap between the end of the motor axon and the effector it innervates is called the (synapse, neuromuscular junction). The transmitter substance released by the motor neurons at their junction with the effectors in both somatic and parasympathetic neurons is (acetylcholine, noradrenalin), whereas the motor neurons of the sympathetic pathways release (acetylcholine, noradrenalin). (pp. 607–12)

Questions 20 to 24 refer to the following diagram showing the effect of transmitter substance on the membrane potential of a neuron.

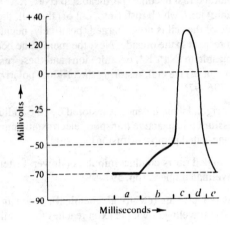

20 In the diagram, what is the resting potential?

   a +50 millivolts   c –50 millivolts
   b 0 millivolts   d –70 millivolts   (p. 597)

21 In the diagram, what is the threshold potential?

   a +50 millivolts   c –50 millivolts
   b 0 millivolts   d –70 millivolts   (p. 597)

22 During which interval (*a, b, c, d,* or *e* in the diagram) would the transmitter substance be active? (p. 597)

23 During which intervals would sodium ions be entering the neuron? (p. 597)

24 During which interval would potassium ions be leaving the neuron? (p. 597)

25 Suppose you pricked your finger with a pin. On the accompanying diagram, draw in and label the components of a reflex arc that originates in the finger. Include the sensory neuron, interneurons, motor neuron, receptor, and effector. Label the parts of each neuron (axon, dendrites, cell body). (p. 608)

Testing knowledge and understanding

Choose the one best answer.

26 Functions of the glial cells include all the following *except*

   a forming myelin sheaths on the axons of peripheral nervous system neurons.
   b providing neurons with nutrients.
   c providing, during development, a framework along which neurons migrate.
   d absorbing substances secreted by neurons.
                                                    (pp. 588–89)

27 Animal control centers tend to be concentrated in the head region; this is known as

   a autonomic control.
   b corticalization.
   c cephalization.
   d rationalization.
                                                    (p. 593)

28 One evolutionary advance found in the annelids (such as earthworms) but *not* in the flatworms is

   a longitudinal nerve cords.
   b prominent ganglia with connecting nerves.
   c a nerve net.
   d one-way conduction.
   e radial symmetry.
                                                    (pp. 593–94)

29 Myelin sheaths

   a are found on all axons.
   b are found on all dendrites.
   c speed up conduction.
   d are made of protein.
   e wrap around all sensory neurons.
                                                    (pp. 597–98)

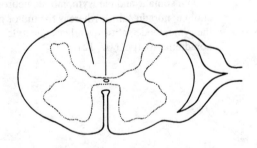

30 Which of the following structures in nerve cells is specialized for receiving and integrating synaptic inputs from a large number of nerve cells?

   a the axon
   b myelin sheath
   c nodes of Ranvier
   d neuroglia
   e dendrites
   (p. 588)

31 Which of the following statements most accurately describes what is happening at the point indicated by the X on the accompanying diagram of the oscilloscope trace of an action potential?

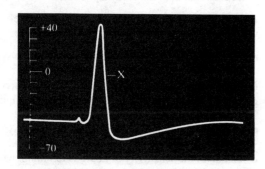

   a K⁺ ions are flowing out of the cell.
   b K⁺ ions are flowing into the cell.
   c Na⁺ ions are flowing into the cell.
   d Na⁺ ions are flowing out of the cell.
   e None of the above.
   (p. 597)

32 Which one of the following statements is *false* concerning the axon in a resting (nonconducting) stage?

   a The membrane is relatively impermeable to sodium ions.
   b The membrane has a positive charge on the outside and a negative charge on the inside.
   c The membrane is highly permeable to large negatively charged organic ions and allows them to leak out.
   d Potassium ions are in higher concentration inside the axon than outside.
   e Sodium ions are in higher concentration outside the axon than inside.
   (pp. 596–97)

33 Which one of the following events occurs *first* when a neuron is stimulated?

   a ATP is hydrolyzed.
   b Na⁺ channels open and Na⁺ rushes inside.
   c K⁺ channels open and K⁺ rushes inside.
   d The sodium-potassium pump exchanges Na⁺ and K⁺.
   e Negatively charged organic ions rush outside.
   (p. 596)

34 Of the following activities, which is the *second* event to occur in the depolarization of a nerve cell?

   a Na⁺ channels open and Na⁺ rushes inside.
   b Na⁺ channels open and Na⁺ rushes outside.
   c K⁺ channels open and K⁺ rushes inside.
   d K⁺ channels open and K⁺ rushes outside.
   e Negatively charged ions rush outside.
   (p. 597)

35 You have set up two electrical recording instruments that will record any electric changes that occur. One is placed at each end of a single neuron 10 cm long. If at the first pair of electrodes the magnitude of the action potential is recorded at 100 millivolts, what will be the magnitude of the action potential when it reaches the second pair of electrodes at the other end?

   a 10 millivolts
   b less than 100 millivolts but cannot be more exact
   c 100 millivolts
   d more than 100 millivolts but cannot be more exact
   e 1,000 millivolts
   (pp. 594–95)

36 On stimulation of a neuron, the influx of sodium ions is the result of

   a active transport.
   b diffusion.
   (p. 596)

37 When a nerve cell is stimulated, the magnitude of the potential difference across the plasma membrane

   a increases.
   b decreases.
   (p. 596)

38 Which one of the following statements concerning the sodium-potassium pump is *false*?

   a The protein complex has a binding site for sodium and a separate binding site for potassium.
   b Sodium is transported out of the cell and potassium is transported into the cell.
   c The pump transports sodium either into or out of the cell.
   d The energy for the pump is provided by the hydrolysis of ATP.
   e The protein complex has two different possible conformations.
   (p. 599)

39 Suppose an intracellular electrode records a resting potential for a given neuron of −70 millivolts. Which one of the recordings below best portrays what will happen if an inhibitory neurotransmitter substance is applied to the neuron? (The arrows indicate the time of application.) (p. 603)

a
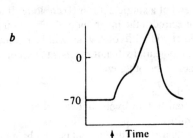

b
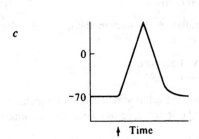

c
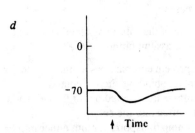

d
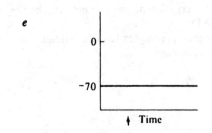

e

40 Suppose you administer a drug that efficiently removes calcium. The effect of this drug on the nervous system would be to

 a prevent action potentials from traveling along axons.
 b prevent transmitter substance in the synaptic cleft from depolarizing the postsynaptic membrane.
 c inhibit the sodium-potassium pump.
 d prevent synaptic vesicles from discharging their contents into the synaptic cleft.
 e prevent transmitter substance from causing the depolarization of the presynaptic membrane. (p. 603)

41 A drug that produces which one of the following effects would probably be *least* effective as a tranquilizer?

 a interferes with the uptake of noradrenalin into synaptic vesicles
 b interferes with the synthesis of noradrenalin in nerve cells
 c prevents the release of noradrenalin from synaptic vesicles
 d blocks receptor sites for noradrenalin on postsynaptic membranes
 e enhances the sensitivity of neurons to noradrenalin (p. 603)

42 The drug lithium carbonate is effective in the treatment of manic-depressive individuals. Its action is to reduce the release of noradrenalin and enhance its reuptake. The effect of ingesting this drug would be to

 a decrease the firing of the postsynaptic neurons at noradrenalin synapses.
 b enhance the sensitivity of the postsynaptic neurons to noradrenalin.
 c block receptor sites for noradrenalin on postsynaptic membranes.
 d stimulate uncontrolled firing of the postsynaptic neurons at noradrenalin synapses.
 e slow down the rate of action-potential movement in the presynaptic neuron. (p. 606)

43 Integration in the nervous system depends on the addition of _____ from different sources.

 a action potentials   d synaptic vesicles
 b specific ions   e resting potentials
 c excitatory and inhibitory postsynaptic potentials (p. 605)

44 Many poisons and drugs act at synapses. One kind of snake venom binds irreversibly with acetylcholine and thus, in an acetylcholine synapse, would be expected to

 a prevent the postsynaptic membrane from depolarizing.
 b cause the postsynaptic membrane to depolarize only once.
 c cause the postsynaptic membrane to depolarize many times in succession.
 d produce muscular spasms.
 e prevent cholinesterase production. (p. 606)

45 The diagram below depicts a typical vertebrate reflex arc involving a single receptor (R), a muscle (M), and three neurons. Assume that you apply an electrical stimulus at

point X. What is the most distant point from X at which you would expect to detect a signal? (p. 603)

46 What is meant by a reflex arc in the nervous system?

   a an inherited behavior pattern that functions through a certain neural pathway
   b a functional unit consisting of a receptor, neural pathways, and an effector
   c peripheral nerves, spinal cord, and brain
   d a homeostatic system of sensory nerves, synapses, and motor nerves
   e automatic responses of the central nervous system
   (pp. 608–609)

47 A nerve

   a is the same as a neuron.
   b is a pathway within the somatic nervous system only.
   c is a bundle of neuron fibers.
   d usually contains only motor neurons.
   e usually contains only sensory neurons. (p. 608)

48 Where A stands for axon, D for dendrite, S for synapse, and CB for cell body, a typical sequence of structures between a receptor and an effector is

   a D-CB-A-S-D-CB-A.
   b A-D-CB-S-A-D-CB.
   c D-CB-A-S-A-CB-D.
   d D-A-S-CB-D-A-CB.
   e A-CB-D-S-D-CB-A. (pp. 608–9)

49 A certain neuron is encountered during dissection of the muscles in the leg of a pig. It is part of the somatic (voluntary) nervous system. Which one of the following observations indicates that it is a motor neuron, not a sensory neuron?

   a Its cell body is located inside the spinal cord.
   b It exhibits an all-or-none response to stimuli.
   c The end of its axon secretes a neurotransmitter.
   d Its axon is myelinated. (p. 608)

50 Which one of the following statements most closely fits the knee-jerk reflex arc?

   a It involves more than two synaptic junctions, but no brain control.
   b It involves one synaptic junction and no brain control.
   c It normally involves two synaptic junctions and can be modified by the brain.
   d It normally involves one synaptic junction and can be modified by the brain.
   e The stretch receptor connects with the brain stem and sends impulses to the dorsal root ganglion motor neurons. (p. 609)

51 The rate of the human heartbeat is controlled by which of the following? (Pick the most complete and specific answer.)

   a sympathetic nervous system
   b parasympathetic nervous system
   c somatic nervous system
   d autonomic nervous system
   e peripheral nervous system (pp. 610–11)

52 The major function of the autonomic nervous system is to

   a transmit impulses from the brain to the central nervous system.
   b regulate and control the peripheral nervous system.
   c control the contraction of skeletal muscles.
   d innervate the internal organs.
   e coordinate nerve impulses. (pp. 610–11)

53 A true *difference* between the somatic and the autonomic nervous system is that

   a acetylcholine is found as a neurotransmitter only in the somatic system.
   b only the somatic system has motor neurons.
   c only the autonomic system connects to the central nervous system.
   d only the somatic system can cause muscle contraction.
   e the somatic system innervates skeletal muscle whereas the autonomic system innervates smooth muscle.
   (pp. 608–11)

54 A certain neuron in a cat is located entirely outside the central nervous system. The synaptic vesicles of its axon release a transmitter substance that is not destroyed by acetylcholinesterase. This neuron is

   a a sensory neuron of the somatic (voluntary) portion of the nervous system.
   b a motor neuron of the somatic system.
   c a second motor neuron of the sympathetic system.
   d a first motor neuron of the parasympathetic system.
   e a second motor neuron of the parasympathetic system.
   (pp. 610–11)

55 All of the following are functions of the autonomic nervous system *except*

   a causing an endocrine gland to secrete hormones.
   b contracting certain blood vessels and dilating others.
   c causing changes in the heart rate.
   d causing the muscles of the intestinal wall to contract.
   e causing the muscles of the arm to contract in response to a pinprick. (p. 610–11)

56 All of the following are functions of the sympathetic nervous system *except*

   a dilation of intestinal capillaries.
   b acceleration of the rate of heartbeat.
   c dilation of capillaries in some skeletal muscle.
   d erection of hairs on the skin.
   e decrease of peristalsis of the digestive tract.
   (pp. 610–11)

57 The control of many complex rhythmic behaviors is based on

   a chains of reflexes.
   b motor programs.
   c habituation.
   d input from stretch receptors alone.
   e autonomic control. (pp. 613–14)

For further thought

1. As you are crossing a street, a car suddenly swerves toward you. Explain how the nervous system enables you to meet this emergency.

2. Certain drugs act as metabolic poisons and prevent the synthesis of ATP. Suppose you treated a neuron with one of these drugs. Would the neuron conduct an impulse? Explain.

3. Cocaine is a drug that inhibits the active reuptake of dopamine into the brain's presynaptic neurons. It also binds inside sodium channels and blocks them. Would you classify cocaine as a stimulant or a depressant? What effect might cocaine have on a person's heart rate, blood pressure, and mental awareness?

4. Practitioners of voodoo in the Caribbean islands were once reputed to feed pufferfish to their enemies. The effect was dramatic: the individuals appeared to be dead and were mourned and buried. According to legend, they awoke days later as "zombies," who could be easily enslaved by their poisoner. Pufferfish is now known to contain a poisonous substance, tetrodotoxin (TTX), that blocks voltage-sensitive sodium channels in both nerve and muscle. Explain the effects of TTX on the individual.

5. Tell how the following drugs affect the nervous system: amphetamine, nicotine, chlorpromazine, LSD, Valium, and novocaine.

ANSWERS

Testing recall

1. sensory reception, conduction, response
2. increases
3. brain, spinal cord
4. a diffuse nerve net
5. two ventral nerve cords, simple sense organs at the anterior end
6. double, solid, ventral
7. limited
8. dendrites, axons, many, one, sensory neurons, interneurons
9. neuroglia, myelin sheath
10. blood vessels, digestive tract, respiratory system, reproductive system
11. two, sympathetic, parasympathetic
12. negative, Na^+, into, positively, K^+, out of
13. sodium-potassium pump, active transport, electrostatic attraction, diffusion
14. faster
15. chemical, Ca^{++}, transmitter chemicals, slower
16. acetylcholine
17. reduces, easier, more polarized, IPSP
18. integration
19. neuromuscular junction, acetylcholine, noradrenalin
20. d
21. c
22. b
23. b, c
24. d, e
25. See the diagram below.

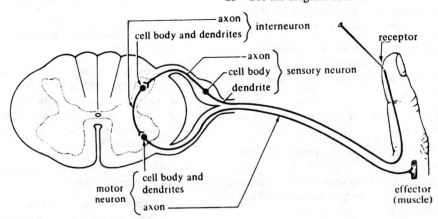

Testing knowledge and understanding

26	a	34	d	42	a	50	d
27	c	35	c	43	c	51	d
28	b	36	b	44	a	52	d
29	c	37	b	45	e	53	e
30	e	38	c	46	b	54	c
31	a	39	d	47	c	55	e
32	c	40	d	48	a	56	a
33	b	41	e	49	a	57	b

Chapter 29

SENSORY PERCEPTION AND PROCESSING

A GENERAL GUIDE TO THE READING

This chapter continues your study of organisms' responses to environmental stimuli by exploring the major senses and the structure and function of the vertebrate brain. Though the nervous systems of other animals are discussed, the major emphasis is on the sense organs and brain of the human being. It is, after all, our brain that sets us apart from other animals. As you read Chapter 29 in your text, you will want to concentrate on the following topics.

1. Receptor function. Try to relate the material on receptor function (pp. 617–20) to the material presented in Chapter 28, particularly in the sections on reflex arcs (pp. 608–609) and on nerve-impulse transmission (especially pp. 596 and 597). Studying Figures 29.1 to 29.5 (pp. 619–21), 29.16 (p. 627), and 29.22 (p. 630) carefully will give you a good understanding of the various receptor types.

2. Vision. Vision is one of the two senses treated in some detail in this text; the other is hearing. The discussion of refraction and accommodation (pp. 625–26) is particularly relevant and interesting. Figures 29.9 (p. 623), 29.13 (p. 625), and 29.17 (p. 628) are helpful in understanding the anatomy of the different types of eyes.

3. Hearing. You may want to devote extra time to this important topic (pp. 629–37).

4. Evolution of the vertebrate brain. Pay particular attention to the section on this topic (pp. 627–46), note the changes at each stage from fish through amphibian and reptile to mammal.

5. The mammalian forebrain. The text (pp. 638–40) presents a fascinating account of the main divisions of this vital organ; here you will learn about the basic structure and functioning of the human brain.

KEY CONCEPTS

1. Each type of sensory receptor functions as a transducer; it converts the energy that constitutes the particular stimulus to which it is attuned into neural stimuli. (p. 618)

2. Regardless of specialization, each type of receptor translates the stimulus into a change in membrane polarization; this is achieved through the opening or closing of specific ion channels in the membrane. (p. 618)

3. The receptors of taste and smell are chemoreceptors; they are sensitive to solutions of different kinds of chemicals which can bind to them by weak bonds. (p. 620)

4 The mechanism of light detection is similar in all animals that respond to light; light energy produces changes in a light-sensitive pigment. (p. 622)

5 Sensory hair cells in the inner ear are responsible for detecting vibrations, acceleration, and position changes that affect equilibrium. (pp. 630–31)

6 Sensation depends on the structure of the brain; the interpretation of incoming action potentials is based on the destination in the brain of the axons carrying them, not on the actual external stimulus or on some special quality of the impulses themselves. (pp. 618–19)

7 The major evolutionary change in the vertebrate brain has been the steady increase in the size and importance of the cerebrum, with a corresponding decrease in the relative size and importance of the midbrain. (pp. 638–39)

8 In the human brain, the cerebral cortex has taken over many of the functions of other parts of the brain and become dominant. (pp. 641–43)

CROSSLINKING CONCEPTS

1 An understanding of receptor cell function must be based on your knowledge of gated channels (Chapter 5) and nerve-impulse transmission (Chapter 28).

2 As in most other chapters, evolution is a prominent theme in your study of vertebrate brains. Comparative embryology and comparative anatomy (Chapter 1) offer insights into the development of the brain's three major sections—forebrain, midbrain, and hindbrain. Evolutionary emphasis is also given to dominance of the cerebral cortex and and its eventual complex folding, which effectively increased surface area.

3 You were first introduced to the hypothalamus in the context of the endocrine system (Chapter 25). In this chapter, you see that another function of the hypothalamus is working as a control center for visceral and emotional responses of the body.

OBJECTIVES

1 Explain how environmental changes lead to changes in membrane polarization, and how sensory cells translate the strength of a stimulus and the temporal pattern of the stimuli into neural stimuli. (pp. 617–19)

2 Describe the pain, touch, and deep-pressure receptors of the skin and the stretch receptor of muscle, and explain how a stimulus can lead to a generator potential and to an action potential in the nerve fiber. (pp. 619–20)

3 Contrast the receptors for taste and for smell with respect to location, basic function, type of receptor cell, and possible mode of action. (pp. 620–22)

4 Using Figures 29.9 (p. 623), 29.10 (p. 624), 29.12 and 29.13 (p. 625), compare the structure of the compound eye with that of the camera eye. In doing so, list similarities and differences between the two kinds of eyes, give the advantages and disadvantages of each, and briefly describe the functioning of the two kinds of eyes. (pp. 623–26)

5 Describe, using Figures 29.12 and 29.13 (p. 625), 29.16 (p. 627), and 29.17 (p. 628), the structure and function of the human eye. Contrast the structure, location, and function of the two principal types of visual receptor cells. (pp. 624–28)

6 Using Figure 29.17 (p. 628), point out some of the connections between receptor cells and other types of neurons in the human retina. Explain the significance of this complex wiring arrangement. (p. 628)

7 Discuss the process of refraction and accommodation in the human eye. In doing so, specify the two parts of the eye responsible for the refraction of light, specify the part of the eye that functions in accommodation, and describe the role of the suspensory ligament in accommodation. Contrast the shape assumed by the lens when near objects are viewed with its shape when distant objects are viewed. (pp. 625–26)

8 Using Figure 29.24 (p. 632), identify each of the following, if shown, and give a function of each: outer ear, middle ear, inner ear, organ of Corti, cochlear canal, vestibular canal, semicircular canals, utriculus, sacculus, tympanic membrane, Eustachian tube, oval window, round window, basilar membrane, and tectorial membrane. Indicate which of the listed parts are involved in hearing, which in maintaining equilibrium, and which in detecting acceleration. (pp. 631–34)

9 Describe the processes by which sound vibrations in the air are transmitted to the hair cells of the organ of Corti. Then explain how we distinguish different pitches. (pp. 634–37)

10 Explain how the inner ear determines the position of the head with respect to gravity and how it detects rotational acceleration. (pp. 631–34)

11 Using Figures 29.33 (p. 638) and 29.34 (p. 639), discuss the evolution of the vertebrate brain, pointing out the changes at each stage from fish through amphibian and reptile to mammal. In doing so, identify and give a major function of each of the following: cerebrum, cerebellum, medulla, thalamus, hypothalamus, optic lobe (mid-brain), and olfactory bulb. (pp. 638–41)

12 Using Figure 29.35 (p. 640), compare the areas of the cerebral cortex devoted to sensory and motor functions and to association in the cat, the monkey, and the human being. (pp. 639–40)

13 List one major function for each of the following parts of the mammalian forebrain: thalamus, reticular formation, hypothalamus, and limbic system. (pp. 638–41)

14 Explain how the two cerebral hemispheres of the human brain differ in function. Describe the functions of Wernicke's area and Broca's area in the brain. What symptoms occur in people who have lesions in these parts of the brain? (pp. 641–43)

15 Contrast short-term and long-term memory, and explain the role of the hippocampus in information storage and retrieval. (p. 644)

KEY TERMS

sensory transduction (p. 617)
chemoreceptor (p. 618)
photoreceptor (p. 618)
mechanoreceptor (p. 618)
thermoreceptor (p. 618)
electroreceptor (p. 618)
generator potential (p. 618)
taste bud (p. 620)
olfaction (p. 621)
eye cup (p. 622)
compound eye (p. 623)
ommatidium (p. 623)
camera eye (p. 624)
sclera (p. 624)
cornea (p. 624)
choroid (p. 624)
ciliary body (p. 624)
iris (p. 624)
pupil (p. 624)
lens (p. 625)
suspensory ligament (p. 625)
retina (p. 624)
refraction (p. 625)
accommodation (p. 626)
presbyopia (p. 626)
rod cell (p. 627)
cone cell (p. 627)
rhodopsin (p. 627)
kinocilium (p. 630)
lateral line system (p. 630)
outer ear (p. 631)
middle ear (p. 631)
inner ear (p. 631)
tympanic membrane (p. 631)
Eustachian tube (p. 631)
oval window (p. 631)
round window (p. 631)
semicircular canals (p. 631)
utriculus (p. 631)
sacculus (p. 631)
otoliths (p. 631)
statolith (p. 633)
cochlea (p. 634)
organ of Corti (p. 634)
tectorial membrane (p. 634)
cephalization (p. 638)
forebrain (p. 638)
midbrain (p. 638)
hindbrain (p. 638)
medulla oblongata (p. 638)
cerebellum (p. 638)
cerebrum (p. 638)
pons (p. 638)
optic lobes (p. 638)
cerebrum (p. 638)
thalamus (p. 638)
hypothalamus (p. 638)
cerebral cortex (p. 638)
neopallium (p. 639)
association area (p. 639)
reticular system (p. 640)
limbic system (p. 641)
corpus callosum (p. 641)
angular gyrus (p. 643)
Wernicke's area (p. 643)
Broca's area (p. 643)
short-term memory, or STM (p. 644)
long-term memory, or LTM (p. 644)
hippocampus (p. 644)
beta amyloid proteins (p. 644)

SUMMARY

Receptor cell function Specialized receptor cells are an animal's principal means of gaining information about its environment. They function as *transducers*; they convert the energy of a stimulus into the electrochemical energy of a nerve impulse. Each type of receptor is maximally responsive to just one kind of stimulus. Regardless of specialization, each type of receptor translates the stimulus into a change in membrane polarization, brought about by the opening or closing of specific gated channels in the membrane.

Stimulation of a sensory receptor produces a local depolarization *(generator potential)* of the receptor cell membrane. When the generator potential reaches threshold level, it triggers an action potential in the sensory neuron. Each receptor sends impulses to a particular part of the brain. It is the part of the brain to which the impulses go, not the stimulus, the receptor, or the message itself, that determines the quality and location of the sensation.

Touch The skin contains sensory receptors for touch, pressure, heat, cold, and pain. These receive information from the outside environment. Receptors inside the body receive information about the condition of the body itself.

Chemicals The receptors of taste and smell are *chemoreceptors*; they are sensitive to solutions of different kinds of chemicals, which can bind to them by weak bonds and thereby open sodium ions gates. The receptor cells for the four taste senses are located in taste buds on different areas of the tongue. The sensations we experience are produced by a blending of these four basic sensations. The receptor cells for smell are true neurons; they are located in the upper part of the nasal passages. These are specialized to detect odors from distant sources.

Light Almost all animals respond to light stimuli. Most multicellular animals have evolved specialized light receptor cells containing a pigment that undergoes a chemical change when exposed to light.

The light receptors of many invertebrates simply detect the presence of light, and perhaps differences in intensity. More complex eyes usually include a lens capable of concentrating light on the receptor cells. Lenses made possible the evolution of image-forming eyes.

There are two basic types of image-forming eyes—*camera-type eyes* (in some molluscs and vertebrates) and *compound eyes* (in insects and crustaceans). A compound eye is made up of many closely packed functional units called *ommatidia*, each of which acts as a separate receptor. Image formation depends on the light pattern falling on the surface of the compound eye. Since each ommatidium points in a slightly different direction, each is stimulated by light coming from different points. A camera-type eye uses a single-lens system to focus light on the many receptor cells that make up the *retina*.

In humans, the light rays coming into the eye are focused by the *cornea* and *lens* onto the light-sensitive *retina*, which contains the specialized receptor cells, the *rods* and *cones*. The sensitive rod cells function in dim light, the cones in bright light. The cones enable us to detect color.

Well-defined image vision depends on precise focusing of the incoming light on the retina by the cornea and lens. The shape of the lens is alterable; it makes possible adjustments in the focus (*accomodation*) depending on whether the object being viewed is close or distant.

The light-sensitive pigment in the rods (*rhodopsin*) is converted into a different form when struck by light and is regenerated in the dark. The pigment conversion leads to a change in membrane permeability and an impulse is generated in associated neurons. Cone vision is more complex; there are three types of cones, each containing a different pigment and each sensitive to different wavelengths of light.

Equilibrium Receptors of the senses of both equilibrium and hearing are *mechanoreceptors*—specialized hair cells. The upper portion of the inner ear consists of three *semicircular canals* and a large vestibule that connects them to the cochlea. The sensory hair cells lining the *utriculus* and *sacculus* (the two cavities of the vestibule) send information to the brain about the position of the head relative to gravity. The three semicircular canals are concerned with sensing changes in the speed or direction of rotation.

Sound Receptors for the sense of hearing are specialized for detection of vibrations. In humans, vibrations in the air pass down the *auditory canal* of the outer ear, strike the *tympanic membrane*, and cause it to vibrate. The vibrations are increased in force as they are transmitted by three small bones across the middle ear to the *oval window*. The resulting movement of the oval window produces movement of the fluid in the canals of the *cochlea*, causing the membrane on which the hair cells of the *organ of Corti* are located to move up and down and rub the hair cells against the overlying *tectorial membrane*. The stimulus to the hair cells is passed into the associated sensory neurons, which carry impulses to the auditory centers of the brain.

Evolution of the brain The brains of invertebrate animals are much smaller in relation to the size of their bodies than those of vertebrates, and their dominance over the rest of the nervous system is usually less pronounced. The brains of primitive vertebrates consist of three regions, the *forebrain, midbrain,* and *hindbrain*. These have been much modified in the evolution of the vertebrates.

Very early in its evolution the forebrain was divided into the *cerebrum* and the more posterior *thalamus* and *hypothalamus*. The midbrain became specialized as the *optic lobes*, and the hindbrain was modified to form the *medulla oblongata,* the *pons*, and the *cerebellum*. The hindbrain has changed little over the course of evolution, the medulla continues to be a control center for some autonomic and nervous pathways, and the cerebellum is still concerned with equilibrium and muscular coordination. The major evolutionary change has been the enormous increase in the size and relative importance of the cerebrum, with a corresponding decrease in the midbrain.

In certain advanced reptiles, a new area of the cerebral cortex, the *neocortex*, evolved. In mammals the neocortex expanded to cover the surface and dominate the other parts of the brain.

In certain advanced reptiles, a new area of the cerebral cortex, the *neopallium*, evolved. In mammals the neopallium expanded to cover the surface and dominate the other parts of the brain.

The mammalian brain The mammalian forebrain consists of the thalamus, hypothalamus, and cortex. The thalamus contains part of the important reticular system, a network of neurons that runs through the medulla, midbrain, and thalamus. The *reticular system* receives inputs from "wiretaps" on all incoming and outgoing brain communication channels. It activates the brain on receipt of stimuli and is an indispensable filter that lets only a few of the major sensory inputs reach the brain's higher centers.

Besides serving as a crucial link between the nervous and endocrine systems, the *hypothalamus* is also the control center for the visceral functions of the body and a major integrating region for emotional responses. The *limbic system*, a functionally related set of structures that form a ring around the anterior end of the brainstem, is also involved; both the limbic system and hypothalamus appear to form the neural basis for such emotions as rage, fear, aggression, motivation, and sexual behavior.

The proportion of the cerebral cortex taken up by purely motor and sensory areas is smaller in humans than in other animals; the association areas occupy the greatest proportion of the cortex. The area of the cortex devoted to each body part is proportional to the importance of that part's sensory or motor activities.

The right and left cerebral hemispheres differ in function. In almost all humans, the left hemisphere is specialized for verbal and analytical ability, whereas the right hemisphere is specialized for visual and spatial relations. For example, the language areas of the brain are located in discrete, well-defined areas in the left hemisphere. Wernicke's area has been identified as an area for both written and spoken language, while Broca's area has been identified as a site for grammatical refinement.

There are different types of memory in humans—short-term memory and long-term memory (permanent storage). The region of the brain known as the *hippocampus* is involved in both short-term memory and the conversion of short-term memory into long-term memory.

CONCEPT MAP

Map 29.1

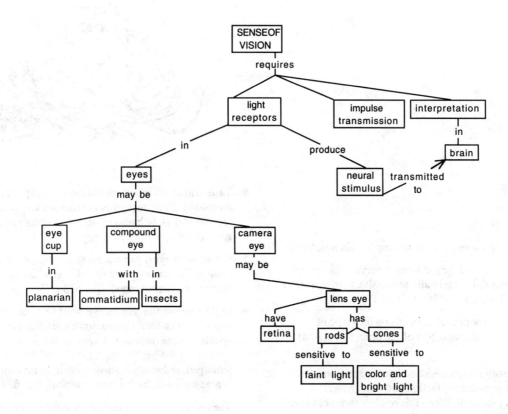

KEY DIAGRAM

The following coloring exercise is designed to help you learn the different parts of the brain and their relative sizes in the brains of different vertebrates. The diagrams below show the brains of the codfish, frog, alligator, and horse. Color the names and the appropriate structures on each of the diagrams, using a different color for each. (Fig. 29.34, p. 639)

CEREBRUM

OPTIC LOBE
(except on horse)

CEREBELLUM

MEDULLA

OLFACTORY BULB

OPTIC NERVE

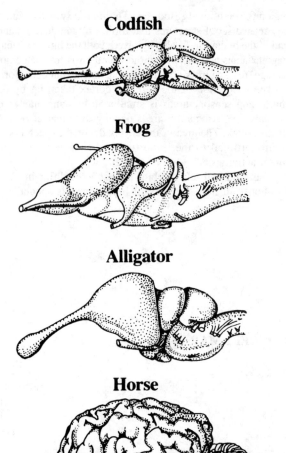

QUESTIONS

Testing recall

Select the correct term or terms to complete each statement.

1. Sensory receptors may be (portions of nerve cells, specialized cells); most are maximally responsive to (one, many) kind(s) of stimulus. (p. 617)

2. The sensation experienced depends on the (type of stimulus, neural message, part of the brain stimulated). (p. 618)

3. When a sensory receptor is stimulated, a small local depolarization called the (action potential, generator potential) is produced. When this reaches threshold level, it triggers a(n) (action potential, generator potential) in the nerve fiber. The (generator potential, action potential) increases in direct relation to the increase in intensity of the stimulus. (p. 618)

4. Taste and smell receptors are (touch receptors, chemoreceptors). Taste receptors are (neurons, specialized receptor cells), whereas olfactory receptors are (neurons, specialized receptor cells). (pp. 620–21)

5. (Compound eyes, camera eyes) are capable of forming images. Camera eyes are found in (molluscs, insects, crustaceans, vertebrates). (pp. 623–25)

6. Light entering the human eye must first pass through the transparent (sclera, cornea, iris); it then passes through an opening in the (sclera, iris, retina). When the circular muscles of the iris contract, the size of the pupil is (enlarged, reduced). Focusing the light rays on the retina is accomplished by the (lens, cornea). (pp. 624–25)

7. The rod cells are more numerous in the (center, periphery) of the retina and function in (dim, bright) light. They are responsible for (color, black-and-white) vision. (p. 627)

8 Within the retina there (is one set, are several sets) of synapses that enable the eye to modify information. (p. 628)

9 Focusing the light rays on the retina is accomplished by the (lens, cornea). When distant objects are viewed, the tension on the suspensory ligament is (high, relaxed) and the lens is (thicker, thinner). (p. 625)

10 Low-frequency sounds stimulate the hair cells near the (base, tip) of the cochlea. (p. 634)

11 The hair cells of the semicircular canals function in sensing (the position of the head, rotational acceleration). (p. 634)

12 Below are listed nine parts of the human body involved in hearing. Arrange them in the order in which they function in sound reception.

   a auditory centers in the brain
   b auditory canal
   c cochlear fluid
   d middle ear bones
   e organ of Corti
   f oval window
   g pinna
   h sensory neurons
   i tympanic membrane (pp. 631–37)

Match each function and description below with the associated part or parts of the brain.

   a cerebellum
   b cerebrum
   c hypothalamus
   d medulla oblongata
   e optic lobes
   f thalamus
   g reticular system

13 part of the forebrain (p. 638)

14 part of the midbrain (p. 638)

15 part of the hindbrain (p. 638)

16 a major sensory-integration center in lower vertebrates (p. 640)

17 filter for incoming sensory information (p. 640)

18 area concerned with balance, equilibrium, and muscular coordination (p. 638)

19 important in regulating the endocrine system (p. 641)

20 site of sensory cortex (p. 639)

21 vital centers for breathing and heart rate (p. 638)

22 associated with memory and learning (p. 644)

23 centers for thirst and hunger (p. 641)

24 concerned with smell in lower vertebrates (p. 638)

Testing knowledge and understanding

Choose the one best answer.

25 Skin has receptors for all the following *except*

   a pressure.   *d* pain.
   b temperature.   *e* UV light.
   c touch.   (pp. 618–20)

26 The sensory receptors for smell in the human nasal cavity are

   a closely associated with sensory neurons.
   b similar to light receptors.
   c chemoreceptors.
   d hair cells sensitive to mechanical distortion.
                                 (p. 621)

27 The light receptors of the human eye are located in the

   a choroid.   *d* cornea.
   b sclera.   *e* retina.
   c lens.   (p. 627)

28 The rods of the human eye

   a are concentrated in the center of the retina.
   b are sensitive to different colors of light.
   c function only in bright light.
   d contain rhodopsin.
   e are or do two of the above (p. 627)

29 Light entering the normal human eye is

   a bent by the cornea.
   b bent by the lens.
   c focused behind the eyeball.
   d focused on the pupil.
   e *a* and *b*.   (p. 625)

30 Refraction and accommodation are two fundamental aspects of image formation in the vertebrate eye. When an object being viewed is very close, which of the following performs the refraction necessary to focus the image on the retina?

   a a maximally stretched lens alone
   b a relaxed lens alone
   c the cornea, together with a maximally stretched lens
   d the cornea, together with a relaxed lens
   e the cornea alone   (pp. 625–26)

31 It is easier to see very dim objects (such as certain stars) by using peripheral vision because

   a images are not focused at night.
   b the rods are on the periphery of the retina.
   c small eye movements are necessary for vision.
   d the pupil enlarges at night.
   e the image is formed upside-down on the retina.
                                 (p. 627)

226 • CHAPTER 29

32 The photoreceptor molecule rhodopsin is

 a composed of a protein and a nonprotein prosthetic group called retinal.
 b a light-sensitive protein without a nonprotein prosthetic group.
 c a derivative of vitamin A that is light-sensitive.
 d a component of the cone cell membrane.
 e found only in insects. (p. 627)

33 Rod cells within the retina of a human eye

 a give rise to more sharply defined images than do cone cells.
 b are more sensitive in dim light than are cone cells.
 c are more abundant toward the center of the retina.
 d connect directly to the optic region of the brain with a single intervening interneuron.
 e and or do two of the above. (p. 627)

Below is a diagram of the inner ear. Use the letters on the diagram to answer the following five questions. More than one answer may be correct.

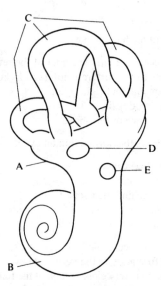

34 senses a change in the direction of motion (p. 634)

35 contains the organ of Corti (p. 634)

36 connects to stapes (p. 634)

37 contains sensory hair cells (pp. 631–34)

38 sense receptors for hearing located here (p. 634)

39 The sensory cells for balance and equilibrium are located in the

 a cochlea.
 b cerebellum.
 c semicircular canals.
 d sacculus and utriculus.
 e organ of Corti
 d sacculus and utriculus.
 e organ of Corti. (p. 631)

40 Which one of the following groups consists of structures all derived from the forebrain?

 a cerebral cortex, thalamus, cerebellum
 b hypothalamus, cerebellum, medulla
 c midbrain, cerebrum, hypothalamus
 d thalamus, hypothalamus, cerebral cortex
 e cerebral cortex, medulla, cerebellum (p. 638)

41 Over the course of vertebrate evolution the position of the chief control center has shifted from the

 a hindbrain to the midbrain.
 b midbrain to the forebrain.
 c cerebellum to the cerebrum.
 d cerebrum to the cerebellum.
 e hindbrain to the cerebrum. (p. 638)

42 Function maps of the surface of the human cerebral cortex reveal a particular relationship between the sensory and motor capabilities of various body parts and the size of the areas devoted to them. In general, which one of the following areas of the cortex surface would you expect to be the smallest?

 a sensory area for the lips
 b sensory area for the hands
 c sensory area for the tongue
 d motor area for the muscles of the fingers
 e motor area for the muscles of the ankle (pp. 641–43)

43 A portion of the brain especially concerned with filtering incoming sensory information and allowing only some of it to reach centers of consciousness is the

 a reticular system.   d cerebellum.
 b hypothalamus.   e cerebrum.
 c medulla.   (p. 640)

44 Which one of the following portions of the brain is incorrectly matched with a function?

 a reticular system—filtering incoming stimuli
 b medulla—control of breathing
 c hypothalamus—control of emotions
 d frontal lobes of the cerebrum—hearing
 e cerebellum—motor coordination (pp. 639–43)

45 The area of the brain that acts as the center for behavioral drives in higher vertebrates is the

 a cerebral cortex.   d midbrain.
 b hypothalamus.   e thalamus.
 c cerebellum.   (p. 641)

46 A person who has a stroke affecting the visual area on the left side of the visual cortex may be rendered

 a completely blind.
 b unable to see on the left side.
 c unable to see on the right side.
 d unable to interpret visual information from the left side. (p. 642)

47 A group of brain areas located in the brainstem that are related functionally in giving rise to feeling and emotions is known as the

   a reticular formation.   d hypothalamus.
   b limbic system.   e midbrain.
   c neocortex.   (p. 641)

48 The part of the human brain responsible for both spoken and written language is

   a Wernicke's area.   d hypothalamus.
   b Broca's area.   e medulla.
   c reticular system   (p. 643)

For further thought

1 Distinguish between the sense of smell and the sense of taste. How does the sense of smell contribute to the sense of taste?

2 When you walk into a dark room from bright sunlight, you can see very little. However, after some time your vision returns. Explain the initial lack of vision and why the vision returns.

3 A person in an automobile accident sustained severe head injuries. A brain scan showed damage to the central portion of the cerebrum on the right side. What symptoms would you expect this person to show?

4 Exposure to loud noises, such as from an overly loud Walkman or an overly amplified guitar, causes the basilar membrane to move violently, and shear off some of the projecting hairs of the hair cells of the organ of Corti. The hairs cannot regenerate. The movement of the membrane is most violent at the base of the cochlea. What would be the effect on the sense of hearing?

ANSWERS

Testing recall

1   portions of nerve cells, specialized cells, one
2   part of brain stimulated
3   generator potential, action potential, generator potential
4   chemoreceptors, specialized receptor cells, neurons
5   compound eyes, camera eyes, molluscs, vertebrates, pinhole, dim, little
6   cornea, iris, reduced, lens, cornea
7   periphery, dim, black-and-white
8   are several sets
9   lens, cornea, high, thinner
10 tip
11 rotational acceleration
12 *g, b, i, d, f, c, e, h, a*
13 *b, c, f, g*
14 *e, g*
15 *a, d, g*

16	*f*	19	*c*	21	*d*	23	*c*
17	*g*	20	*b*	22	*b*	24	*b*
18	*a*						

Testing knowledge and understanding

25	*e*	31	*b*	37	*A, B, C*	43	*a*
26	*c*	32	*a*	38	*B*	44	*d*
27	*e*	33	*b*	39	*d*	45	*b*
28	*d*	34	*C*	40	*d*	46	*c*
39	*e*	35	*B*	41	*b*	47	*b*
30	*d*	36	*D*	42	*e*	48	*a*

Chapter 30

EFFECTORS AND ANIMAL LOCOMOTION

A GENERAL GUIDE TO THE READING

This is the final chapter of the six that deal with the reaction of organisms to stimuli. The focus of this chapter is on the role of muscles and skeletons in producing movement. Here, as in other chapters, a variety of animals are discussed, but special attention is given to vertebrates: in this instance to the human skeleton and muscles. As you read Chapter 30 in your text, you will want to concentrate on the following topics.

1. Skeletons. Many students fail to recognize the importance of a skeleton in effecting movement. Muscles can contract only if they have a firm base or skeleton against which they can pull. Try to keep this in mind as you read pages 650 to 653, which describe the three basic skeletal types.

2. Vertebrate muscle. The section on the three types of vertebrate muscle (pp. 654–59) should be studied carefully. You will want to pay particular attention to the material on skeletal muscle, since this is prerequisite to understanding the later discussion of the mechanisms of skeletal muscle contraction.

3. Muscle contraction. The sections that discuss the molecular basis of contraction and its control (pp. 659–66), are difficult, and you may need to read them over several times. Pay close attention to the figures that accompany the text and to Table 30.3 (p. 666); these are an essential part of the material, and will help you learn it.

KEY CONCEPTS

1. The underlying mechanism of effectors of motion depends on either microfilaments or microtubules; for example, muscular movement is produced by the action of microfilaments. (p. 650)

2. The principal effectors of movement in higher animals are the muscle cells—elongated cells that are specialized for contraction. (p. 650)

3. A skeleton plays an important role in effecting movement; it provides the mechanical resistance against which the muscles can act. (pp. 650–53)

4. Individual muscle fibers resemble individual nerve cells in that they fire only if a stimulus is of threshold intensity, duration, or rate. (pp. 656–57)

5. Muscle contraction involves a ratchetlike mechanism in which actin and myosin microfilaments slide past one another. ATP provides the energy for this movement. (pp. 660–64)

6. The stimulatory transmitter substance acetylcholine released by the neuronal axon causes a depolarization of the muscle membrane. If the depolarization is above threshold, an action potential is triggered over the surface of the fiber by the same combination of voltage-gated channels that is seen in nerve-cell membranes. The action potential activates the contraction process propagating the movement of Ca^{++} ions. (pp. 664–66)

CROSSLINKING CONCEPTS

1. Your study of microfilaments in Chapter 6 gave you a general understanding of actin and myosin; this prepared you for the detailed information about muscle function given in this chapter.

2. Antagonistic control is a concept that was first introduced in the study of endocrine system hormones and was returned to in the study of nervous system control. In this chapter, with the study of paired muscles, you again see its importance.

3. In Chapter 28, you were introduced to the somatic and autonomic nervous systems. Skeletal and smooth muscles are stimulated separately and distinctly by these two systems.

4. The role played by ATP and the concept of oxygen debt will take on more meaning if you acquired a solid understanding of aerobic respiration from Chapter 8.

5. The importance of calcium has been seen in its role in mineral nutrition (Chapter 17), its regulation by parathyroid hormone and calcitonin (Chapter 25), and its role in nerve-impulse transmission across synapses (Chapter 28).

OBJECTIVES

1. Describe the locomotion characteristic of an earthworm. In doing so, discuss the roles of the longitudinal and the circular muscles in effecting movement, and explain the importance of the hydrostatic skeleton. Finally, show how segmentation is an advantage to an animal like an earthworm. (p. 651)

2. Distinguish between an exoskeleton and an endoskeleton, and list the advantages and limitations of each. Using Figure 30.4 (p. 652), explain how an appendage can be flexed by an insect and by a human. (pp. 652–53)

3. Using Figure 30.5 (p. 653), point out a Haversian system, a Haversian canal, the spaces where the bone cells are located, and the intercellular matrix. (p. 653)

4. Give the difference between the two terms in each of the following pairs: spongy bone—compact bone, appendicular skeleton—axial skeleton, and tendon—ligament. (pp. 653–54)

5. Using Figure 30.7 (p. 656), identify smooth, skeletal (striated), and cardiac muscle. Contrast the three types of muscle with respect to the shape of the cell, the presence or absence of multiple nuclei in a cell, the presence or absence of striations, and the source of innervation (the somatic or the autonomic nervous system). (pp. 654–56)

6. For each of the following tissues, indicate whether the muscle cells are predominantly striated or smooth: iris of the eye, wall of an artery, leg muscle, abdominal muscle, tongue, wall of the small intestine, face muscle, and wall of the esophagus. (pp. 654–56)

7. Using 30.12 (p. 659) explain the terms simple twitch, tetanus, summation, and fatigue; compare the extent of shortening of a muscle undergoing simple twitches with that of a muscle in tetanus. Give the causes of fatigue in a muscle. (p. 659)

8. Using Figures 30.14 (p. 661) and 30.15 (p. 662), identify a sarcomere, a bundle of muscle fibers, a myofibril, a Z line, an I band, an A band, an H zone, a thick filament, and a thin filament. (pp. 661–62)

9. Explain the sliding-filament theory of skeletal muscle contraction. In doing so, indicate the contribution to muscular contraction of each of the following: actin filament, myosin filament, myosin heads, regulatory proteins, Ca^{++}, ATP, and creatine phosphate. You will find the examination of Figures 30.16 to 30.22 (pp. 663–66) helpful in meeting this objective. (pp. 662–66)

10. Using Figure 30.20 (p. 664), identify a sarcomere, thick filaments, thin filaments, transverse tubules, the sarcoplasmic reticulum, and the cell membrane. Describe the relationship between transverse tubules and the cell membrane of a muscle cell, and explain the role of the sarcoplasmic reticulum with respect to calcium ions and muscle contraction. (pp. 664–66)

11. Outline the events occurring between the time a nerve impulse reaches a neuromuscular junction and the time the muscle fiber contracts. You will find Table 3 (p. 666) useful in meeting this objective. (pp. 664–66)

KEY TERMS

effector (p. 649)
hydrostatic skeleton (p. 651)
segmentation (p. 651)
setae (p. 651)
exoskeleton (p. 652)
endoskeleton (p. 652)
spongy bone (p. 653)
compact bone (p. 653)
Haversian system (p. 653)
Haversian canal (p. 653)
canaliculi (p. 653)
osteoporosis (p. 653)
axial skeleton (p. 654)
appendicular skeleton (p. 654)

ligament (p. 654)
tendon (p. 654)
skeletal (voluntary, or striated) muscle (p. 654)
striations (p. 655)
smooth muscle (p. 655)
cardiac muscle (p. 656)
simple twitch (p. 658)
summation (p. 659)
tetanus (p. 659)
fatigue (p. 659)
isotonic contraction (p. 659)
isometric contraction (p. 659)
tonus (p. 660)
creatine phosphate (p. 660)
myoglobin (p. 660)
oxygen debt (p. 660)
myofibril (p. 661)
I band (p. 662)
A band (p. 662)
H zone (p. 662)
Z line (p. 662)
sarcomere (p. 662)
thick filament (p. 662)
thin filament (p. 662)
actin (p. 662)
myosin (p. 662)
sliding-filament theory (p. 662)
sarcoplasmic reticulum (p. 664)
T system (p. 664)
regulatory proteins (p. 665)

SUMMARY

Although plants do move by differential growth or turgor changes, the most elaborate mechanisms for producing movement are found among animals. Movement in animals depends on either microtubules or microfilaments. The *effectors* are the parts of the organism that do things, that carry out the organism's response to a stimulus. The principal effectors of movement in higher animals are the muscle cells. Muscles require some type of skeleton to act as a mechanical support.

Hydrostatic skeleton Movement in many invertebrate animals is produced by the alternating contraction of the longitudinal and circular muscles against the incompressible fluids in the body cavity. The fluids function as a *hydrostatic skeleton*. The hydrostatic skeleton of the earthworm is particularly efficient; the body cavity is segmented, and each segment has its own muscles. Consequently, the worm is capable of localized movement, and burrows very effectively.

Hard, jointed skeleton Both arthropods and vertebrates have evolved paired locomotory appendages, a hard, jointed skeleton, and an elaborate musculature. Arthropods have an *exoskeleton*, a hard body covering with all muscles and organs inside it. Besides providing support, the exoskeleton functions as protective armor and as a waxy barrier to prevent water loss. Periodic molting of the exoskeleton is necessary for growth. An exoskeleton limits the ultimate size of an organism.

Vertebrates have an *endoskeleton*, an internal framework with the muscles outside. It is composed of bone and/or cartilage. *Cartilage* is firm, but not as hard or brittle as bone. It is found wherever firmness combined with flexibility is needed. *Bone* has a hard, relatively rigid matrix and provides structural support and protection.

Vertebrate skeletons are divided into two components, the *axial* skeleton and the *appendicular* skeleton. Some bones are connected by immovable joints; others are held together at movable joints by *ligaments*. Skeletal muscles, attached to the bones by *tendons*, contract and bend the skeleton at these joints. The movable bones behave as a lever system with the fulcrum at the joint. The action of any specific muscle depends on the exact positions of its attachments and on the type of joint between them. The muscles operate in antagonistic and synergistic groups.

Muscle types Vertebrates possess three different types of muscle: skeletal, smooth, and cardiac. The abundant *skeletal* muscle is responsible for most voluntary movement. Each muscle fiber is long and cylindrical, contains many nuclei, and is crossed by light and dark bands called *striations*. The fibers are usually bound together into bundles, and the bundles into muscles. Skeletal muscle is innervated by the somatic nervous system.

Smooth muscle forms the muscle layers in the walls of the viscera and the blood vessels. The spindle-shaped cells interlace to form sheets of tissue that are innervated by the autonomic nervous system. Smooth muscle is primarily responsible for movements in response to internal changes, whereas skeletal muscle is concerned with making adjustments to the external environment. These differences in action are reflected in many differences in the physiological characteristics of these two muscle types.

Cardiac muscle is found only in the heart. It shows some characteristics of skeletal muscle and some of smooth muscle. Its fibers are striated, but it is innervated by the autonomic nervous system and acts as smooth muscle.

The physiology of skeletal muscle activity Individual skeletal-muscle fibers contract only if they receive a stimulus of threshold intensity and duration. In vertebrates, individual muscle fibers seem to exhibit the all-or-none property, but muscles give a graded response. When a single threshold stimulus is applied to a muscle, a *simple twitch* occurs. When frequent stimuli are applied to a muscle, the muscle does not have time to relax between contractions and the contractions add together (*summation*).. If the stimuli are very frequent, the muscle may not relax at all between successive stimulations; the resulting strong sustained reaction is called *tetanus*. If the frequent stimulation continues, the muscle may fatigue and be unable to sustain the contraction.

Molecular basis of contraction The energy for muscle contraction comes from ATP. *Creatine phosphate* are, which is stored in muscle, can be used to generate ATP for immediate contraction. For continued contraction, ATP must be produced from the complete oxidation of glucose and/or fatty acids to

CO_2 and water. During strenuous muscular activity, the energy demands are greater than can be met by respiration, because oxygen cannot be gotten to the tissues fast enough. Consequently, lactic acid fermentation takes place, and the muscle incurs an *oxygen debt*. The panting that an animal exhibits, even after activity is over, supplies the oxygen needed to reconvert the lactic acid into pyruvic acid and to metabolize it.

Analysis of muscle shows that its contractile components are the proteins *actin* and *myosin*. Within each *myofibril* the thick myosin filaments and thin actin filaments are interdigitated, with the myosin located exclusively in the A bands, and the actin primarily in the I bands but extending some distance into the A bands. The contractile unit (between the Z lines) is called the *sarcomere*. According to the sliding-filament theory, cross bridges from the myosin filaments hook onto the actin at specialized receptor sites, and bend and pull the actin along the myosin. The necessary energy comes from the hydrolysis of ATP by the myosin cross-bridges. As the filaments slide past each other, the width of the H zone and I band decreases.

Electrochemical control of contraction The membrane of the resting muscle fiber is polarized, with the outside charged positively in relation to the inside. Acetylcholine molecules released by the neuronal axon reduce this polarization, and so trigger an action potential that sweeps across the muscle fiber. When the action potential penetrates into the interior of the fiber via the *T system*, calcium ions are released by the *sarcoplasmic reticulum*. The Ca^{++} ions stimulate contraction by binding to regulatory proteins on the actin and causing a conformational change, which exposes the myosin-binding sites of actin. Myosin heads bind to actin and bend in a power stroke, and the filaments to slide past one another.

The cycle of attachment, power stroke, and recovery flip can continue as long as there is nervous stimulation (leading to release of calcium ions) and sufficient ATP (to detach the myosin heads from actin and recock them).

CONCEPT MAP

Map 30.1

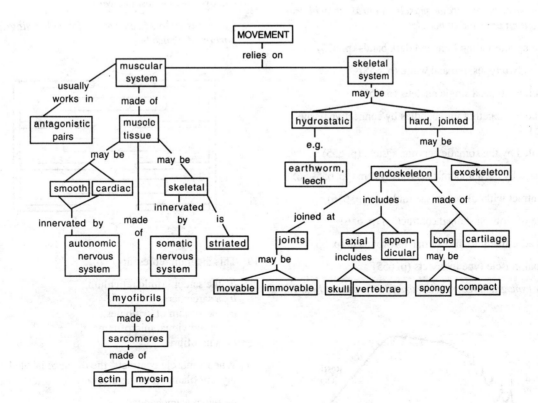

QUESTIONS

Testing recall

The next two questions refer to the two bones in the drawing.

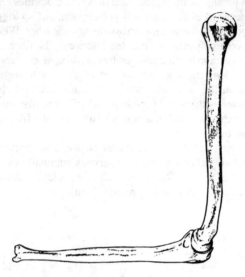

1. Draw in and label ligaments, two antagonistic muscles (showing points of attachment to the bones), and tendons. (p. 652)

2. Do these bones form part of an endoskeleton, exoskeleton, or hydrostatic skeleton? (p. 652)

For items 3 to 12, indicate whether the phrase is true of skeletal, smooth, or cardiac muscle. A phrase may be true of more than one kind of muscle.

3. crossed by alternating light and dark bands (p. 655)
4. occurs in two types, red and white (p. 655)
5. each cell contains a single nucleus (p. 655)
6. fibers bound together into bundles by connective tissue (p. 655)
7. innervated by the somatic nervous system (p. 655)
8. innervated by the autonomic nervous system (pp. 655–56)
9. can contract without nervous stimulation (p. 656)
10. capable of slow, sustained contraction (p. 656)
11. effects adjustment to the external environment (p. 654)
12. principal muscle type in insects (p. 655)

Use the kymograph record below to answer questions 13 to 16.

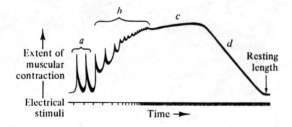

13. Referring to the diagram, state what is happening to the muscle in areas *a*, *b*, *c*, and *d*. (p. 659)
14. Why do several stimuli in rapid succession produce more contraction than a single stimulus of the same strength? (p. 659)
15. Are you using simple twitches or tetanic contractions when you write the answer to this question? (p. 659)
16. What causes fatigue? (p. 659)
17. Which of the following substances or processes are used by the cell to generate ATP for muscle contraction? (More than one may be correct.)

    a glycolysis   *d* creatine phosphate
    b Krebs cycle   *e* myosin
    c cell respiration   (pp. 660–61)

18. Place the following seven events in the contraction of a muscle fiber in order.

    a Transmitter substance diffuses across synaptic cleft at neuromuscular junction.
    b Actin and myosin slide past each other.
    c Permeability of the muscle membrane is altered.
    d Impulse arrives at the junction between the motor neuron and the muscle fiber.
    e Sarcoplasmic reticulum releases stored calcium ions.
    f Action potential travels along membrane of the muscle cells and into the tubules.
    g Calcium ions bind to regulatory proteins (pp. 664–66)

Choose the one best answer.

Questions 19 to 24 refer to the following diagram of a portion of a muscle.

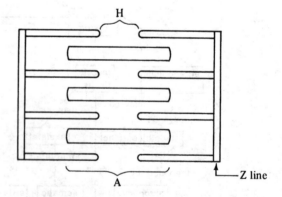

19. This diagram represents

    a the sarcoplasmic reticulum.
    b a sarcomere.
    c the system of T tubules.
    d an individual microfilament.
    e a myofibril. (p. 662)

20. When a muscle contracts, the distance labeled A (the myosin filament) in the diagram

    a remains unchanged.
    b decreases.
    c increases. (p. 663)

21 When a muscle contracts, the distance labeled H in the diagram

   a remains unchanged. *b* decreases. *c* increases.
   (p. 663)

22 What would be the main substance remaining if you dissolved away the thin filaments?

   a actin   *b* myosin   (p. 662)

23 What would be the main substance remaining if you dissolved away the thick filaments?

   a actin   *b* myosin   (p. 662)

24 The energy for the movement of the cross-bridges is supplied by

   a ATP.
   b calcium ions.
   c lactic acid.
   d actin.
   e myosin.   (p. 660)

Testing knowledge and understanding

Choose the one best answer.

25 Which of the following statements concerning the skeletal system is *false*?

   a Some bones are held together by ligaments.
   b Bones help protect some internal organs.
   c Bones contain both nerve and blood cells.
   d Aside from nerve and blood cells there are no living cells in bone.
   e Many muscles are attached to bone.   (p. 653)

26 An important advantage of an endoskeleton over an exoskeleton is that the former

   a provides more structural support.
   b provides a structure against which muscles can work.
   c helps prevent water loss.
   d does not impose a severe limitation on overall size.
   e helps determine the shape of the animal.   (p. 652)

27 Vertebrate skeletal muscle is usually

   a under control of the sympathetic nervous system.
   b under control of the parasympathetic nervous system.
   c under control of the somatic (voluntary) nervous system.
   d dually innervated by sympathetic and parasympathetic neurons.
   e dually innervated by somatic and parasympathetic neurons.   (p. 655)

28 If a single excised frog muscle is stimulated electrically and the responses are recorded, which one of the following *cannot* be demonstrated?

   a threshold
   b tetanus
   c summation
   d fatigue
   e antagonism   (pp. 658–59)

29 A sustained muscular contraction is

   a fatigue.
   b a twitch.
   c tetanus.
   d oxygen debt.
   e the latent period.   (p. 659)

30 Muscular fatigue may occur with an accumulation of

   a ATP.
   b lactic acid.
   c oxyhemoglobin.
   d myoglobin.
   e glucose.   (p. 659)

31 Regeneration of ATP in the muscle cell utilizes stores of

   a myosin.
   b actin.
   c creatine phosphate.
   d NAD.
   e pyruvic acid.   (p. 660)

32 The energy levels in the muscle are directly restored after muscle contraction by

   a cell respiration.
   b lactic acid formation.
   c ADP hydrolysis.
   d protein breakdown.
   e oxygen entry.   (p. 660)

33 A muscle has built up an oxygen debt. When there is enough oxygen for aerobic respiration to resume, all the following will occur *except* that

   a lactic acid will be converted to pyruvic acid.
   b O_2 will be used up.
   c acetyl-CoA will be converted to CO_2 and H_2O.
   d pyruvic acid will be converted to acetyl-CoA.
   e an excess of NADH will accumulate.   (pp. 660–61)

34 Once an impulse is transmitted across the neuromuscular junction by acetylcholine, continued transmission due to the same impulse is prevented by the breakdown of acetylcholine by

   a ATPase.
   b phosphatase.
   c acetylcholinesterase.
   d the sodium-potassium pump.
   e acetylase.
   (see Chapter 28, p. 603)

35 The function of the T tubules in muscle contraction is to

   a carry the impulse into the myofibrils of the muscle cell.
   b release calcium ions.
   c release sodium ions.
   d split ATP.
   e take up calcium ions.   (pp. 664–65)

36 At the start of a muscle contraction, calcium ions are released from

   a actin.
   b the T tubule.
   c the motor neuron.
   d the sarcoplasmic reticulum.
   e myosin   (p. 664)

37 The role of calcium ions in muscle contraction is to

   a facilitate the binding of ATP.
   b trigger depolarization of the membrane of muscle fibers.
   c bind to a regulatory protein associated with actin, and so allow cross-bridges with myosin to form.
   d promote release of vesicles containing transmitter molecules.   (p. 665)

38 The reason that an A band within a sarcomere appears darker than adjacent I bands is that

   a myofibrils are narrower in diameter at I bands.
   b multiple nuclei tend to cluster within the A bands.
   c the cell membrane is more opaque near A bands.
   d A bands contain both actin and myosin, whereas I bands do not. (p. 662)

39 During muscle contraction one molecule of ATP is hydrolyzed each time

   a a calcium ion is moved out of the sarcoplasmic reticulum into the cytoplasm.
   b myosin binds to regulatory proteins.
   c actin binds to regulatory proteins.
   d a muscle fiber contracts by 30 percent of its length.
   e a cross-bridge between an actin molecule and a myosin molecule is made and broken. (p. 665)

40 Which of the following chemicals are necessary to sustain a muscle contraction?

   a actin, myosin, calcium ions
   b actin, myosin, ADP
   c actin, ATP, myoglobin
   d myosin, calcium phosphate, ATP
   e myosin, calcium ions, myoglobin (p. 665)

41 Which of the following statements is *false*?

   a In skeletal muscles, there is a regular alternation of light-colored I bands and dark-colored A bands.
   b It is now thought that the A bands correspond to the lengths of thick myosin filaments.
   c When a muscle contracts, the A bands become much shorter, whereas the I bands remain roughly the same length.
   d It is thought that cross-bridges from the myosin filaments, acting like ratchets, provide the pull that slides the filaments together during contraction.
   e The direct stimulant for contraction is thought to be release of intracellular calcium from cisternae of the sarcoplasmic reticulum. (p. 663)

42 When a skeletal muscle contracts, all the following occur except

   a the H zone almost disappears.
   b the Z lines move closer together.
   c the I bands diminish markedly in width.
   d ATP is hydrolyzed to ADP + Pi.
   e the A bands diminish markedly in width. (p. 662)

For further thought

1 The earthworm has exploited the hydrostatic skeleton to its fullest. How might the earthworm's movements differ if the worm were not segmented?

2 Why does a person pant and sweat after vigorous exercise?

3 Explain what happens when rigor mortis sets in after death.

ANSWERS

Testing recall

1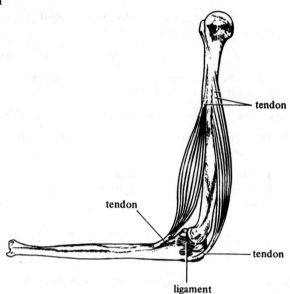

2 endoskeleton
3 skeletal, cardiac
4 skeletal
5 smooth, cardiac
6 skeletal
7 skeletal
8 smooth, cardiac
9 smooth, cardiac
10 smooth
11 skeletal
12 skeletal

13 a two single twitches: each with latent period, contraction, relaxation
    b summation: muscle does not fully relax between contractions
    c tetanus: muscle does not relax at all between contractions; slow, sustained contraction
    d fatigue: the result of the buildup of lactic acid

14 The muscle does not fully relax between contractions.
15 tetanic contractions
16 buildup of lactic acid
17 a, b, c, d
18 d, a, c, f, e, g, b
19 b   21 b   23 a
20 a   22 b   24 a

Testing knowledge and understanding

25 d   30 b   35 a   39 e
26 d   31 c   36 d   40 a
27 c   32 a   37 c   41 c
28 e   33 e   38 d   42 e
29 c   34 c

Chapter 31

EVOLUTION: ADAPTATION

This chapter and the next provide a detailed discussion of evolution, a theme first introduced in Chapter 1 and woven through many subsequent chapters. You may therefore find it helpful to begin your study of evolution by reviewing pages 7 to 13, with particular attention to the summary on page 13. The subject matter in Chapters 31 and 32 is easily understood and very interesting. As you read this chapter in your text, you will want to concentrate on the following topics.

1. Sources of genetic variation. Pay careful attention to the discussion of this topic (pp. 670–71). Make sure you understand that evolution can act only through variation that is inherited. These heritable variations come from mutations and from recombination of existing alleles. In eucaryotes, much major evolutionary change is based on gene duplication followed by the independent evolution of the spare copies of the gene. Also, learn the difference between the theory of evolution by natural selection and the theory of evolution by the inheritance of acquired characteristics.

2. The gene pool. The essential point on pages 675 to 676 is that evolution involves the gene pool of a population. A population can evolve; an individual cannot. The term gene pool refers to the total of all the genes in a population.

3. The Hardy-Weinberg Law. This important law (p. 675) sets up the conditions under which evolution would not occur. The fact that these conditions cannot all be met provides an indirect demonstration that all populations are evolving. Pay careful attention to the five conditions for genetic equilibrium, and study the forces that drive evolution when these conditions are not met. Notice, too, that the law can be used not only to predict the allelic and genotypic frequencies in successive generations in a population that is not evolving but also to calculate allelic frequencies when only phenotypic frequencies can be measured directly. Details on how this is done are presented in the box on page 677, and a study guide for solving population-genetics problems is included in this chapter of the *Study Guide*.

4. The role of natural selection. Although the text concentrates on natural selection as being the most important force in bringing about evolution, it is important for you to realize that evolution does not depend exclusively on any one mechanism. For perspective, read carefully the three paragraphs on giraffes on page 674 before reading pages 680 to 687. There will be more discussion of this topic in Chapter 32. There are a wealth of examples provided in this section that will help clarify the concept of natural selection for you.

5. Sexual selection. Sexual selection operates on traits used exclusively to attract and keep mates. Be sure to distinguish sexual selection from natural selection. (p. 687)

6. Adaptation. Begin your study of this topic by learning the exact meanings of "adaptation" and "fitness" in evolutionary biology; the definitions of these terms on page 688 are probably different from those you have learned previously. The text (pp. 687–99) provides a host of examples of different adaptations. As you read about them, try to see how each increases fitness.

KEY CONCEPTS

1. The modern theory of evolution is based on five concepts: (1) more offspring are produced than the habitat can support,

(2) the characteristics of living things differ between individuals of the same species, (3) many differences are the result of heritable genetic differences, (4) some differences affect how well adapted an organism is, and (5) some differences in adaptedness are reflected in the number of offspring successfully reared.

2 The evolutionary raw material is genetic variation in a population. Mutations produce new alleles, which are sources of variation on which natural selection may act. Once a variety of alleles is in existence, recombination through sexual reproduction and crossing over during meiosis are the primary sources of genetic variation on which natural selection may act. Natural selection can act on genetic variation only when it is expressed as phenotypic variation. (pp. 670–74)

3 Eucaryotic gene evolution depends in part on gene duplication followed by changes in base sequence that give rise to functionally different proteins. (p. 670–73)

4 Evolutionary change means change in allelic frequencies (and hence in genotypic ratios) in populations of organisms, not in individuals. (pp. 675–76)

5 By natural selection, we mean nonrandom reproduction, or, more specifically, reproduction that is to some degree correlated with genotype. (p. 679)

6 Evolutionary change is not automatic; it occurs only when something disturbs the genetic equilibrium. Mutation pressure and selection pressure are always disturbing this equilibrium, and migration and genetic drift may also do so. (pp. 676–79)

7 Many characteristics are both advantageous and disadvantageous; the evolutionary fate of such characteristics depends on the algebraic sum of the separate selection pressures. (pp. 685–86)

8 Sexual selection affects traits used exclusively to attract and keep mates. The two principal forms of sexual selection involve (1) contests between members of one sex (usually males) for access to the other and (2) male features that function only to attract females (female choice). (pp. 686–87)

9 Fitness refers to an individual's (or allele's or genotype's) probable genetic contribution to succeeding generations. Adaptations are genetically determined characteristics that increase an organism's fitness. (p. 688)

10 An adaptation need not be 100 percent effective to give the individuals possessing it significantly greater chances of surviving to reproduce. (pp. 687–88)

OBJECTIVES

1 Discuss sexual recombination as a major source of variation upon which natural selection may act. (p. 670)

2 Explain the role of gene duplication in the evolution of eucaryotic organisms. In doing so, cite examples of gene duplication and describe three ways in which it can occur. (pp. 670–73)

3 Explain why changes that affect only somatic cells, such as variations produced by practice, education, diet, or medical treatment, cannot bring about evolution. Contrast Lamarck's theory of evolution by the inheritance of acquired characteristics with Darwin's and Wallace's theory of evolution by natural selection. (p. 674)

4 Explain the concept of the gene pool. Given the frequency of two alleles, calculate the ratios of the genotypes produced by them, using a Punnett square or the algebraic formula given on page 677. Your instructor may pose additional problems for you to solve. (pp. 675–78)

5 State the Hardy-Weinberg Law, and discuss its five conditions for the maintenance of genetic equilibrium, describing the forces that prevent these conditions from being fulfilled. In doing so, use the terms mutational equilibrium, genetic drift, mutation pressure, gene flow, selection pressure, and random reproduction. (pp. 675–80)

6 Explain how natural selection on a phenotype in one generation can affect the genotype of the next generation. Be able to calculate how changes in allelic frequencies in the parental generation can alter allelic frequencies in the offspring. (Fig. 31.9, p. 681, may be helpful. See also the "Study Guide for Population-Genetics Problems" in this chapter.) (pp. 680–81)

7 Contrast the roles of selection and mutation in directing evolutionary change. Indicate other factors that may be important in evolution. (p. 680)

8 Using Figure 31.13 (p. 684), contrast directional selection, stabilizing selection, and disruptive selection. Indicate which type of selection is involved in the experiment shown in Figure 31.12 (p. 683). (pp. 683–85)

9 Using an example, explain how a characteristic can have both positive and negative effects, and indicate what determines whether or not a trait will increase or decrease in the population. Include a discussion of heterozygote surperiority, and show how this condition affects the fate of certain alleles. (pp. 685–88)

10 Distinguish between selection that affects physical survival and sexual selection. Name, describe, and give an example of two types of sexual selection. (pp. 686–87)

11 Define adaptation and fitness in their evolutionary sense, and explain, using examples, how such phenomena as flower structure, cryptic appearance, warning coloration, and mimicry are adaptive. Distinguish between Batesian and Müllerian mimicry and give an example of each. (pp. 688–95)

12 List the three types of symbiotic interactions, and show how they differ from one another. State whether their differences indicate distinct classes or a continuum of differences. Distinguish between external and internal parasites, and discuss two adaptations of internal parasites. Explain why most well-adapted parasites do not generally kill or seriously harm their host species. (pp. 695–99)

KEY TERMS

phyletic evolution (p. 669)
branching evolution (p. 669)
population (p. 669)
recombination (p. 670)
gene duplication (p. 670)
evolution by natural selection (p. 674)
evolution by the inheritance of acquired characteristics (p. 674)
population genetics (p. 675)
Müllerian mimicry (p. 694)
symbiosis (p. 695)
commensalism (p. 695)
mutualism (p. 695)
parasitism (p. 695)
parasitoid (p. 697)
degenerate (p. 697)
specialization (p. 697)
red-queen model (p. 698)

SUMMARY

Modern evolutionary theory is based on five concepts: (1) more offspring are produced than the habitat can support, (2) the characteristics of living things differ between individuals of the same species, (3) many differences are the result of heritable genetic differences, (4) some differences affect how well adapted an organism is, and (5) some differences in adaptedness are reflected in the number of offspring successfully reared. Evolution is the change in the genetic makeup of a population in successive generations. There are always variations among members of a population; if there is selection against certain variants and for other variants, the overall makeup of the population may change with time.

Variation New genes may arise by duplication of preexisting genes; the duplicate genes may then undergo small-scale mutations that give rise to functionally different products. Gene duplication can occur in several ways, including nondisjunction, chromosomal breakage and fusion, transposition, and reverse transcription.

Mutations also produce new alleles of existing genes. The processes of meiosis, crossing over, and recombination at fertilization provide almost endless variation in the population. Together, these processes provide the genetic variability on which natural selection can act to produce evolutionary change.

Natural selection can act on genetic variation only when it is expressed phenotypically. Nongenetic variations and variations produced by somatic mutations are not evolutionary raw material. Somatic mutations do not affect the genotype of the germ cells. Lamarck's hypothesis of acquired characteristics is not tenable.

Population genetics Population genetics is based on the concept of the *gene pool*, the sum total of all the genes possessed by all the individuals in the population. The frequencies of the various alleles of a given gene are used to characterize the gene pool. Evolution is a change in the allelic frequencies within gene pools.

Evolutionary change is not automatic; it occurs only when something disturbs the genetic equilibrium. According to the Hardy-Weinberg Law, both allelic frequencies and genotypic ratios remain constant from generation to generation in sexually reproducing populations if the following conditions for stability are met: large population (so there will be no *genetic drift*), no mutation (or else mutational equilibrium), no migration (i.e., no *gene flow* into or out of the population), and totally random mating and reproductive success (no *selection pressure*). In reality, many populations are large, and some populations exist without migration, but the conditions of no mutation and random reproduction are never met in any population. Mutations are always happening and mutational equilibrium is rare; hence *mutation pressure* can cause slow shifts in allelic frequencies. Reproduction is never totally random; no aspect of reproduction is completely devoid of correlation with genotype. Nonrandom reproduction, or natural selection, is the universal rule. Thus the Hardy-Weinberg Law describes the conditions under which there would be no evolution, and since these conditions cannot be met in nature, it follows that evolution is always occurring. Evolutionary change is a fundamental characteristic of the life of all populations.

Natural selection The majority of inherited characteristics are controlled by multiple genes with multiple alleles, and their expression is influenced by environment. These characteristics tend to vary continuously over a wide range; when graphed, the frequencies often form a bell-shaped curve rather than showing the two or three phenotypes characteristic of Mendelian inheritance. All populations are subject to *selection pressure*, which disturbs the genetic equilibrium. Even very slight selection pressures can lead to major changes in allelic frequencies over time. Changing environmental conditions can exert *directional selection*, which causes the population to evolve along a particular functional line.

Natural selection can be creative. Even in the absence of new mutation, selection can produce new phenotypes by combining old genes in new ways. Mutation, by contrast, is not usually a major directing force in evolution; its principal evolutionary role is to provide new variations on which future selections can act.

Sometimes a population is subject to two or more opposing directional selection pressures (such as selection for the extremes). Such *disruptive selection* may divide the population into distinct groups.

Natural selection may also play a conservative role; it acts to preserve favorable gene combinations by eliminating less adaptive new combinations created by recombination and mutation. Another way to look at this is selection against the extremes. This sort of selection has been termed *stabilizing selection*.

Many characteristics have both advantageous and disadvantageous effects, and a single allele may influence multiple characteristics (*pleiotropy*). The evolutionary fate of such characteristics or alleles depends on whether the sum of all the various positive selection pressures acting on them is greater or less than the sum of the negative selection pressures. Sometimes the effects of a given allele are more advantageous in the heterozygotes than in the homozygous condition (*heterozygote superiority*).

Sexual selection Darwin distinguished between selection that affects physical survival and selection that operates on traits used exclusively to attract and keep mates (*sexual selection*). Male contests are over access to reproductively ready females; these can take the form of fights for dominance or for territory. Characteristics such as large size, offensive weapons, and defensive structures influence these contests. Other uniquely male features exist only to attract females (*female choice*).

Adaptations Adaptations are genetically controlled characteristics that enhance an organism's *fitness*—chances of perpetuating its genes in succeeding generations. Adaptations can be structural, physiological, or behavioral; genetically simple or complex; highly specific or general.

The flowering plants depend on pollinators (e.g., wind, birds, insects) to carry pollen from the male to the female parts of the plant. The flowers are adapted in shape, structure, color, and odor to their particular pollinating agent. The plants and their pollinators have evolved together, each becoming more finely tuned to the other's peculiarities. Such evolutionary interaction is called *coevolution*.

Cryptic (concealing) appearance is an adaptation that helps animals escape predation. Kettlewell's experiments on the light and dark forms of pepper moths (*Biston betularia*) showed clearly that those moths that most closely resemble their background have the best chance of escaping predation. This is an example of *polymorphism*—the occurrence in a population of two more forms of a genetically determined character.

Some animals that are disagreeable to predators (because of sting, bad taste, or smell) have evolved *warning coloration*. They benefit by being gaudily colored and conspicuous because predators can easily learn to recognize and avoid them after one or two unpleasant encounters. The avoidance of aposematic insects may not depend solely on learning; this avoidance response may evolve in predators.

Another protective adaptation is *mimicry*: members of different species resemble (mimic) one another. There are two types of mimicry. In *Batesian mimicry*, an unprotected species resembles a distasteful species; the mimicry is based on deception. *Müllerian mimicry* involves the evolution of similar appearances by two or more distasteful species. This type of mimicry is advantageous to both because predators learn more easily to avoid both prey species.

Symbiosis Many organisms have evolved adaptations for *symbiosis*—for living together. There are three types of symbiosis: commensalism, mutualism, and parasitism. However, there are no sharp boundaries between them; they grade into one another.

Commensalism is a relationship in which one species benefits while the other neither benefits nor is harmed. The advantages the commensal species receives from its host frequently include shelter, support, transport, and/or food. Often it is difficult to determine whether a relationship is commensalism or *mutualism*, in which both species benefit.

Parasitism is a symbiotic relationship in which one species benefits while the other species is harmed. External parasites live on the outer surface of their host, and internal parasites live inside the host's body. Internal parasitism is usually marked by more extreme specializations than external parasitism; these include structural *degeneracy* (evolutionary loss of structures), resistant body walls, a complex life cycle, and large reproductive potential. Over time the host and the parasite undergo coevolution, and eventually reach a dynamic balance in which both can survive without serious damage. Most long-established host-parasite relationships are balanced ones.

CONCEPT MAP

Map 31.1

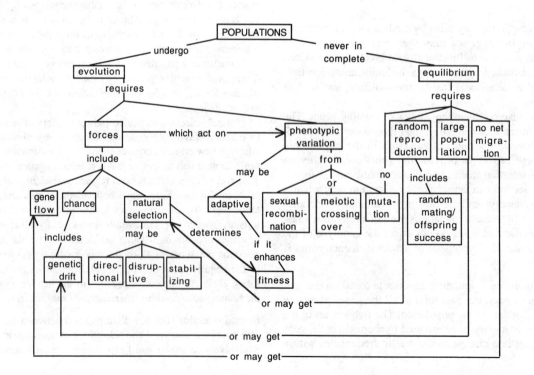

STUDY GUIDE FOR POPULATION-GENETICS PROBLEMS

The best way to gain an understanding of population genetics is to work with it. The following information is intended to get you started on problems; additional problems are provided at the end of this section. Before you begin, review page 677 in your text.

1. Allelic frequencies. We can express the frequency of alleles in a population numerically.

   Example 1: Suppose we have 100 alleles of a particular gene in a population, with 70 of these being allele A and 30 being allele a:

   Frequency A = 70/100 = 0.7, or 70%

   Frequency a = 30/100 = 0.3, or 30%

   Example 2: Calculating allelic frequencies in alleles showing intermediate inheritance. In cats, tail length is determined by a single pair of alleles. Long-tailed cats have the genotype I^L/I^L, short-tailed I^L/I^l and no-tailed I^l/I^l. In a given population, 49 cats had long tails, 42 had short tails, and 9 had no tails. What are the frequencies of the I^L and I^l alleles?

   Solution

   The total number of individuals is 49 + 42 + 9 = 100. Since each cat has two alleles, the total number of alleles in the population is 2 × 100 = 200.

   The number of I^L alleles = (2 × 49) + 42
   = 140/200 = 0.7

   The number of I^l alleles = (2 × 9) + 42
   = 6/140 = 0.3

   Note that the frequency of the two alleles adds up to 1.

2. The expression $p + q = 1$. In population genetics the symbol p represents the frequency of one allele; q, the frequency of the other. Thus in the preceding example, $p = 0.7$ and $q = 0.3$. We can then use the expression $p + q = 1$ to represent the relationship between the two alleles; since all individuals in this population have either allele A or allele a, the two frequencies must add up to 1. When we know the frequency of the alleles we can calculate the frequency of the various genotypes they will produce:

	0.7 A	0.3 a
0.7 A	0.49 A/A	0.21 A/a
0.3 a	0.21 A/a	0.09 a/a

   The Punnett square tells us that 49 percent of the population will be A/A; 21 plus 21, or 42 percent, will be A/a; and 9 percent will be a/a.

3. The Hardy-Weinberg equilibrium. The same results can be obtained algebraically by means of the formula

   $$p^2 + 2pq + q^2 = 1$$

   Here, p^2 is the frequency in the offspring of the homozygous dominant genotype, $2pq$ is the frequency of the heterozygous genotype, and q^2 is the frequency of the homozygous recessive genotype.

   $(0.7)^2 + 2(0.7)(0.3) + (0.3)^2 = 1$

   $0.49 + 0.42 + 0.09 = 1$

   As shown in your text (p. 676), the allelic frequencies in the offspring generation are the same as they were in the parent generation. The population is therefore said to be in genetic equilibrium.

4. Solving problems. We can use the two equations presented above to solve problems.

   a. In a rabbit population 10 percent of the alleles for coat color are for albino (b), and 90 percent are for black (B). What percentage of the rabbits are heterozygous if the Hardy-Weinberg assumptions hold true?

   Solution

   Frequency of B = 90% = 0.9 = p
   Frequency of b = 10% = 0.1 = q
   $p^2 + 2pq + q^2 = 1$
   Heterozygotes = $2pq$ = 2(0.9)(0.1) = 0.18
   18 percent of the rabbits are heterozygous.

   b. The frequency of the recessive allele b in a population is 0.2. Then a sudden catastrophe exerts a selective pressure, which reduces the frequency to 0.16 in a single generation. What will be the frequencies of all genotypes (BB, Bb, bb) in the next generation?

   Solution

   Frequency of b = 0.16 = q
   $p + q = 1$
   $p + 0.16 = 1$
   $p = 0.84$

   $p^2 + 2pq + q^2 = 1$
   $(0.84)^2 + 2(0.84)(0.16) + (0.16)^2 = 1$

   Genetic frequencies in the next generation (rounded):
   $p^2 = 0.70 = B/B$
   $2pq = 0.27 = B/b$
   $q^2 = 0.03 = b/b$

   c. Allele A, for unattached earlobes, is dominant over allele a, for attached earlobes. If 36 percent of a population has attached earlobes, what are the frequencies of allele A and allele a in this population?

   Solution

   $p^2 + 2pq + q^2 = 1$
   Frequency of attached earlobes (a/a) = 0.36 = q^2
   $\sqrt{0.36} = q$
   0.6 = q = frequency of allele a
   $p + q = 1$
   $p + 0.6 = 1$
   $p = 0.4$ = frequency of allele A

d Allele *T*, for the ability to taste a chemical called PTC, is dominant over allele *t*, for the absence of this ability. In a population of 1,000 individuals, 750 are tasters and 250 are nontasters. What are the frequencies of allele *T* and allele *t* in this population?

Solution

$$p^2 + 2pq + q^2 = 1$$

Frequency of non-tasters $(t/t) = 250/1,000 = q^2$

$$0.25 = q^2$$
$$\sqrt{0.25} = q$$
$$0.5 = q = \text{frequency of allele } t$$
$$p + q = 1$$
$$p + 0.5 = 1$$
$$p = 0.5 = \text{frequency of allele } T$$

Population-Genetics Problems

1 The frequency of two alleles in a gene pool is 0.1 (*A*) and 0.9 (*a*). What is the percentage in the population of heterozygous individuals? Of homozygous recessives? Assume that the population is in Hardy-Weinberg equilibrium.

2 Allele *W*, for white wool, is dominant over allele *w*, for black wool. In a sample of 900 sheep, 891 are white and 9 are black. Estimate the allelic frequencies in this sample, assuming that the population is in equilibrium.

3 In a population that is in Hardy-Weinberg equilibrium, the frequency of the recessive homozygote genotype of a certain trait is 0.09. What is the percentage of individuals homozygous for the dominant allele?

4 In a population that is in Hardy-Weinberg equilibrium, 36 percent of the individuals are recessive homozygotes for a certain trait. For the same trait, what is the percentage in this population of homozygous dominant individuals? Of heterozygous individuals?

5 Allele *T*, for the ability to taste a particular chemical, is dominant over allele *t*, for the inability to taste it. At Cornell University, out of 400 surveyed students, 64 were found to be nontasters. What is the percentage of heterozygous students? Assume that the population is in equilibrium.

6 In humans, Rh-positive individuals have the Rh antigen on their red blood cells, whereas Rh-negative individuals do not. Assume that the Rh-positive phenotype is produced by a dominant gene *Rh*, and the Rh-negative phenotype is produced by its recessive allele *rh*. In a population that is in Hardy-Weinberg equilibrium, if 84 percent of the individuals are Rh-positive, what are the frequencies of the *Rh* allele and the *rh* allele at this locus?

7 In corn, yellow kernel color is governed by a dominant allele; white, by its recessive allele. A random sample of 1,000 kernels from a population that is in equilibrium reveals that 910 are yellow and 90 are white. What are the frequencies of the yellow and white alleles in this population? What is the percentage of heterozygotes in this population?

8 A rare disease, which is due to a recessive allele that is lethal when homozygous, occurs with a frequency of one in a million. How many individuals in a town of 14,000 can be expected to carry this allele?

Answers to Population-Genetics Problems

1 18 percent heterozygous individuals, 81 percent homozygous recessives

2 0.9 white, 0.1 black

3 49 percent

4 16 percent homozygous dominant individuals, 48 percent heterozygous

5 48 percent

6 *Rh* 0.6, *rh* 0.4

7 yellow 0.7, white 0.3, 42 percent heterozygotes

8 about 28 individuals

QUESTIONS

Testing recall

Determine whether each of the following statements is true or false. If false, correct the statement.

1 For evolution to occur in a population, reproduction must be totally random. (p. 679)

2 Genetic drift has more evolutionary impact on small populations than on large ones. (p. 678)

3 Somatic mutation does not provide raw material for evolution. (p. 674)

4 Many genes are simultaneously subject to both positive and negative selection pressures. (p. 685)

5 Polymorphism is the occurrence in a population of two or more distinct forms of a genetically determined character. (p. 693)

6 Cryptically colored insects are usually distasteful to predators. (p. 694)

7 Industrial pollution caused the English pepper moth (*Biston betularia*) to absorb coal dust and become dark. (p. 693)

8 Two unrelated beetle species that have the same range look very much alike. Both produce a noxious chemical that makes them taste bad to potential predatory birds. This is probably an example of Müllerian mimicry. (p. 694)

9 In a large population with random mating, mutational equilibrium, and no migration, a disadvantageous allele will gradually disappear. (pp. 675–80)

10 Selection of heterozygotes can result in an increase in the frequency of lethal recessive alleles. (p. 686)

11 Mutations are more significant for evolutionary change than recombination of existing alleles. (p. 670)

12 Sexual selection may result in characteristics that natural selection selects against. (pp. 686–87)

The following important terms are sometimes confused by students. Distinguish between the terms within each group.

13 adapted—preadapted (pp. 682–88)

14 directional selection—disruptive selection—stabilizing selection (pp. 683–85)

15 Batesian mimicry—Müllerian mimicry (pp. 693–95)

16 cryptic appearance—warning coloration (pp. 691–93)

17 symbiosis—mutualism—commensalism—parasitism (pp. 695–97)

18 parasite—predator—parasitoid (pp. 697–98)

Testing knowledge and understanding

Choose the one best answer.

Questions 19 to 22 refer to the following situation.

Two Siamese cats and three Persians survive a shipwreck and are carried on driftwood to a previously uninhabited tropical island. All five cats have normal ears, but one carries the recessive allele for folded ears (his genotype is *F/f*).

19 What are the frequencies of allele *F* and allele *f* in the population of this island?

  a $F = 0.8; f = 0.2$
  b $F = 0.9; f = 0.1$
  c $F = 0.1; f = 0.9$
  d $F = 0.2; f = 0.8$
  e none of the above (p. 675)

20 If you assume a Hardy-Weinberg equilibrium for the ear alleles (admittedly very improbable), about how many cats would you expect to have folded ears when the island population reaches 20,000?

  a 0
  b 20
  c 200
  d 1,000
  e 2,000 (p. 676)

21 No new cats arrive on the island, and no one leaves this tropical paradise. If the number of folded-eared individuals differs greatly from the answer to the previous question, this could be due to a combination of which factors?

  a gene flow and genetic drift
  b gene flow and natural selection
  c genetic drift and natural selection
  d gene flow and nonrandom mating (pp. 675–80)

22 Over time, the population reaches 20,000. Twenty percent of this island population has long hair, but the population that the two Siamese had come from had only 1 percent long-haired individuals. This is most likely an example of

  a genetic drift.
  b natural selection.
  c selective mutation.
  d gene flow.
  e nonrandom mating. (p. 676)

23 In a population that is in Hardy-Weinberg equilibrium, there are two alleles for a certain gene, *A* and *a*. The frequency of *A* is 80 percent. What percentage of the population will be heterozygous for this gene?

  a 4 percent
  b 16 percent
  c 32 percent
  d 64 percent
  e 100 percent (p. 677)

24 According to the Hardy-Weinberg Law, all the following are conditions for genetic equilibrium *except*

  a mutations must not occur or must be at mutational equilibrium.
  b there must be no immigration or emigration.
  c reproduction must be nonrandom.
  d the population must be large. (p. 676)

25 A large population of a certain species of freshwater fish lives in South America. No close relatives of this species are known. Suppose you could somehow cause all mutations to cease in this population and prevent all immigration into the population and emigration from it. Which one of the following statements *best* expresses the probable future of the population?

  a All evolution will promptly cease because without mutation there will be no raw material for evolution.
  b The population will begin to deteriorate after four or five generations because of excessive inbreeding that will result from the absence of immigration and emigration.
  c The population will continue to evolve for a long time as selection acts on the variability produced by recombination.
  d Major evolutionary changes will continue in the population because of genetic drift.
  e Though the population will cease to evolve, it may survive for a long time if the environment remains constant. (pp. 675–80)

26 In the peppered moth the allele for light color is recessive and the allele for dark color is dominant. In a population that is in Hardy-Weinberg equilibrium, if there are 640 light moths and 360 dark moths, how many of the moths are heterozygous?

  a 96 percent
  b 64 percent
  c 48 percent
  d 32 percent
  e 8 percent (p. 677)

27 Natural selection can best be defined as

   a survival of the best-adapted individuals.
   b nonrandom reproduction.
   c differential population growth rates.
   d enhanced survival of those individuals best adapted to attract mates.
   e the elimination of the weak by the strong. (pp. 680–87)

28 The theory of natural selection postulates that

   a in each generation, all the individuals well adapted for their environment live longer and produce more progeny than the less well-adapted individuals.
   b the deaths of individuals occur completely at random with respect to the physical environment.
   c the deaths of individuals occur completely at random with respect to their genotypes.
   d to at least a small extent the survival and reproductive success of individuals depend on the extent to which they are genetically adapted to their environment.
   e most deaths of individual organisms occur soon after fertilization, as a result of hereditary deficiencies. (pp. 680–87)

29 A city was intensively sprayed with DDT in 1953 in an effort to control houseflies. The number of flies was immediately greatly reduced. Each year thereafter the city was again sprayed with DDT, but the flies gradually increased in numbers until 10 years later they were almost as abundant as they were when the control program began. Which one of the following is the most likely explanation?

   a Flies from other areas moved in and replaced the ones killed by the DDT.
   b The few flies that were affected by DDT but survived developed antibodies to DDT, which they passed on to their descendants.
   c The DDT caused new mutations to occur in the surviving flies, and this resulted in resistance to DDT.
   d The DDT killed susceptible flies, but the few that were naturally resistant lived and reproduced and their offspring repopulated the area. (pp. 680–87)

30 Careful measurements have shown that duck eggs of average size are more likely than smaller or larger eggs to produce viable ducklings. This is an example of

   a disruptive selection.   d polymorphism.
   b stabilizing selection.   e convergence.
   c directional selection.   (pp. 684–85)

31 Flowers pollinated by hummingbirds are often red and usually odorless; hummingbirds see red easily but have a poor sense of smell. This is an example of

   a mimicry.   d stabilizing selection.
   b polymorphism.   e convergence.
   c coevolution.   (p. 688)

32 In England the peppered moth has two color forms. The prevalence in an area of one or the other form is determined by the amount of soot polluting the area. Both forms of the moth exhibit

   a cryptic appearance.   d warning coloration.
   b Batesian mimicry.   e coevolution.
   c Müllerian mimicry.   (pp. 691–93)

33 The nasty-tasting viceroy butterfly looks like the equally distasteful monarch butterfly. This is an example of

   a cryptic coloration.   d coevolution.
   b cryptic appearance.   e Müllerian mimicry.
   c Batesian mimicry.   (p. 694)

34 Many mosses live on the trunks and branches of trees. They flourish but do not seem to harm the trees in any way, since they are occupying only the bark surface, which is dead tissue. The interaction between these two organisms could *best* be described as

   a symbiosis.   d parasitism.
   b commensalism.   e predation.
   c mutualism.   (pp. 695–96)

35 *Pseudomyrmex ants* live in an intimate association with the *Acacia* plant, a Central American plant with very large thorns. The queen ant bores a hole in one of the large thorns, hollows it out, and lays her eggs. As the colony grows, more thorns are excavated and inhabited. The ants continuously patrol the plant and remove herbivorous insects and competing vegetation. The relationship between the ants and the *Acacia* could *best* be described as

   a symbiosis.   d predation.
   b mutualism.   e commensalism.
   c parasitism.   (p. 695)

36 The relationship between a farmer and her flock of egg-producing hens could *best* be described as

   a predation.   d mutualism.
   b parasitism.   e competition.
   c commensalism.   (p. 695)

For further thought

Answers to Problems 1–4 are included at the end of the Answers section of this chapter.

1 In a certain African tribe, 4 percent of the population is born with sickle-cell anemia. What is the percentage in

the tribe of individuals who enjoy the selective advantage of the sickle-cell gene (increased resistance to malaria)?

2 In the United States, approximately one child in 10,000 is born with PKU (phenylketonuria), a syndrome that affects individuals homozygous for the recessive allele. What is the frequency of this allele in the population? What is the frequency of the normal allele? What is the percentage in the population of carriers of this trait?

3 Artificial selection can alter the distribution of phenotypes in a population. Suppose the distribution of phenotypes within a population looks like this:

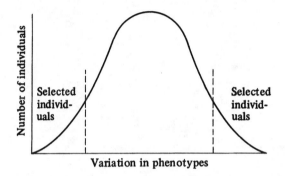

Disruptive selection is then practiced: only those organisms at both extremes of the distribution are chosen for breeding; those in the central portion are rejected. Draw a curve showing the expected phenotypic distribution of the F_1.

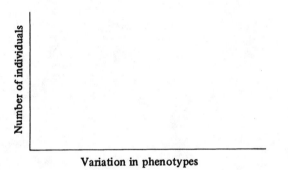

4 The introduction of DDT to control mosquitoes in the 1940s brought rapid control over the mosquitoes carrying the malarial pathogen. Today, resistance to DDT has developed in many mosquito populations and is reducing the effectiveness of malarial control programs. Explain how it was possible for the mosquitoes to evolve DDT resistance in such a short time.

5 Children born with Tay-Sachs disease die within two or three years of birth. The disease is inherited as an autosomal recessive. Considering the strong selection pressure against individuals with the disease, explain why the allele persists in the population.

ANSWERS

Testing recall

1 false—nonrandom
2 true
3 true
4 true
5 true
6 false—brightly colored or tasty
7 false—to exhibit an increase in the frequency of the dark forms
8 true
9 false—no change in frequency
10 true
11 false—less significant
12 true
13 *Adapted*, the more general of the terms, refers to an organism's possession of any genetically controlled characteristic that increases its fitness to survive in the existing environment. *Preadapted* refers to an organism's possession of genes that increase its fitness to survive if the environment changes in certain respects; the organism is adapted to a new environment before it encounters that environment.
14 *Directional selection* acts against individuals exhibiting one extreme of a character: it causes the population to evolve along a particular line. *Disruptive selection*, which occurs when two opposing selection pressures are present, favors individuals exhibiting both extremes of a character, and thereby divides the population into two distinct types. *Stabilizing selection* acts against individuals exhibiting characters that are too different from the mean condition, and thus maintains stability in the population.
15 In *Batesian mimicry*, a palatable species mimics in appearance some unpalatable species' warning coloration; in *Müllerian mimicry* two or more unpalatable species mimic each other in appearance.
16 Animals with a *cryptic appearance* tend to blend into their surroundings so as to be undetectable; those with *warning coloration* have evolved colors and patterns that contrast boldly with their environment. The latter are often distasteful to predators.
17 *Symbiosis* means "living together." *Mutualism* (in which both species benefit from the association), *commensalism* (in which one species benefits and the other neither benefits nor is harmed), and *parasitism* (in which one species benefits at the expense of the other) are types of symbiosis.

18 A *parasite* passes much of its life on or in the body of the living host from which it obtains its food, whereas a *predator* eats its prey very quickly and goes its way. Parasites generally do not kill their hosts; *parasitoids* do.

Testing knowledge and understanding

19	b	24	c	29	d	33	e
20	c	25	c	30	b	34	b
21	c	26	d	31	c	35	b
22	a	27	b	32	a	36	d
23	c	28	d				

For further thought

1 32 percent are heterozygous and resistant to malaria.
2 The frequency of the PKU allele is about 1 percent; the normal allele frequency is 99 percent. About 1.98 percent of the population are carriers (heterozygotes).

Chapter 32

EVOLUTION: SPECIATION AND PHYLOGENY

A GENERAL GUIDE TO THE READING

This chapter applies the evolutionary concepts presented in Chapter 31 to the process of speciation—the process by which an ancestral species may split and give rise to two or more different descendant species. The splitting process has been exceedingly frequent over millions of years and is responsible for the immense diversity of life today. This chapter is a very important one, and as you read it you will want to concentrate on the following.

1. Species and speciation. You will want first to learn the definition of "species" (p. 704) and then concentrate on how speciation occurs (pp. 707–13). An understanding of the role of isolation, both extrinsic and intrinsic, is key. This material is crucial; when you finish the chapter, you should have a thorough understanding of divergent speciation.

2. Sympatric speciation. You will want to understand what is meant by this term, and how sympatric speciation differs from divergent evolution of allopatric species. Also you will want to distinguish between chromosomal and nonchromosomal sympatric speciation.

3. Adaptive radiation. Darwin's finches are used to illustrate the process of adaptive radiation and to provide a model by which the evolution of the enormous diversity of species living today can be understood. This section synthesizes many of the ideas that are introduced early in the chapter.

4. Competition. The relative importance of competition and chance (through genetic drift or natural catastrophes) to the evolution of populations is described on pages 718 to 721.

5. Gradualism versus punctuated equilibrium. The current controversy over the mechanism of evolution is described on pages 721 to 725. Figure 32.24 (p. 725) provides a good example of the two hypotheses in action. When you read this section, note that the so-called instantaneous speciation is hardly that; speciation may take thousands of years. Take note of the Burgess fauna, an excellent case of relatively rapid evolution. (pp. 725–26)

6. Phylogeny. You will want to understand some of the problems faced by the systematist (pp. 726–31) and to learn the classification hierarchy (Table 32.3, p. 734). A help in remembering the order of phylogenetic categories is the mnemonic sentence "*K*indly *p*rofessors *c*annot *o*ften *f*ail *g*ood *s*tudents," in which the first letter of each word is the first letter of a category.

KEY CONCEPTS

1. In the modern view, a species is a genetically distinctive group of natural populations that share a common gene pool and that are reproductively isolated from all other such groups. (pp. 703–5)

2. Divergent speciation usually begins when two sets of populations are separated geographically by external barriers; as the two population systems evolve independently, they accumulate differences that lead in time to the development of intrinsic isolating mechanisms. (pp. 707–12)

246 • CHAPTER 32

3 Sympatric speciation does not involve geographical isolation. Species that arise by polyploidy are genetically distinctive and are reproductively isolated from the parent species. Habitat preference and host-specificity, and certain behavioral isolating mechanisms can also produce reproductive isolation, which leads to speciation. (pp. 712–13)

4 Adaptive radiation via divergent evolution—the evolutionary splitting of lineages into many separate lineages—has occurred very frequently; it is responsible for the immense variety of living things. (pp. 713–16)

5 Although competition is believed to be the predominant force leading to speciation over time, the effects of chance through rare environmental crises such as natural catastrophes may be important in the evolution of certain populations. (pp. 718–21)

6 Two opposing hypotheses of evolutionary change have been proposed: gradualism, which states that speciation occurs by the gradual accumulation of small changes over long periods of time, and punctuated equilibrium, which states that most evolution is marked by long periods of equilibrium punctuated by short periods of rapid change in small populations, resulting in speciation. (pp. 721–25)

7 The phylogeny of a group of organisms represents the evolutionary history of that group. Biologists attempt to reconstruct evolutionary history by using a systematic approach to discover the relationship among species and then fitting these relationships into an orderly scheme. (pp. 726–34)

CROSSLINKING CONCEPTS

1 Scientific method. The discussion of different classification strategies illustrates the roles of objectivity, subjectivity, and intuition and the importance of technological developments in determining the process of scientific research (Chapter 1).

2 We know there is variation within species and that sources of that variation include mutation and recombination (Chapters 9 and 10). The significance of that variation is again emphasized in this chapter because it plays an important role in adaptive radiation. Also, note the importance of nondisjunction (Chapter 14) in creating polyploidy, a form of chromosomal sympatric speciation.

3 A study of evolution necessarily introduces concepts pertaining to the environment. The effect of geographic isolation illustrates the role of the environment in creating allopatric species. The specialization of the Galapagos finches toward different food sources illustrates the fact that organisms tend to carry on a particular function within their environment; this idea eventually leads to the creation of niches. These concepts will be developed more fully when you study ecology (Chapter 33).

OBJECTIVES

1 Discuss intraspecific geographic variation; in doing so, distinguish among population, deme, subspecies, and cline. (pp. 703–707)

2 Give a definition of a species and, using Figure 32.9 (p. 710), explain the geographic-isolation model of divergent speciation. In doing so, be sure to take into account the roles of mutation, recombination, natural selection, and the gene pool and to distinguish between extrinsic and intrinsic isolating mechanisms. Then briefly describe 10 intrinsic isolating mechanisms. Tell whether the isolating mechanism prevents mating, production of hybrids, or perpetuation of hybrids. (pp. 707–12)

3 Explain why the biological species concept is difficult to apply to asexual organisms, fossil organisms, and allopatric populations. (pp. 704–705)

4 Describe speciation by polyploidy, and indicate why plant breeders frequently take advantage of the process. (pp. 712–13)

5 Differentiate between sympatric speciation and allopatric speciation. Give two examples of nonchromosomal sympatric speciation. (pp. 712–13)

6 Explain what is meant by adaptive radiation and discuss the evidence for this phenomenon, using the Galapagos finches as an example. Be sure to discuss the contributions of extrinsic and intrinsic isolation, competition, character displacement, and species recognition. (pp. 714–18)

7 Contrast the role of competition with the effects of sudden changes such as natural catastrophes on the evolution of natural populations and explain what is meant by an evolutionary bottleneck. Indicate whether competition or sudden changes are believed to be the predominant force leading to speciation over time. (pp. 718–21)

8 Compare the hypothesis of punctuated equilibrium with that of gradualism. Using Figure 32.24 (p. 725), explain how the giraffe evolved from the pre-okapi, using the two different hypotheses. (pp. 721–25)

9 Describe the sources and types of information used by systematists to determine phylogenetic relationships. Include information on classical evolutionary taxonomy, phenetics, and cladistics. (pp. 726–34)

10 Distinguish between divergent and convergent evolution and give examples of each. Indicate why convergence poses a problem for the systematist. Explain the difference between homologous and analogous similarities. (pp. 727–28)

11 List in order the categories of the classification hierarchy used today. Then explain how a species is named. (pp. 732–34)

KEY TERMS

population (p. 703)
deme (p. 703)
species (p. 704)
biological species concept (p. 704)
intraspecific variation (p. 706)
cline (p. 706)
subspecies (p. 707)
speciation (p. 707)
phyletic evolution or anagenesis (p. 707)
branching evolution, or cladogenesis, or divergent speciation (p. 707)
allopatric (p. 708)
founder effect (p. 708)
intrinsic reproductive isolation (p. 709)
sympatric (p. 709)
ecogeographic isolation (p. 709)
habitat isolation (p. 710)
temporal isolation (p. 710)
behavioral isolation (p. 710)
mechanical isolation (p. 711)
gametic isolation (p. 711)
developmental isolation (p. 711)
hybrid inviability (p. 711)
hybrid sterility (p. 711)
hybrid vigor (p. 711)
selective hybrid elimination (p. 712)
polyploidy (p. 712)
sympatric speciation (p. 712)
allopolyploidy (p. 712)
macroevolution (p. 713)
adaptive radiation (p. 713)
character displacement (p. 716)
niche (p. 718)
niche rule (p. 718)
evolutionary bottleneck (p. 719)
punctuated equilibrium (p. 724)
gradualism (p. 724)
Burgess fauna (p. 725)
phylogeny (p. 727)
convergence (p. 727)
homologous (p. 727)
analogous (p. 727)
molecular clock (p. 731)
classical evolutionary phylogeny (p. 731)
phenetics (p. 732)
cladistics (p. 732)
ancestral (p. 732)
derived (p. 732)
shared derived characters (p. 732)

SUMMARY

Species A *population* of sexually reproducing organisms is a group of individuals that share a common gene pool. A *deme* is a small local population; demes are usually temporary units of population that intergrade with other such units.

The existence of discrete clusters of living things that can be called species has long been recognized, but the concept of what a species is has changed many times. According to the *biological species concept*, a *species* is a genetically distinctive group of natural populations (demes) that share a common gene pool and that are *reproductively isolated* from all other such groups. This definition of a species is hard to apply to asexual organisms, fossil organisms, and allopatric populations.

The vast majority of plant and animal species show geographic variation, most of which probably reflects differences in selection pressures resulting from local environmental conditions. Gradual variation correlated with geography is called a *cline*. An abrupt shift in a character in a geographically variable species may occur; in this case the resulting populations are designated as subspecies. These are groups of natural populations within a species that differ genetically and that are partly isolated from each other reproductively because they have different ranges.

Divergent speciation In the process of *divergent speciation*, one ancestral species gives rise to two or more descendant species, that grow increasingly unlike as they evolve. The initiating factor is usually geographic separation; if a population is divided by some physical or ecological barrier, the separated (*allopatric*) populations will no longer be able to exchange genes. In time, the populations will evolve in different directions because they start out with different gene frequencies, they experience different mutations, and they are exposed to different environmental selection pressures. Eventually, the populations may become genetically so different that they develop *intrinsic isolating mechanisms*—biological characteristics that prevent effective interbreeding should they again become *sympatric*.

Many intrinsic isolating mechanisms act by preventing effective mating. The two populations may be so specialized ecologically that they cannot become sympatric, or they may occupy different habitats; they may breed at different times, be behaviorally isolated, or be physically unable to mate. Even if mating occurs, fertilization may not take place or the embryo may not survive. If hybrids are born, they may be inviable, sterile, or selected against.

Sympatric speciation Speciation that does not involve geographic isolation is called *sympatric speciation*. There are two types of sympatric speciation. One type, *polyploidy*, is genetic in origin. Polyploidy, which is rare in animals but common in plants, is the occurrence in cell nuclei of more than two complete sets of chromosomes, usually as a result of nondisjunction in meiosis or some other abnormal event in cell division. In *allopolyploidy* there is a multiplication of the number of chromosomes in a hybrid between two species.

Sympatric speciation can also occur without a genetic basis. Reproductive isolation can be achieved by means other than geographic isolation; such factors as habitat preference and host-specificity may create the reproductive isolation necessary for speciation.

Adaptive radiation Divergent evolution, or the evolutionary splitting of species into many separate descendant species, results in *adaptive radiation*. The rate of evolutionary divergence is not always constant; when conditions change rapidly and organisms have new evolutionary opportunities available to them, they may undergo a rapid evolutionary burst. Adaptive radiation like that of Darwin's finches on the Galapagos

Islands helps account for the tremendous diversity among living things on the earth today. Competition, leading to *character displacement*, is thought to be the predominant force in the speciation process. According to Gause's competitive exclusion principle, or *niche rule*, no two species occupying the same niche can long coexist. However, the assumption that competition is the major mode of speciation is being challenged by the view that chance crises may have been important in the evolution of certain populations. Crises can be caused by any environmental factor that severely affects a population so as to cause major changes in allelic frequencies (called an *evolutionary bottleneck*). As the population passes through the evolutionary bottleneck caused by extraordinary environmental conditions, one character or another may gain ascendancy and rapid evolutionary change results.

Two opposing hypotheses of evolutionary change have been proposed: *gradualism*, which states that speciation occurs by the gradual accumulation of small changes over long periods of time, and *punctuated equilibrium*, which states that most evolution is marked by long periods of equilibrium followed by short bursts of rapid change.

The concept of phylogeny One of the tasks of *systematics* is to discover the relationships among species and to trace the ancestors from which they are descended. Because so many kinds of evidence must be weighed, reconstructing the evolutionary history, or *phylogeny*, of any group of organisms entails an element of speculation.

One problem in interpreting the similarities among organisms is *convergence*. Organisms that are not closely related may come to resemble each other because they occupy similar habitats and adopt similar environmental roles. Systematists must try to determine whether the similar features are *homologous* (inherited from a common ancestor) or merely *analogous* (similar in function and often in appearance but of different evolutionary origin) before speculating about the degree of relationship between the organisms.

The morphology of the adult and embryo, combined when possible with information from the fossil record, has traditionally been the basic source of data for reconstructing the evolutionary history of organisms. More recently, techniques have been developed to compare the DNA, mRNA, and proteins of different speices, and these provide additional information on relatedness.

There are three major approaches to systematics. *Classical evolutionary taxonomy* depends more than any other on experience and subjective judgment; it uses as many independent characters of the species in question as possible to construct phylogenies. *Phenetics* (which weighs all characteristics equally) and *cladistics* (which focuses on *shared derived characters*) are new approches that attempt to use more objective methods in classification.

Classification The classification system used today attempts to encode information about the organism's evolutionary history. It uses a hierarchy of categories in which each category (taxon) is a collective unit containing one or more groups from the next-lower level. The principal categories are kingdom, phylum, class, order, family, genus, and species. In the modern system of nomenclature each species is given a name consisting of two Latin words; the first names the genus to which the species belongs, the second the individual species designation.

CONCEPT MAP

Map 32.1

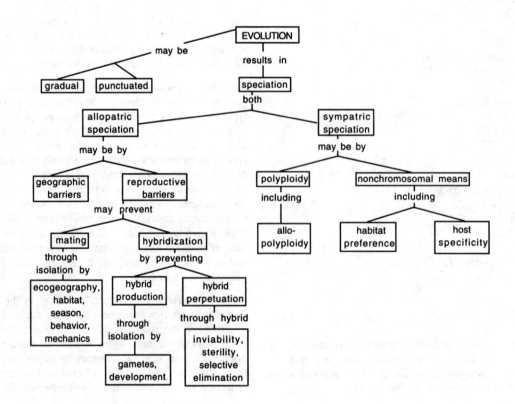

QUESTIONS

Testing recall

Determine whether each of the following statements is true or false; if false, correct the statement.

1 By definition, no hybridization between two closely related species is possible. (p. 704)

2 When two or more closely related species are sympatric, differences in their courtship displays often play an important role in preventing hybridization between them. (p. 710)

3 Character displacement is more likely to be observed in two related bird species from different islands than in two such species from the same island. (p. 716)

4 Two species can never produce viable hybrids. (pp. 711–12)

5 Two subspecies of the same species cannot be sympatric for long without fusing. (pp. 712–13)

6 Speciation by polyploidy is much more common in animals than in plants. (pp. 712–13)

7 The catastrophic effects of severe environmental crises may be an important force leading to speciation in certain populations. (pp. 719–21)

8 The hypothesis of adaptive radiation states that most evolution is marked by long periods of equilibrium followed by short bursts of rapid change. (pp. 721–25)

9 The wings of birds and the wings of butterflies are examples of convergently evolved structures. (p. 727)

10 The modern system of scientific naming of plants and animals dates from the work of the Swedish naturalist Carolus Linnaeus. (p. 733)

The following important terms are sometimes confused by students. Distinguish between the terms within each of the following groups.

11 species—subspecies (pp. 704–707)

12 deme—cline (pp. 703–706)

13 genetic drift—founder effect (p. 708)

14 divergent speciation—sympatric speciation (pp. 707–12)

15 extrinsic isolating mechanism—intrinsic isolating mechanism (p. 709)

16 sympatric—allopatric (pp. 708–709)

17 homologous—analogous (p. 727)

18 convergence—divergence (p. 727)

19 phenetics—cladistics (p. 732)

Testing knowledge and understanding

Choose the one best answer.

20 Which of the following pairs represents two species?

a German shepherd, poodle
b honeybee worker, queen honeybee
c monarch butterfly, viceroy butterfly
d housefly larva, housefly adult
e Eskimo, Peruvian Indian   (pp. 704–705)

21 In North America there is a progressive decrease in the size of song sparrows from the north to the south of their range. This is an example of

a a deme.   d behavioral isolation.
b interspecific variation.   e a cline.
c the founder effect.   (pp. 706–707)

22 Population A occurs only in Africa. The closely related population B occurs only in South America. Only a few minor structural differences between A and B can be found. Population C, which is closely related to both A and B and is sympatric with B, differs more noticeably from B in several characters. In an attempt to determine whether A and B belong to the same or to different species, individuals from the two populations are mated in the laboratory; they produce viable offspring. Populations A and B *must* be regarded as

a members of the same species and subspecies.
b members of the same species but of different subspecies.
c members of different species.
d uncertainly related, because insufficient evidence is available to settle the matter.   (pp. 709–12)

23 Gradual variation of a characteristic correlated with geography is an example of

a a cline.   d genetic drift.
b polymorphism.   e gene flow.
c phenotypic variation.   (p. 706)

24 Which one of the following isolating mechanisms is generally considered most important in *initiating* the process of divergent speciation?

a geographic isolation   d seasonal isolation
b behavioral isolation   e hybrid sterility
c developmental isolation   (p. 708)

25 You carefully study populations of two very similar meadow mice, one from the Northeast and one from Texas. You want to know whether the populations belong to the same species or to two different species. You could *most* confidently decide this if you could

a show that the ranges of the two mice overlap without hybridization occurring.
b bring the two types of mice into the laboratory and show that they can cross and produce viable offspring.
c demonstrate that the natural ranges of the two types of mice are entirely allopatric.
d show that there are statistically significant structural differences between the two types.
e show that the Northeastern mice live in wetter habitats than the Texan mice.   (pp. 708–12)

26. A biologist studies a large field in which hundreds of plants, all belonging to a single species, are growing. Plants of this species occur nowhere else in the state, and no other species of the same genus are known to science. The biologist demonstrates that all the plants in the field are probably infertile. Five years later the biologist again visits the field and notices a few plants that look slightly different from the others. On conducting cross-pollination experiments, she discovers that these unusual plants can cross with one another but not with the normal plants; they appear to be members of a second species closely related to the original one. The *most likely* explanation is that

   a a new mutant gene arose in some members of the original population, causing it to split into two species.
   b genetic drift in this relatively small population led to speciation.
   c habitat differences in two different parts of the field led to divergent adaptation that resulted in speciation.
   d the unusual plants are polyploids.
   e the original species first evolved polymorphism, and then each form became a separate species. (pp. 712–13)

27. One plant species has a diploid number of 8. A second plant species has a diploid number of 10. These two plant species hybridize and a single fertile hybrid is produced. This new species produces *gametes* that probably have how many chromosomes?

   a 4   *d* 18
   b 5   *e* 36
   c 9   (pp. 712–13)

28. All the following are intrinsic isolating mechanisms *except* one, which is an extrinsic isolating mechanism. Which one is the extrinsic isolating mechanism?

   a developmental isolation
   b geographical isolation
   c habitat isolation
   d behavioral isolation
   e mechanical isolation   (pp. 709–12)

29. All the following are examples of intrinsic reproductive isolation *except*

   a two populations separated by a mountain range.
   b two populations with different habitats that keep potential mates from each population separate.
   c populations that breed during different seasons of the year.
   d organisms that can mate but fail to produce viable zygotes.
   e two populations with very different courtship displays. (pp. 709–12)

30. New plant species may arise in one generation as a result of

   a seasonal isolation.   *d* divergent speciation.
   b behavioral isolation. *e* the founder effect.
   c polyploidy.   (pp. 712–13)

31. An East Coast mouse and a West Coast mouse are related, but distinguishable. When the coast-to-coast railroad was completed, a considerable overlap of range of the two groups of mice developed. In the overlap area, the two types of mice evolved further: one lived primarily from small seeds, and the other primarily from insects. The concept that best describes this scenario is

   a adaptive radiation.
   b convergent evolution.
   c character displacement.
   d evolutionary bottleneck.
   e shared derived characters.   (p. 716)

32. The evolution of the Galapagos finches from a common ancestral form is thought to have occurred by the mechanism of

   a convergence.   *d* adaptive radiation.
   b coevolution.   *e* sympatric speciation.
   c parallel evolution.   (pp. 714–17)

33. All the following factors may be involved in sympatric speciation *except*

   a polyploidy.   *d* habitat isolation.
   b host-specificity.   *e* geological isolation.
   c sexual imprinting.   (pp. 712–13)

Questions 34 and 35 refer to the following evolutionary tree. Each letter in the tree represents a different species; C, D, E, and F live in different continents.

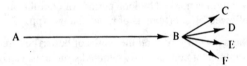

34. The evolution of species C, D, E, and F from B looks like an example of

   a adaptive radiation.   *c* convergent evolution.
   b parallel evolution.   *d* polyploid speciation. (p. 713)

35. Species D and F are
   a sympatric.   *b* allopatric.   (p. 708)

36. Natural catastrophes that cause major changes in allelic frequencies in a population are referred to as a(n)

   a evolutionary bottleneck.
   b geographic isolation.
   c extrinsic isolating mechanism.
   d intrinsic isolating mechanism.
   e adaptive radiation.   (p. 719)

37. The term punctuated equilibrium refers to the concept that

   a there have been long periods in the history of the earth when no new species evolved.
   b species tend to be static for long periods and then sudden "speciation events" occur.
   c the number of species on the earth remains at equilibrium.
   d evolution requires geographic isolation.
   e species with small body size evolve more rapidly than those with larger body size. (pp. 721–25)

38 The Australian mole is actually a marsupial rather than a placental mammal like the North American or European mole. The two animals are similar in appearance because

   a they evolved from a molelike common ancestor.
   b there are practically no placental mammals in Australia.
   c the selection pressures on both were similar.
   d they have undergone a long period of coevolution.
   e marsupials and placental mammals are closely related.
   (p. 727)

39 Convergent evolution between two species would be most likely to occur as a result of

   a a series of identical mutations occurring in both species.
   b hybridization between the two species.
   c interbreeding by both species with members of a third species.
   d exposure of both species to similar selection pressures.
   e genetic drift between the species. (p. 727)

40 Which one of the following is a correct hierarchical sequence of taxonomic groups?

   a class—order—family—genus
   b order—class—family—genus—species
   c family—order—species—genus
   d class—phylum—order—family (pp. 732–33)

41 The scientific name of the human species is properly written

   a homo sapiens.   d Homo Sapiens.
   b homo sapiens.   e Homo sapiens.
   c Homo Sapiens.   (pp. 733–34)

For further thought

1 The apple maggot, *Rhagoletis pomonella*, parasitizes apple trees. The adult flies generally mate on the fruit tree, the maggots parasitize, and the female lays her eggs there. Recently, some apple maggots were found on cherry trees in an area where cherry orchards and apple orchards were close to one another. Today the two populations are genetically and behaviorally distinctive: cherry-preferring maggots emerge in early summer, when the cherries ripen, whereas the apple-preferring maggots emerge in late summer, when the apples ripen. Hybridization between the two populations is possible. Can these two populations be considered as two distinct species? If so, what type of speciation is going on here? Justify your answers to both these questions.

2 Discuss the ways in which one species may give rise to two or more new species. Be sure to explain the role of mutation, recombination, gene pool, extrinsic isolating mechanisms, and intrinsic isolating mechanisms and give specific examples. Discuss the role that natural selection plays in this process.

3 On the basis of your understanding of the biological species concept and its limitations, give at least one explanation for the disagreements among scientists about the classification of species.

4 Explain why convergent evolution poses such difficulties for systematists.

5 Your text states, "Much research may be necessary before they [some species] can be fitted into the classification system with any degree of certainty, and their assignment to genus or family (or even order) may have to be changed as more is learned about them." What does this statement imply about the nature of scientific knowledge?

ANSWERS

Testing recall

1 false—hybridization may occur, but the hybrids are often invisible or sterile
2 true
3 false—more likely in species from the same island
4 false—sometimes; the hybrids may be sterile
5 true
6 false—more common in plants
7 true
8 false—punctuated equilibrium
9 true
10 true
11 A *species* is a genetically distinctive group of natural populations that share a common gene pool and are reproductively isolated from all other such groups. *Subspecies* are groups of natural populations within a species that differ genetically and are partly isolated from each other genetically because they have different ranges.
12 A *deme* is any small local population of a species. A *cline* is a gradual variation in the character of a species correlated with geography.
13 *Genetic drift* is evolutionary change that is caused by chance. *Founder effect* is a type of genetic drift; it occurs when a new population is formed by a small number of individuals whose allelic frequencies differ from those of the parental population.
14 *Divergent speciation* requires geographic isolation for speciation; *sympatric speciation* does not require geographic isolation but relies on some other mechanism, such as polyploidy or some intrinsic isolating mechanism, to achieve reproductive isolation for speciation.
15 An *extrinsic isolating mechanism* is a physical or geographic barrier that divides a population and prevents interbreeding between the two segments. An *intrinsic isolating mechanism* is any biological characteristic that prevents two populations from interbreeding effectively.
16 *Sympatric* means occupying the same range, whereas *allopatric* means occupying different ranges.
17 When characters in different organisms are similar because they are inherited from a common ancestor, they are *homologous*; when characters in different organisms are similar in function but of different evolutionary origin, they are *analogous*.
18 *Convergence* occurs when organisms that are not closely related become more similar because of independent

adaptations to similar environmental situations; *divergence* occurs when organisms become more dissimilar over time.

19 *Phenetics* is a system of classifying organisms that weights all characteristics equally; *cladistics* is a system of classifying organisms that focuses on shared derived characters.

Testing knowledge and understanding

20	c	26	d	32	d	37	b
21	e	27	c	33	d	38	c
22	d	28	b	34	a	39	d
23	a	29	a	35	b	40	a
24	a	30	c	36	a	41	e
25	a	31	c				

Chapter 33

ECOLOGY

A GENERAL GUIDE TO THE READING

This chapter and the next examine the interactions among organisms, and their relationships to the chemical, biological, and physical environment. The chapter will explore the interactions between individuals within a population, between species within a community, and between the community and the environment. Ecology differs from other areas of biology in that the emphasis is on these interactions and relationships rather than on the individual organism or part of the organism. Ecologists are fond of saying, "You cannot do just one thing." Elimination of one species can cause a ripple effect through the whole ecosystem. As you read Chapter 33 in your text, try to appreciate the complexity of the interrelationships among organisms. You will want to focus on the following.

1. Basic terminology. Terms used frequently are defined on pages 737 to 738; you need to learn them before continuing your reading.

2. Population growth. You will want to be familiar with the various types of population growth curves, as shown in Figures 33.6 (p. 741), 33.7 (p. 741), and 33.10 (p. 743).

3. Population regulation. In the section on this topic (pp. 745–48), the differences between *r*-selected and *K*-selected species are important, as are the differences between density-dependent and density-independent types of limits on population growth. Pay particular attention, too, to the concept of niche and the competitive exclusion principle. (pp. 749–50)

4. Ecological succession. In reading this material (pp. 752–58), pay close attention to the causes of succession and the trends in succession. A helpful summary of the trends is presented on pages 756–57.

KEY CONCEPTS

1. Ecology is the study of interactions between organisms and their environment. The various ecosystems are linked by biological, chemical, and physical processes. (p. 737)

2. Many natural populations show an initial period of exponential growth at low densities, followed by a deceleration in growth at higher densities, and an eventual leveling off as the density approaches the carrying capacity of the environment. Such a population is primarily limited by influences that provide feedback control because it depends at least partially on the density of the population itself. (p. 741)

3. In some populations density grows exponentially, but then falls precipitously before reaching the carrying capacity of the environment; this is a result of density-independent limiting factors. (pp. 745–48)

4. In natural populations, growth is subject to a number of limitations; the regulation of population density can occur in different ways in different populations. (pp. 745–52)

5. The niche is the function and position of an organism in the ecosystem. The more the niches of two different species overlap, the more intense the competition between them will be. (pp. 750–51)

6 A species does not exist as an isolated entity; it is always interacting in a variety of ways with other species. This association of interacting populations forms a community which has its own characteristic structure and complex array of interrelationships. (p. 752)

7 The trend of most ecological successions is toward a more diverse, complex, and longer-lasting community. (pp. 752–57)

8 Most ecological successions eventually reach a climax stage that is more stable than the stages preceding it. The more complex organization of the climax community, and its larger organic structure and more balanced metabolism, enable it to buffer its own physical environment to such an extent that it can perpetuate itself as long as the environment remains essentially the same. (p. 757)

CROSSLINKING CONCEPTS

1 At this point in your studies, you should have a good picture of the levels of biological organization and how they relate to different chapters. From smallest to largest, levels include chemical, cell, tissue, organ, system, organism, population, community, ecosystem, and biosphere. Individual biologists often choose to work at some particular level that commands their interest.

2 You have already studied the Galapagos finches in relation to speciation and evolution (Chapter 32). This chapter offers an opportunity to broaden your understanding of the concept of a niche and the role it plays in competition. You will want to pay close attention to what may happen when species with similar niches occupy the same territory.

3 Your understanding of the endocrine system (Chapter 25) will deepen your understanding of the physiological basis of density-dependent limiting factors. You will particularly want to consider the linkage between hormones and crowding.

OBJECTIVES

1 Define population, community, ecosystem, and biosphere, and indicate how each is related to the others. (pp. 737–38)

2 Distinguish between uniform, clumped, and random distributions. Indicate the conditions under which each occurs, and which is the most common. (pp. 739–40)

3 Draw an exponential growth curve. Write the equation for the curve, and define all the terms used in the equation. (pp. 740–41)

4 Draw an S-shaped (logistic) curve. Label the carrying capacity, the inflection point, the portion of the curve showing an accelerating rate of population growth, and the portion showing a decelerating rate. Compare this curve with the exponential growth curve. Then explain what is meant by zero population growth, and describe how this condition is reached. (pp. 741–42)

5 Draw an exponential growth curve with sudden crash, and list factors that might cause the crash. Distinguish between density-dependent and density-independent limitations on population growth, and indicate which type of limitation regulates growth in populations that show, respectively, an exponential growth curve with sudden crash and an S-shaped growth curve. (p. 742)

6 On a single graph draw type I, type II, and type III survivorship curves; then state which curve is most common in natural populations. (pp. 743–44)

7 Distinguish between a r-selected species and a K-selected species with respect to number of offspring, type of survivorship curve, type of environment (stable or unstable), and type of growth curve (S-shaped or boom-and-bust). (p. 746)

8 Using examples, discuss the ways in which parasitism, predation, intraspecific competition, emigration, and physiological and behavioral mechanisms can act as density-dependent limitations on population growth. Explain, using an example, how destroying the balance between predator and prey in a community can upset the ecology of an area. (pp. 748–52)

9 Carefully define the concept of ecological niche, and explain its significance with respect to the competitive exclusion principle and the principle of limiting similarity. Specify the four possible results of intense interspecific competition. (pp. 749–50)

10 Describe the process of ecological succession, and indicate why the species in a given area change over time. Distinguish between primary and secondary successions, and give an example of each. Also, summarize the trends seen in many successions, and explain what is meant by a climax community. (pp. 752–58)

11 Describe the effect of human intervention in biological communities. (pp. 758–60)

KEY TERMS

ecology (p. 737)
population (p. 737)
community (p. 737)
ecosystem (p. 737)
biosphere (p. 738)
population density (p. 739)
uniform distribution (p. 739)
random distribution (p. 739)

clumped distribution (p. 739)
exponential growth curve (p. 740)
intrinsic rate of increase (p. 741)
logistic or S-shaped growth curve (p. 741)
carrying capacity (p. 741)
density-dependent limitation (p. 741)
zero population growth (p. 742)
density-independent limitation (p. 742)
survivorship curve (p. 743)
r-selected species (p. 745)
K-selected species (p. 746)
intraspecific competition (p. 749)
interspecific competition (p. 749)
niche (p. 749)
principle of limiting similarity (p. 750)
succession (p. 752)
pioneer community (p. 753)
humus (p. 753)
primary succession (p. 754)
secondary succession (p. 754)
climax community (p. 757)
monoculture (p. 758)

SUMMARY

Ecology is the study of the interaction between organisms and their environment. There are three higher levels of organization: *populations*, groups of individuals belonging to the same species; *communities*, units composed of all the populations living in a given area; and *ecosystems*, the sum total of the communities and their physical environment considered together. The various ecosystems are linked to one another by biological, chemical, and physical processes. The entire earth is a true ecosystem; this global ecosystem is called the *biosphere*.

Populations as units of structure and function The distribution of individuals within an area may be uniform, random, or clumped. *Uniform* and *random* distributions are relatively rare and occur only where environmental conditions are fairly uniform. A uniform distribution results from intense competition or antagonism between individuals; a random distribution occurs when there is no competition, antagonism, or tendency to aggregate. *Clumping* is the most common distribution because environmental conditions are seldom uniform, reproductive patterns favor clumping, and animal behavior patterns often lead to congregation.

Population growth All organisms have the potential for explosive growth; under ideal conditions their growth curve would be *exponential*. The equation for an exponential growth curve is $I = rN$, where I is the rate of increase of the population, r is the intrinsic rate of increase of the population (average birth rate – average death rate), and N the number of individuals in the population at a given moment. If r is positive, the population will grow at an ever-accelerating rate.

The exponential growth of many real populations begins to level off as the density approaches the *carrying capacity* (K) of the environment. Such a growth curve is called an *S-shaped* or *logistic growth curve* and results from a changing ratio between births and deaths. After the curve has leveled off, births and deaths are in balance and the population has *zero population growth*. This occurs because environmental limitations become increasingly effective in slowing population growth as the population density rises (i.e., the limitations are *density-dependent*).

The populations of many small short-lived animals, or those living in variable environments, go through a period of exponential growth, followed by a sudden crash (boom-and-bust curve). The crash occurs before the populations reach the carrying capacity; it is due to a *density-independent* limitation such as weather or other physical environmental factors. The operation of such a limitation does not depend on the density of the organisms.

In addition to the birth rate and death rate, the potential life span, the average life expectancy, the average age of reproductive maturity, and the age distribution are important determinants of the makeup of a population. Determining the mortality rates for the various age groups in the population gives a *survivorship curve*. A type I curve is one in which all the organisms live to old age and die quickly, a type II curve shows a constant mortality rate at all ages, and a type III curve is typical of populations where the mortality among the young is very high, but those who survive the early stages tend to live for a long time. In nature, high mortality among the young is the rule.

Population regulation The regulation of population density in organisms with boom-and-bust curves is primarily due to the work of density-independent limitations. These organisms produce large numbers of small, quickly maturing offspring. Since they have evolved high intrinsic rates of increase (i.e., high r), they are called r-*selected species*.

Organisms with S-shaped growth curves, whose population limitation is primarily density-dependent, are called K-*selected species*; the maximum density for population stability is determined largely by the environment's carrying capacity. The fitness of these organisms depends on their ability to exploit the limited environmental resources efficiently. k-selected species tend to be large, longer-lived organisms producing only a few slow-maturing young.

Both *parasitism* and *predation* usually influence the prey (or host) species in a density-dependent manner. In general, the density of predators or parasites fluctuates in direct proportion to the changes in the density of the prey.

Intraspecific competition is one of the chief density-dependent limiting factors. As population density rises, competition for limited environmental resources becomes increasingly intense and acts as a brake on population growth.

Interspecific competition can also act as a density-dependent limitation. The more the niches of the species overlap, the more intense the interspecific competition. *Niche* refers to the functional role and position of an organism in an ecosystem. Every aspect of an organism's existence helps define that organism's niche. Two species cannot for long simultaneously occupy the same niche in the same place. The more similar two niches are, the more likely it is that both species will be competing for a limited resource. According to the *principle of limiting similarity*—a modification of the competitive exclusion principle—there is a limit to the amount of niche overlap compatible with coexistence. Extinction of one of the competing species, range restriction, character displacement, or alternation of dominance between periods of stability and crisis will usually be the outcome of intense interspecific competition.

In some animals, crowding induces physiological and behavioral changes that result in increased emigration from the crowded region. Crowding may also result in decreased resistance to disease and in hormonal changes that reduce the reproductive rate. The importance of this mechanism in nature is unclear.

The biotic community The species composing a community interact with each other and with the physical environment. The biotic community they form can be considered a unit of life, with its own characteristic structure and functional interrelationships.

Ecological succession is an orderly process of community change involving the replacement, over time, of the dominant species within a given area by other species. Succession results in part from the modification of the habitat by the organisms themselves or by physiographic changes, and in part from differences in dispersal and growth patterns among species. The species that predominate in the early stages tend to be *r*-selected speceis that are rapid growers and easily dispersed.

Successions in different places and at different times are not identical. The sequence of changes in *primary successions* (those occurring in newly formed habitats) is longer and slower than in *secondary successions* (those occurring in areas where previous communities were destroyed). The trend of most successions is toward a more complex and longer-lasting ecosystem in which less energy is wasted and hence a greater biomass can be supported without further increase in the supply of energy.

If no disruptive factor interferes, most successions reach a stage that is more stable than those that preceded it—the *climax community*. It will persist as long as the climate, physiography, and other environmental factors remain the same.

Human activities have often increased community instability by decreasing species diversity and structural complexity. A prime example is modern agriculture, with its emphasis on *monocultures* (the planting of a single crop species in a large field) and the use of chemicals to control insect infestations and outbreaks of disease.

CONCEPT MAP

Map 33.1

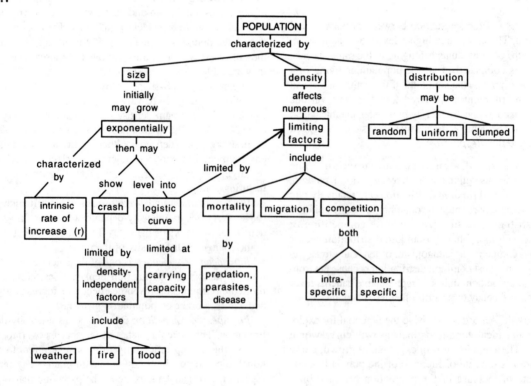

QUESTIONS

Testing recall

Match each of the descriptive phrases listed below with the term it best defines. Use each answer only once.

- a abiotic
- b biosphere
- c climax community
- d community
- e competition
- f ecological niche
- g ecosystem
- h K
- i mortality rate
- j population
- k r
- l succession

1 global ecosystem (p. 738)

2 all organisms in a given place at a given time (p. 737)

3 functional role of an organism in its community (p. 749)

4 progressive change in the plant and animal life of an area (pp. 752–53)

5 intrinsic growth rate of a population (p. 741)

6 a major determiner of population density (p. 740)

7 carrying capacity of the environment (p. 741)

8 group of individuals belonging to the same species (p. 737)

9 stable stage of succession (p. 757)

10 density-dependent limitation on population growth (p. 749)

11 sum total of the physical features and organisms in a given area (p. 737)

Indicate whether the following statements are true or false, and correct the false statements.

12 Clumped spacing of members of a population is more common than uniform or random spacing. (p. 739)

13 A population with more offspring per generation will always have a faster growth rate (r) than a similar population with fewer offspring per generation. (pp. 740–42)

14 The house fly, which has a short life span and produces a large number of eggs, could be considered a K-selected species. (pp. 745–46)

15 Predation usually acts as a density-independent limiting factor on populations. (pp. 748–49)

16 The term niche refers to an organism's limiting physical environmental factors, such as temperature, humidity, and soil pH. (p. 749)

17 Interspecific competition between two closely related species living in the same area may lead to the extinction of one of the competing species. (p. 750)

18 The species composition of the community in a given area changes markedly during the course of ecological succession. (p. 752)

19 Energy utilization is usually less efficient in a climax community than in a pioneer community. (p. 757)

20 All succession in a given large climatic area will eventually converge to the same climax type; there is only one type of climax community for a given region. (p. 757)

21 Human intervention generally makes communities more complex. (p. 758)

Testing knowledge and understanding

22 All the following are density-dependent factors that limit population growth *except*

- a intraspecific competition.
- b interspecific competition.
- c disease.
- d fire.
- e predation. (p. 749)

23 The *best* definition of a niche would be

- a an organism's habitat.
- b the ways in which an organism uses its environment to make a living.
- c the relationship of an organism to other species.
- d the physical limiting factors of an organism.
- e the factor to which an organism can adapt least. (p. 749)

24 Which of the following statements about r-selected species is *false*?

- a They generally inhabit unstable environments.
- b They are good competitors for limited resources.
- c They have a type III survivorship curve.
- d They are usually limited by density-independent factors.
- e They have many, quickly maturing offspring. (p. 745)

Questions 25 to 29 refer to the following situation.

A New York State farmer stocked his farm pond with 580 bluegill fish fingerlings. Bluegills usually reproduce first as yearlings and regularly thereafter. The farmer recorded the number of fish each year for the next 10 years. He obtained the following data:

Year	Number of fish
Stocked	580
1	600
2	750
3	1200
4	1400
5	1460
6	1440
7	1450
8	1460
9	1450
10	1460

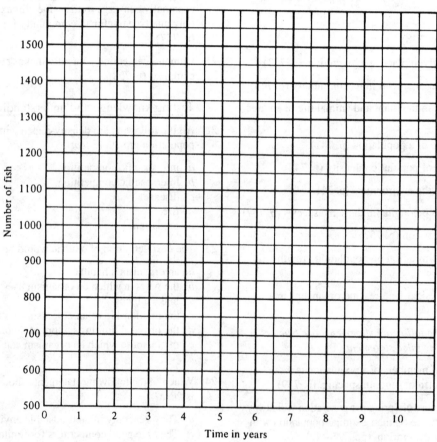

25 Plot these data on the graph.

26 What kind of growth curve is seen? (pp. 741–42)

27 Would you consider the bluegill a *K*-selected species or a *r*-selected species? (p. 746)

28 Mark the point on the curve where the rate of increase is greatest. (p. 741)

29 What factors might be involved in slowing the population growth from the fourth year on? (pp. 749–50)

Choose the one best answer.

30 To an ecologist, an "interbreeding group of individuals that occupies a specific geographic area" is a(n)

  a population.
  b species.
  c community.
  d gene pool.
  e ecosystem. (p. 737)

31 The graph below, showing the total number of persons in the United States in each census, 1790 to 1970, indicates that the population

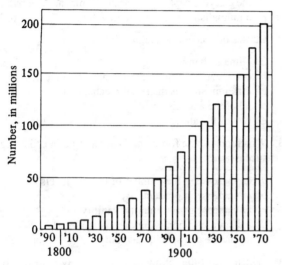

  a was growing exponentially.
  b had reached carrying capacity in 1970.
  c was at zero population growth in 1970.
  d exhibited an S-shaped growth curve.
  e none of the above (p. 741)

32 In the curve below, at which stage of population growth is the birth rate equal to the death rate? (p. 741)

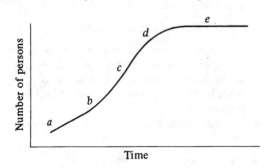

33 Which one of the following populations will show the greatest population increase in the next year assuming exponential growth?

  a population A, with 200,000 individuals, when $r = 0.020$
  b population B, with 500,000 individuals, when $r = 0.040$
  c population C, with 2 million individuals, when $r = 0.008$
  d population D, with 10 million individuals, when $r = 0.002$
  e population E, with 30 million individuals, when $r = 0.001$ (pp. 740–41)

34 Suppose that you study the growth of a population. You find that it increases in size exponentially and then reaches steady state. Which of the following factors is highly *unlikely* to have been a factor in limiting the population?

  a predation
  b parasitism
  c fire
  d competition
  e emigration (p. 748)

35 Which one of the following would be most likely to act as a limiting factor for a population of insects showing a boom-and-bust growth curve?

  a predation
  b competition
  c disease
  d a period of unfavorable weather
  e physiological changes induced by crowding (p. 742)

36 A certain fish species lives in large lakes. It feeds on small insects and other invertebrates. Spawning occurs in late spring, when males and females congregate in shallow water and engage in brief courtship displays, after which they release gametes in great clouds. The adult fish then swim back to deeper water. Young fish become sexually mature when they are three years old.

Which of the following survivorship curves is most likely to apply to this species? (p. 743)

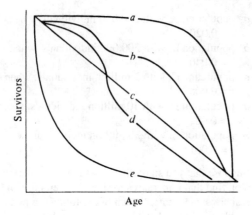

37 All the following characteristics are typical of a *r*-selected species *except*

   a it lives in an unpredictable environment.
   b it has a high reproductive rate.
   c it is small in size.
   d it has a long life span.
   e it exhibits little or no parental care of young. (p. 745)

38 Which one of the following statements is *false*?

   a Theoretically, the ecological niche of any given species is determined by an almost infinitely large number of different factors.
   b Linked fluctuations in the density of predator and prey species indicate that the major limiting factor for the predator is the availability of its food.
   c The reproductive potential of any species of plant or animal usually equals the actual reproductive rate.
   d Severe interspecific competition often leads to extinction, range restriction, or evolutionary divergence. (p. 740)

39 Which of the following statements is *correct*?

   a Two species may not live in the same habitat.
   b The more dissimilar the niches of two species, the stronger is their competition.
   c No two species can occupy exactly the same niche in the same geographical range.
   d No two species may occupy the same ecosystem.
                                              (p. 750)

40 When two species in similar niches compete for a limited resource,

   a one species may become extinct.
   b their habitats may diverge.

   c the species may alternate dominance between normal periods and crisis periods.
   d character displacement may occur.
   e all of the above.   (p. 750)

41 If two sympatric species occupy very similar niches under natural conditions, one would expect

   a the species to hybridize.
   b intense interspecific competition to occur.
   c extensive interspecific cooperation to occur.
   d the carrying capacity of the environment to be reduced.
   e a mutualistic symbiosis to develop.   (p. 750)

42 Causes of succession include

   a climate change.
   b one species altering the environment for the next species.
   c different species dispersal mechanisms.
   d all of the above.
   e two of the above.   (pp. 752–54)

43 Which one of the following is *not* a trend in ecological succession?

   a an increase in the number of trophic levels
   b an increase in productivity
   c an increase in community stability
   d a decrease in nonliving organic material
   e an increase in species diversity   (pp. 752–57)

44 Which one of the following is *least* likely to be true of an ecological succession?

   a The species composition of the community changes continuously during the succession.
   b The total number of species rises initially, then stabilizes.
   c The total biomass in the ecosystem declines after the initial stages.
   d The total amount of nonliving organic matter in the ecosystem increases.   (pp. 752–57)

45 The climax stage of a biotic succession

   a persists until the environment changes significantly.
   b changes rapidly from time to time, seldom remaining at any stage for more than a decade or so.
   c is the first stage in the reclamation of land from a lake bottom.
   d is a stage in which the dicot plants are always dominant.
                                              (p. 757)

46 Which ecosystem is the most unstable?

   a Sahara desert
   b African grassland
   c Alpine tundra
   d Montana wheat field
   e Costa Rican tropical rain forest   (pp. 758–60)

For further thought

1. The estimated population of Iran in 1990 was 55,650,000. If there were 44 births per 1,000 and 10 deaths per 1,000, what is *r* for this population? What would be the approximate size of the population at the end of 1991? Contrast *r* and the annual population growth of Iran with that of the People's Republic of China, where the estimated population in 1990 was 1,130,000,000, the birth rate was 23 per 1,000, and the death rate was 7 per 1,000. Which country will show a greater population increase in the years 1990 to 1995 if the rates of increase remain unchanged?

2. Look at the human population growth curve shown in question 31. Can our population continue to grow indefinitely? What are some of the limitations on population growth that may occur as the population density increases?

3. There are basically two different philosophical positions to take in regard to population growth. The position often taken by biologists is that the earth has a carrying capacity and that ultimately human population growth must be limited, either by self-regulation or by catastrophe. The position often taken by economists is that there is no reason to be concerned about population growth because we will always have technological advances that extend the carrying capacity. Which position most reflects your observations of the world?

4. From your readings and experience, give examples of the important ecological maxim: "You can never do just one thing."

5. Research the role of migration in the growth and development of the United States. What fraction of the total growth rate of the population is from birth rate? What fraction is from immigration? What projections are being made for this trend? From what you have read about migration as a density-dependent limiting factor, evaluate the implications of our current immigration policy.

6. Your book indicates that human intervention often has the effect of simplifying community structure. Give one example from your experience that illustrates this principle. Do you know of any examples that result in more complexity?

ANSWERS

Testing recall

1. *b*
2. *d*
3. *f*
4. *l*
5. *k*
6. *i*
7. *h*
8. *j*
9. *c*
10. *e*
11. *g*
12. true
13. false—often
14. false—*r*-selected
15. false—density-dependent
16. false—niche includes more than environmental factors; it is the functional role and position of the organism in its ecosystem
17. true
18. true
19. false—more efficient
20. false—species comprising a given community are products of local environmental conditions—there is no single climax community for a region
21. false—more simple

Testing knowledge and understanding

22. *d*
23. *b*
24. *b*
25. The curve should be similar to the one in Figure 33.7 (p. 741).
26. S-shaped
27. *K*-selected
28. The rate of increase is greatest at the inflection point (Fig. 33.7, p. 741).
29. density-dependent factors, especially competition
30. *a*
31. *a*
32. *e*
33. *e*
34. *c*
35. *d*
36. *e*
37. *d*
38. *c*
39. *c*
40. *e*
41. *b*
42. *e*
43. *d*
44. *c*
45. *a*
46. *d*

Chapter 34

ECOSYSTEMS AND BIOGEOGRAPHY

A GENERAL GUIDE TO THE READING

This last chapter, the second on the topic of ecology, considers some of the ways in which the movement of energy and materials binds together the community and the physical environment as a functioning system. First it looks at the movement of energy and nutrients through ecosystems, and then it goes on to consider how physical forces create regional variations in climate and thus influence the dispersal of species and the distribution of communities on earth. As you read this chapter, you will want to focus on the following.

1. The flow of energy. Learn the terms presented on pages 763 to 766, and make sure you understand why the flow of energy is always noncyclic.

2. The cycle of materials. At first the four cycles presented on pages 766 to 774 may look forbiddingly detailed, but careful study will show that they share certain general features. In all of them, materials cycle between living and nonliving systems. You will find the cycles easier to learn if for each, you identify the reservoir in the nonliving system, recognize how the chemicals enter and leave the cycle, and understand how they are passed from organism to organism within the living system.

3. Biological magnification. You have doubtless heard for years about the effects of DDT and other persistent chemicals; the discussion in the text shows the reason for the problem. Figure 34.23 (p. 776) is helpful in understanding biological magnification.

4. Soil. Soil is a resource often taken for granted. It is important for you to focus on the soil's key role as a bridge between the inorganic and organic worlds and the variety of ways that humans can cause soil depletion. (pp. 777–81)

5. Biomes. Biomes are a function of climate differences brought about by differences in latitude and altitude (which in turn determine light, temperature, and moisture availability). The major biomes of the world are described (pp. 784–89). The accompanying photographs will help you recognize the characteristics of each.

6. Continental drift. Especially important in the section on biogeography (pp. 794–99) is the material on the concept of continental drift, since it helps explain the present distribution of plants and animals.

KEY CONCEPTS

1. The movement of energy and materials knits a community together and binds it with the physical environment to form a functioning system. (pp. 763–66)

2. Radiant energy from the sun is the ultimate energy source for life on earth. This energy is captured by the producers and passed on to the consumers. (pp. 763–66)

3. Energy is constantly lost from an ecosystem as it is passed along the links of a food chain. Energy flow is always noncyclic. (pp. 763–66)

4. The water and mineral components of the biosphere are used over and over again. They cycle through an ecosystem, and they can be passed around it indefinitely. (pp. 766–74)

5 Soil is essential to plants not only as a substrate, but as a reservoir for water and essential minerals. Thus, the soil is a key link between the inorganic and organic worlds. (pp. 777–81)

6 The sun is the ultimate energy source for life, and the distribution of its energy helps determine the distribution of living things. Because the earth's surface curves away from the path of incident light, areas at different latitudes receive different amounts of sunlight and precipitation. Changes in altitude produce effects similar to changes in latitude. (p. 782)

7 A limited number of major categories of climax formations, called biomes, can be recognized; classification of a region as belonging to a particular biome is determined by which climax community is the most common in the region. (pp. 784–89)

8 The distribution of biomes is a consequence of climate, physiography, and other environmental factors within each geographic area. (pp. 784–89)

9 Understanding the distribution of life today requires understanding the earth's land masses and climate, both present and past. (pp. 794–99)

CROSSLINKING CONCEPTS

1 Your past study of photosynthesis (Chapter 7) and respiration (Chapter 8) will deepen your understanding of the broader concept of matter cycling between the biotic and abiotic worlds. Photosynthesis and respiration are two key processes that link organisms to the water and carbon cycles.

2 The pyramid of productivity, which describes energy flow through the producers and consumers of ecosystems, is a result of the Second Law of Thermodynamics. As you will recall from Chapter 4, this law states that every transfer of energy involves some loss of energy as waste heat. Thus each stage will have less energy than the stage before.

3 Chapter 16 mentioned that the bacteria living in the soil and symbiotically in root nodules of legumes and some other plants are very important because they are able to fix nitrogen. In this chapter, you will see how this fixed nitrogen circulates through the ecosystem. Protein synthesis (Chapter 11) and nitrogenous wastes (Chapter 22) are important examples of how organisms are connected with the nitrogen cycle.

4 In a sense, evolution is ecology over time. Ecology at a particular moment is the interaction of organisms with their environment. It is the environment that determines the fitness of particular organisms. The adaptations that you studied before, in an evolutionary context of being naturally selected, now should take on additional meaning as you see the intimate relationships between organisms and their natural environments.

OBJECTIVES

1 Diagram the flow of energy through an ecosystem, including producers, decomposers, primary consumers, secondary consumers, first trophic level, second trophic level, and third trophic level. State the difference between a food chain and a food web. (pp. 763–65)

2 Distinguish between gross primary productivity and net primary productivity. (pp. 763–64)

3 Explain why the distribution of productivity within an ecosystem can always be represented as a pyramid and why there are seldom more than four or five levels in a food chain. In doing so, specify the percentage of energy present at one trophic level that can usually be passed on to the next, and indicate the reason for the great decrease from level to level in the amount of available energy. (pp. 764–66)

4 Using Figure 34.8 (p. 767), describe the water cycle, specifying the roles of evaporation, transpiration, and rainfall. (pp. 766–68)

5 Using Figure 34.10 (p. 768), describe the carbon cycle. In doing so, explain how carbon enters the living system and how it leaves, indicate the role of microorganisms in the cycle, and identify the reservoir for carbon. In the context of global warming, describe how humans are affecting the carbon cycle. (pp. 768–71)

6 Using Figure 34.16 (p. 772), describe the oxygen cycle. In doing so, explain how the carbon and oxygen cycles are linked and how ozone is formed. Then describe the role of the ozone layer in shielding the earth's surface from harmful radiation and the factors that are leading to its depletion.

7 Using Figure 34.18 (p. 773), describe the nitrogen cycle. In doing so, discuss nitrogen fixation, nitrification, and denitrification; indicate the role of microorganisms in the cycle; and identify the reservoir for nitrogen. (pp. 773–74)

8 Using Figure 34.20 (p. 775), describe the phosphorus cycle. In doing so, explain how phosphorus enters the living system, indicate the role of microorganisms such as bacteria in the cycle, and identify the reservoir for phosphorus. Then describe the effect of increased phosphorus levels in a freshwater ecosystem, and show how they may accelerate eutrophication. (pp. 774–75)

9 Using Figure 34.23 (p. 776), explain the process of biological magnification. (p. 776)

10 Describe how the structure and properties of soil affect the availability of water, oxygen, and minerals to plants. Give three ways that humans can deplete soil fertility. Define acid rain and explain its effect on soil fertility. (pp. 777–81)

11 Explain how the earth's curvature and axis of rotation influence the amount of sunlight reaching a given area, and state how this influences the temperature and precipitation in that area. (p. 782)

12 List the major biomes of the world; for each, specify the principal characteristics of the area, some representative plants and animals, and the climatic factors that influence it. Using Figure 34.44 (p. 789), contrast altitudinal biomes with latitudinal biomes. Describe the trends in species diversity and vertical distribution of species with change in latitude. (pp. 784–89)

13 Distinguish between the terms benthic division and pelagic division, and photic zone and aphotic zone. Compare the coral reef community and the rain forest. (pp. 792–93)

14 Outline the main features of the continental-drift concept, and indicate how it helps explain the present geography of life. (pp. 794–99)

15 Discuss the similarities and the histories of the two island continents, and indicate how their isolation from other continents is reflected in the character of their plants and animals. (pp. 794–99)

16 Name the four main biogeographical regions of the World Continent, and explain why the plants and animals of these regions are more alike than are those of the island continents. Indicate what barriers separate these regions from one another. (pp. 794–99)

KEY TERMS

gross primary productivity (p. 763)
net primary productivity (p. 763)
detritus (p. 763)
food chain (p. 763)
food web (p. 764)
producer (p. 764)
decomposer (p. 764)
primary consumer (p. 764)
secondary consumer (p. 764)
trophic level (p. 764)
top predators (p. 764)
pyramid of productivity (pp. 764–66)
pyramid of biomass (p. 766)
pyramid of numbers (p. 766)
water cycle (p. 766)
aquifer (p. 766)
carbon cycle (p. 768)
greenhouse effect (p. 769)
nitrogen cycle (p. 773)
nitrogen fixation (p. 773)
nodule (p. 774)
nitrification (p. 774)
denitrification (p. 774)
phosphorus cycle (p. 774)
eutrophication (p. 775)
biological magnification (p. 776)
loam (p. 777)
humus (p. 778)

acid rain (p. 778)
salinization (p. 781)
biosphere (p. 781)
biome (p. 784)
tundra (p. 785)
taiga (p. 785)
deciduous forest (p. 786)
tropical rain forest (p. 787)
grassland (p. 788)
desert (p. 788)
continental shelf (p. 792)
continental slope (p. 792)
abyssal plains (p. 792)
ocean ridges (p. 792)
benthic division (p. 792)
pelagic division (p. 792)
plankton (p. 792)
photic zone (p. 792)
aphotic zone (p. 792)
estuary (p. 793)
coral reef (p. 793)
continental drift (p. 794)
Pangaea (p. 794)
Laurasia (p. 794)
Gondwanaland (p. 794)
Australian region (p. 796)
Neotropical region (p. 797)
World Continent (p. 798)
Nearctic region (p. 798)
Palaearctic region (p. 798)
Oriental region (p. 798)
Ethiopian region (p. 798)

SUMMARY

Flow of energy The flow of energy and materials knits a given community together and binds it with the physical environment as a functioning system. Almost all forms of life obtain their high-energy organic nutrients, directly or indirectly, from photosynthesis. The total amount of energy bound into organic matter by photosynthesis is called *gross primary productivity*; *net primary productivity* is the amount left after subtracting the amount the plant uses in respiration. Heterotrophs obtain their energy by consuming green plants, other heterotrophs, or the dead bodies or wastes of other organisms.

The *food chain* is the sequence of organisms, including the producers (autotrophic organisms), *primary consumers* (herbivores), *secondary consumers* (herbivore-eating carnivores), and *decomposers*, through which energy and materials may move in a community. In most communities the food chains are completely intertwined to form a *food web*. The successive levels of nourishment in the food chains are called *trophic levels*. The producers constitute the first trophic level, the primary consumers the second, the herbivore-eating carnivores the third, and so on. Since many species eat a varied diet, trophic levels are not hard-and-fast categories.

The flow of energy through the ecosystem is one-way. At each successive trophic level there is loss of energy from the system; only about 10 percent of the energy at one trophic level is available for the next. The reasons for this loss include lack of complete consumer harvest of available biomass, inability to assimilate all that is consumed, and respiration with its associated dissipation of heat energy. The distribution of productivity within a community can be represented by a *pyramid of productivity*, with the producers at the base and the last consumer level at the apex. In general, the decrease of energy at each successive trophic level means that less biomass can be supported at each level; thus many communities show a *pyramid of biomass*. Some communities also show a *pyramid of numbers*: there are fewer individual herbivores than plants, and fewer carnivores than herbivores.

Cycles of materials The endless cycling of water to earth as rain, back to the atmosphere through evaporation, and back to earth again as rain maintains the freshwater environments and supplies water for life on land. The *water cycle* is also a major factor in modifying temperatures and in transporting chemical nutrients through ecosystems.

Carbon cycles from the inorganic reservoir to living organism and back again. Carbon dioxide in the atmosphere or in water is converted into organic compounds by photosynthesis; the resulting organic compounds may be released as CO_2 by respiration or consumed by animals or decomposers. Eventually the carbon will be released as CO_2 and the cycle will begin again. Human activities, particularly the burning of fossil fuels and the clearing of forests, have increased the CO_2 and methane levels in the atmosphere; eventually this may lead to a change in the earth's temperature since the heat radiated from earth is absorbed by atmospheric CO_2 and radiated back to warm the earth in a *greenhouse effect*.

The oxygen in the atmosphere came from two processes: photosynthesis and the dissociation of water vapor in the upper atmosphere through the action of ionizing radiation. The present level of oxygen in the atmosphere is maintained largely because the oxygen produced during photosynthesis is balanced by that used in respiration. Oxygen in the outer atmosphere is found as ozone (O_3), which shields the surface of the earth from mutagenic and carcinogenic ultraviolet radiation. Various human activities, particularly the use of fluorocarbons, are causing the depletion of the protective ozone layer.

Biological *nitrogen fixation* by microorganisms, some of them living symbiotically in root *nodules*, provides most of the usable nitrogen for the earth's ecosystems. The microorganisms reduce atmospheric N_2 to compounds such as ammonia (NH_3) that can be absorbed by plant roots. Most flowering plants abosrob nitrate (NO_3^-) rather than ammonia. The nitrate is produced from ammonia by *nitrification*, which is carried out by certain groups of bacteria. When the plant dies, nitrogen is returned to the soil as NH_3, which can be recycled. Some bacteria carry out *denitrification*, converting NO_2^-, or NO_3^- into N_2 gas.

Phosphorus also cycles from the inorganic reservoir to living organisms and back again. Unlike the carbon and nitrogen cycles, which have important atmospheric reservoirs, the phosphorus cycle has a primary reservoir in rocks. Phosphate rock dissolves slowly and becomes available to plants, which pass it to animals. Some is excreted by animals; the rest is released from organic compounds when the organism dies. Sewage, detergents, and fertilizer runoff have greatly increased the phosphate levels in the aquatic environment, and thereby accelerated the *eutrophication* (aging) process of lakes.

Modern industry and agriculture have been releasing vast quantities of chemicals into the environment. Some chemicals, such as mercury, are harmless when released, but are made toxic by microorganisms. Some (for example, DDT and PCBs) show *biological magnification*. When ingested, these persistent chemicals are retained in the body and tend to become increasingly concentrated as they are passed up the food chain to the top predator.

The properties of soils—their particle sizes, amount of organic material, and pH, among others—determine how rapidly water and minerals move through the soils and how available to plants they will be. The proportions of clay, silt, and sand particles help determine many of these soil properties. Decaying organic material (*humus*) promotes proper drainage and aeration.

A complex equilibrium exists between the ions free in the soil water and those adsorbed on clay and organic particles. Many factors, acidity in particular, can shift this equilibrium; for example, acid rain, a by-product of air pollution, damages the soil. The plants and animals living in or on the soil also profoundly affect soil structure and chemistry. The cutting down of forests, poor farming practices, and irrigation have ruined many soils.

The biosphere The sun is the ultimate energy source for life, and the distribution of its energy helps determine the distribution of living things. Because the earth's surface curves away from the path of incident light, areas at different latitudes receive different amounts of sunlight and precipitation.

Most biologists recognize a number of major climatic regions called *biomes*. The *tundra* is the northernmost biome of North America, Europe, and Asia. The subsoil is permanently frozen. There are many organisms on the tundra but relatively few species. South of the tundra lies the zone dominated by the coniferous forests, the *taiga*. The number of species living in the taiga is larger than on the tundra.

The biomes south of the taiga show much variation in rainfall and thus more variation in climax communities. The *deciduous forests* predominate in temperate zones with abundant rainfall and long, warm summers. Tropical areas with abundant rainfall are usually covered by *tropical rain forests*, which include some of the most complex communities on earth; the diversity of species is enormous. Huge areas in both the temperate and tropical regions are covered by *grassland* biomes. These occur in areas of low or uneven rainfall. Places where the rainfall is very low form the *desert* biomes. Deserts are subject to the most extreme temperature fluctuations of any biome type.

A series of different biomes can also be found on the slopes of tall mountains. As with latitude, climatic conditions change with altitude, and biotic communities change correspondingly.

Aquatic ecosystems also vary with varying physical conditions. The *benthic division* comprises the ocean bottom with all bottom-dwelling organisms, whereas the *pelagic division* includes the water above the bottom with all the swimming and floating organisms. The upper, well-lighted region of the ocean constitutes the *photic zone*, and the deeper, lightless region the *aphotic zone*. Because of the scarcity fo mineral nutrients, the open ocean is similar in productivity to a desert. The most complex oceanic communities occur in the shallow waters above the continental shelf. Two regions of particularly high productivity and diversity are *estuaries*, which occur where freshwater rivers flow into the ocean, and *coral reefs*, which are found in the tropical zone.

Biogeography To understand the present geography of life, we must combine knowledge of present conditions with evidence from the fossil record and with geological evidence of past configurations of the earth's land masses and their climates. Geological evidence indicates that 225 million years ago all the earth's land masses were combined in a single supercontinent called *Pangaea*. Pangaea broke up into a northern supercontinent called *Laurasia* and a southern one called *Gondwanaland*. Soon Gondwanaland broke up; India drifted to the north and the African–South American mass separated from the Antarctic–Australian mass. Later, each of these masses split. The division of Laurasia into North America and Eurasia was one of the last to occur. As the continents moved, their climates changed and altered the distribution of organisms. In addition, fossil evidence indicates that the earth's climate has undergone many changes over time.

Because of the long isolation of the *Australian region* from the other continents, the plants and animals found there are most unusual; many species, particularly the mammalian fauna, that are common in Australia exist nowhere else. Most of the ecological niches that are filled by placental mammals on other continents are filled by marsupials in Australia. There is a considerable amount of convergent evolution between these two groups.

The South American continent, the *Neotropical region*, has also been an island continent through much of its history, but its nearness to North America and its recent connection via the Central American land bridge led to increased species diversity; many of its plants and animals resemble those of the World Continent (Europe, Asia, Africa, and North America).

The World Continent can be divided mainly into the *Nearctic* (most of North America), *Palaearctic* (Europe, northernmost Africa, northern Asia), *Oriental* (southern Asia), and *Ethiopian* (Africa south of the Sahara). North America and Eurasia have been connected through much of their history by the Siberian land bridge.

CONCEPT MAPS

Map 40.1

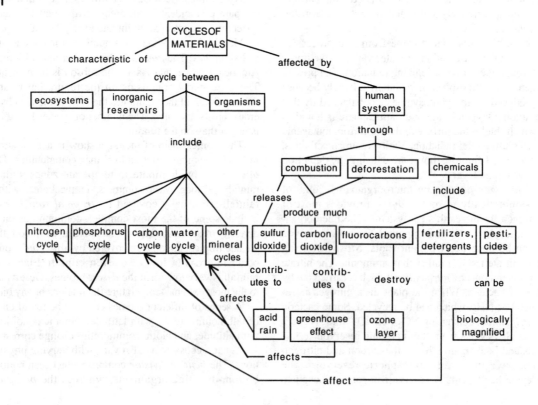

QUESTIONS

Testing recall

Determine whether each of the following statements is true or false. If it is false, correct it.

1 Net productivity is higher than gross productivity. (p. 763)
2 Food chains almost always begin with some sort of unicellular autotrophic organism. (p. 763)
3 The pyramid of productivity holds true for all populations. (p. 766)
4 The pyramid of numbers holds true for all populations. (p. 766)
5 In a normal biological community, there will be less usable energy at the carnivore trophic level than at the herbivore trophic level. (p. 766)
6 In most biological communities there are more top predators than primary consumers. (p. 766)
7 Most flowering plants obtain their nitrogen in the form of nitrate. (p. 773)
8 Carnivores are more likely to have high concentrations of DDT in their tissues than comparable herbivores in the same ecosystem. (p. 776)
9 Modern industry and agriculture have been releasing vast quantities of new or previously rare chemicals into the environment. (p. 776)
10 The subsoil of the taiga is permanently frozen. (p. 785)
11 A temperate deciduous forest usually has fewer plant species than a taiga forest. (p. 786)
12 A deciduous forest has more vertical stratification than a tropical forest. (p. 787)
13 Almost no producers are found in the benthic division in the aphotic zone of the ocean. (p. 792)
14 Many marsupials of Australia have convergently evolved with placentals of other continents. (p. 796)

Testing knowledge and understanding

Choose the one best answer.

15 The diagram below shows a particular food web. Each letter represents a different species. Arrows indicate the

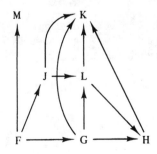

flow of energy and materials. Which of the following would have the greatest total biomass?

a F   d K + M
b J + G   e H
c K   (pp. 764–66)

16 Which one of the species is probably carnivorous?

a J   c L
b M   d G   (p. 765)

17 Which species is a decomposer?

a F   d K
b G   e L
c H   (p. 765)

The arrows in questions 18 to 23 indicate the direction of energy flow. For each question, evaluate the validity of the direction of the arrow. If the arrow points the correct way, the answer is a; if the direction is incorrect, the answer is b.

18 autotroph → heterotroph (p. 764)
19 carnivore → herbivore (p. 764)
20 dead carnivore → decomposer (p. 765)
21 respiration → photosynthesis (p. 764)
22 primary consumer → secondary consumer (p. 765)
23 producer → decomposer (p. 765)
24 The diagram below shows the flow of materials between trophic levels. Which arrow is incorrect? (p. 765)

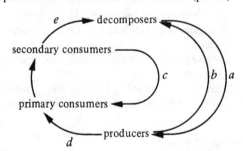

25 When we eat carrot sticks, we are acting as

a producers.   c secondary consumers.
b primary consumers. d tertiary consumers.
                                 (pp. 764–66)

26 In most food chains,

a there are fewer individuals at the top predator level than at the second trophic level.
b there is less usable energy at the herbivore level than at the carnivore level.
c there are few individuals at the decomposer level.
d there is less usable energy at the autotrophic level than at the carnivore level.   (pp. 764–66)

27 Assume that green plants in an ecosystem produce 1,000 grams of dry mass per square meter and there is an efficiency of 10 percent in each transfer of energy from one trophic level to the next. How many total grams of dry mass will be produced by the tertiary consumers per square meter per year?

   a 100
   b 10
   c 1
   d 0.1
   e 700   (pp. 764–66)

28 Of the five elements that are the most important constituents of living things, the only one that requires the action of microorganisms to enter the living system is

   a carbon.
   b hydrogen.
   c oxygen.
   d nitrogen.
   e phosphorus.   (pp. 773–74)

29 Combustion of fossil fuels most directly affects

   a the carbon cycle.
   b the water cycle.
   c the nitrogen cycle.
   d the phosphorus cycle.
   e none of the above.   (p. 768)

30 Important processes of the water cycle include all of the following *except*

   a precipitation.
   b evaporation.
   c rock weathering.
   d percolation.
   e transpiration.   (p. 767)

31 Though free nitrogen (N_2) is abundant in the atmosphere, living organisms are unable to use it until it is "fixed" into certain nitrogen compounds. Which one of the following statements best describes nitrogen fixation?

   a Only flowering plants are able to fix nitrogen.
   b Nitrogen is fixed by both free-living and symbiotic bacteria.
   c Only symbiotic bacteria, not free-living bacteria, fix nitrogen.
   d Most nitrogen is fixed by lightning during thunderstorms.
   e Animals, but not plants, are able to fix nitrogen.   (p. 773)

32 Biological magnification is the

   a increase in size of organisms at higher trophic levels.
   b increase in number of organisms at higher trophic levels.
   c concentration of stable, nonexcretable chemicals in organisms at higher trophic levels.
   d concentration of nitrates and phosphates in a polluted lake.
   e capture of small organisms by larger organisms at higher trophic levels.   (p. 776)

33 DDT has been more of a problem in predatory birds such as the bald eagle than in seed-eating birds. This is because

   a seed-eating birds are smaller.
   b predatory birds eat more than seed-eating birds.
   c DDT is absorbed by animals, but not plants.
   d biological magnification concentrates DDT in higher trophic levels.
   e seed-eating birds are able to metabolize DDT and excrete its by-products.   (pp. 776–77)

34 The cycle most important for modifying temperature fluctuation is

   a the water cycle.
   b the carbon cycle.
   c the nitrogen cycle.
   d the phosphorus cycle.
   e none of the above.   (p. 767)

35 The pollutant that contributes most to acid rain is

   a Sulfur dioxide.
   b Carbon monoxide.
   c ozone.
   d methane.
   e flurocarbons.   (p. 778)

36 The cycle with a major reservoir in rock is

   a the carbon cycle.
   b the nitrogen cycle.
   c the phosphorus cycle.
   d two of the above.
   e all of the above.   (p. 775)

37 The step in the nitrogen cycle that provides the nitrogen source used by carnivores is the formation of

   a ammonia (NH_3).
   b nitrate ions (NO_3^-).
   c nitrate ions (NO_2^-).
   d atmospheric nitrogen (N_2).
   e amino acids.   (p. 773)

38 Water, oxygen, and minerals are most available to plants growing in soils that are called

   a sands.
   b loams.
   c clays.
   d silts.   (p. 777)

39 Soil fertility can be destroyed by

   a cutting down forests.
   b acid rain.
   c overgrazing.
   d overirrigation.
   e all the above.   (pp. 777–81)

40 Large parts of the former Soviet Union are covered in coniferous forests. These regions are cold in winter and warm in summer. This biome is known as

   a tundra.
   b taiga.
   c grassland.
   d deciduous forest.
   e desert.   (p. 785)

41 The climax formation (biome) of much of the state of Kansas is

   a tundra.
   b grassland.
   c tropical rain forest.
   d taiga.
   e deciduous forest.   (p. 788)

42 Which one of the following statements is *false*?

  a Deserts often display extreme daily variations in temperature.
  b Epiphytes are very common in tropical rain forests.
  c The subsoil of the tundra is permanently frozen.
  d The dominant trees of the taiga are birch and maple.
  e Vast numbers of waterfowl nest on the tundra.
  (pp. 785–86)

43 Vermont is in the

  a temperate deciduous biome of the Nearctic region.
  b taiga biome of the Nearctic region.
  c temperate deciduous biome of the Palaearctic region.
  d taiga biome of the Palaearctic region.
  e temperate deciduous biome of the Neotropical region.
  (pp. 786–98)

44 The ecosystems most associated with great diversity are

  a the taiga.   d coral reefs.
  b the deciduous forest. e two of the above.
  c the tropical forest.   (pp. 787–93)

45 If you were climbing a high mountain in Central America, in what order would you cross (1) deciduous forest, (2) taiga, (3) tundra, and (4) tropical rain forest?

  a 4, 3, 2, 1   d 3, 4, 1, 2
  b 4, 3, 1, 2   e 4, 1, 2, 3
  c 3, 4, 2, 1   (p. 789)

46 Distribution of life on the earth is determined by

  a altitude.   d two of the above.
  b latitude.   e all of the above.
  c geological history.   (pp. 793–99)

47 The present distribution of plant and animal life can best be explained by assuming that the earth's land masses have drifted from one place to another. According to this concept of continental drift,

  a South America and Australia were connected through most of geologic time.
  b North America and Europe were connected throughout much of their geologic history.
  c India split away from Asia and drifted southward.
  d the entire Oriental region belonged to the same supercontinent as South America and Africa 135 million years ago.
  e the separation of Europe from Asia occurred early in geologic history.
  (pp. 798–99)

48 When the supercontinent Laurasia broke up, it gave rise to all the following except

  a North America.   d India.
  b Greenland.   e Asia.
  c Europe.   (p. 794)

49 If the living representatives and all known fossils of a very ancient group of organisms were found only in Africa and Australia, on which of the following continents would you *most* expect to find additional fossils of this group?

  a South America   d Asia
  b North America   e Europe
  c Greenland   (pp. 794–99)

50 Which biogeographic regions were island continents?

  a Oriental and Ethiopian
  b Nearctic and Palaearctic
  c Nearctic and Neotropical
  d Australian and Neotropical
  e Australian and Oriental
  (pp. 796–99)

For further thought

1 Suppose that the fox in the food web diagramed in Figure 34.2 (p. 765) in your text feeds only on rabbits, squirrels, mice, and seed-eating birds, all of which feed on plant material. How many square meters of plant material are then required to support the fox if the net primary productivity of the plant material is 8,000 Kcal/m^2/year? Assume that a fox's daily caloric requirement is 800 Kcal and that only 10 percent of the energy at one trophic level can be passed on to the next. If the fox in Figure 34.2 were to feed only on insect-eating birds, how many square meters of plant material would be required?

2 Trace the route that a molecule of CO_2 might follow as it cycled through the ecosystem. (Include at least four organisms in the pathway.)

3 Calcium ions are required nutrients for both plants and animals. The reservoir for calcium is in rocks. Design a calcium cycle, including at least three organisms.

4 As the world's human population continues to grow, the problem of feeding people becomes increasingly difficult. What are some of the environmental consequences of attempting to feed so many people?

5 In 1992, there was a first Earth summit with diplomats from all over the world, representing their countries. Research the positions the United States took in this conference regarding global warming and biodiversity. What role do you feel our country, as one of the world's leading industrialized countries, should play in meetings such as this?

6 There is considerable controversy about the relationships among combustion, global warming, potential effects of global warming, and relevant national policies. Research this area and give your conclusions.

ANSWERS

Testing recall

1. false—lower
2. false—autotroph may be unicellular or multicellular
3. true
4. false—many
5. true
6. false—fewer
7. true
8. true
9. true
10. false—tundra
11. false—more
12. false—less
13. true
14. true

Testing knowledge and understanding

15	a	24	c	33	d	42	d
16	c	25	b	34	a	43	a
17	d	26	a	35	a	44	e
18	a	27	c	36	c	45	e
19	b	28	d	37	e	46	e
20	a	29	a	38	b	47	b
21	b	30	c	39	e	48	d
22	a	31	b	40	b	49	a
23	a	32	c	41	b	50	d

Chapter 35

ANIMAL BEHAVIOR

A GENERAL GUIDE TO THE READING

This chapter focuses on the mechanisms behind the behavior of organisms and applies what you have learned about control and effector mechanisms to the question of what, how, and why animals do what they do. As you read Chapter 35 in your text, you will want to concentrate on the following.

1. Fundamental components of behavior. You will want to pay particular attention to the material on sign stimuli and releasers (pp. 805–807), since understanding it is prerequisite to understanding much of the material that follows in this chapter.

2. Motor programs. Motor programs were previously discussed in Chapter 28. Make sure you understand the difference between a fully innate motor program and a learned motor program (pp. 807–809).

3. Learning and behavior. The relationship between instinctive behavior and learning is considered in some detail (pp. 809–10). The discussion of difficulties in studying learning is especially pertinent. You will need to know the differences between trial-and-error learning and classical conditioning, and the importance of imprinting on an animal's behavior (pp. 811–14).

4. Communication. You will find the discussion of animal communication (pp. 815–21) interesting and informative, and well illustrated by the diagrams and photographs. The dance language of the honeybees is an especially elegant example of social communication. Figure 35.30 (p. 821) will help you learn this material.

5. Orientation and navigation. You will want to understand the distinction between the two types of orientation behavior described on pages 822 to 826: compass-and-timer, and compass-and-map. Note the importance of biological clocks (p. 822) for sun-compass orientation.

6. Social behavior. The discussion of social behavior, or sociobiology (pp. 827–34), shows how spatial factors influence sociality and how altruistic behavior can evolve. Note particularly the fallacy of the notion that behavior can evolve "for the good of the species." To understand the section on mating strategies, you will want to focus on the concept of parental investment. (p. 830)

7. The evolution of behavior. The final segment of the chapter (pp. 834–35) emphasizes the concept that behavior is a biological attribute and evolves just as physical characteristics evolve. Try to keep this fundamental concept in mind as you do your reading.

KEY CONCEPTS

1. Anthropomorphism must not be used in the description and analysis of animal behavior patterns. We cannot assume that because an animal's actions are outwardly similar to our own, they have the same purpose or motivation. (pp. 803–804)

2. Behavior cannot generally be classified as strictly innate or strictly learned; even the most rigidly automatic

behavior depends on the environmental conditions under which it evolved, whereas most learning appears to be guided by innate mechanisms. (p. 805)

3. Releasers can initiate certain critical behavior responses automatically, and thus bypass the time-consuming and error-prone process of learning. (p. 806)

4. Motor programs appear to be all-or-none innate responses to releasers; once begun, they proceed to completion with little or no need for further feedback. On the other hand, learned behavior can become stereotyped and largely automatic; that is, it can take on the characteristics of an innate motor program. (pp. 807–809)

5. An animal may be motivated to do one thing now and another later. The general health of the animal, maturational state, hormonal level, activity of its central nervous system, sensory stimuli, and previous experience that led to learning are all involved in determining an animal's current behavior. (pp. 809–15)

6. Communication probably evolved first for mating and reproduction, and from the mechanism of sexual communication evolved the wide variety of signals that are now used to communicate other useful information. (p. 815)

7. The olfactory sense is immensely important in the lives of many animals and constitutes a basis for effective communication. (pp. 815–17)

8. Sound is an effective means of communication among many animals; it is a particularly useful form of communication in environments where vision is limited. (pp. 817–19)

9. A display is a behavior that has evolved specifically as a signal. Agonistic displays reduce physical combat; reproductive displays help avoid matings between individuals of different species and synchronize the sexual physiology of the male and female. (pp. 819–20)

10. All living things, whether individual cells or whole multicellular organisms, appear to have an internal sense of time—a biological clock. (p. 822)

11. Many animals have navigational abilities that allow them to find their way over great distances through unfamiliar territory. Some birds use a compass-and-timer strategy; others use a compass sense in conjunction with a map sense. (pp. 822–26)

12. Social behavior is strongly influenced by the way the individuals of a species are organized in space, which is in turn strongly dependent on the distribution of the food supply and on the animal's method of exploiting it. (pp. 827–28)

13. Among many social animals, individuals exhibit behavior that may reduce their personal reproductive success while increasing the reproductive success of other members of the species. (pp. 829–30)

14. The evolution of different mating systems in different species is often dependent on parental investment—the cost to the parent in terms of its ability to produce more offspring. (pp. 830–34)

15. Behavior is adaptive: in the course of evolution, natural selection brings about an increase in well-adapted and a decrease in poorly adapted behavior patterns. (pp. 834–35)

CROSSLINKING CONCEPTS

1. Much of animal behavior is based on the hormonal and nervous controls described in the Chapters 25, 26, 28, 29, and 30. The fullest comprehension of this material, therefore, rests on a thorough understanding of the endocrine and nervous systems and their chemical mechanisms. In particular, the concepts of motor programs and interactions of the endocrine and nervous systems will aid your understanding of animal behavior.

OBJECTIVES

1. Define behavior and indicate why the analysis of animal behavior poses some problems for the scientist studying it. (pp. 803–804)

2. Define the terms innate and instinct. (p. 804)

3. Differentiate among taxis, kinesis, and reflex; give an example of each. (p. 805)

4. Describe, citing the experiments of Lorenz and Tinbergen, some of the characteristics of sign stimuli and releasers, and indicate their role in animal behavior. Explain why sign stimuli are biologically adaptive and why they can be disadvantageous as well. Finally, define and give an example of a supernormal stimulus. (pp. 805–807)

5. Differentiate between innate motor programs and learned motor programs. (pp. 807–809)

6. Summarize the current thinking on the relationship between inheritance and learning in animal behavior. (pp. 809–10)

7. State the five major factors that complicate the study of learning. (p. 810)

8. Define the five major types of learning, and give an example of each. Distinguish between parental and sexual imprinting, and explain what is meant by a critical period (sensitive phase). (pp. 811–14)

9 Define a pheromone, and contrast the functions of releaser and primer pheromones. Describe the roles that both classes of pheromones play in the lives of social insects such as ants and bees. (pp. 815–17)

10 Describe the environmental conditions in which sound is a particularly useful form of communication, and indicate four functions of bird song. (pp. 817–18)

11 Explain what is meant by a display, and discuss the importance of reproductive and agonistic displays. In doing so, indicate the significance of the threat and appeasement displays to animals involved in an agonistic encounter. (pp. 818–20)

12 Using Figure 35.29 (p. 820), describe the honeybee's round dance and waggle dance and the information each conveys. Given the position of a food source and the sun in relation to the hive, predict the angle at which a waggle dance would be performed. (pp. 820–21)

13 Explain what is meant by a biological clock and by circadian rhythm, and describe how the environment can reset the clock. (p. 822)

14 Discuss the two strategies used by birds to navigate on long flights, indicating what is meant by a map sense and a compass sense. Then, describe the various cues that migratory birds and homing pigeons use to orient themselves on long flights. (pp. 822–26)

15 Explain how spatial factors may influence social behavior, and differentiate among the terms individual distance, territory, and home range (Fig. 35.43, p. 828 may be helpful). Using Figure 35.42 (p. 828), show how the nature of the food supply affects the type of territory occupied. (pp. 827–28)

16 Describe altruistic behavior and the possible benefits such behavior has for the individual. In doing so, explain the concepts of kin selection of reciprocal altruism. (pp. 829–30)

17 Describe the four different vertebrate mating systems, and explain the role of parental investment in determining the evolution of these systems. (pp. 830–31)

18 Describe the social organization in honeybee hives. Give four ways in which insect societies differ from vertebrate societies. (pp. 831–33)

19 Explain what is meant by a dominance hierarchy in vertebrate social organization, and indicate the advantages of this system. (pp. 833–34)

20 Explain, giving an example, how the biologist uses the evidence from the comparison of species and behavioral analysis to study the evolution of behavior patterns. (pp. 834–35)

KEY TERMS

ethology (p. 803)
anthropomorphism (p. 803)
innate (p. 804)
instinct (p. 804)
learned behavior (p. 804)
taxis (p. 805)
kinesis (p. 805)
reflex (p. 805)
sign stimulus (p. 805)
releaser (p. 806)
pheromone (p. 806)
supernormal stimulus (p. 807)
fixed-action pattern (p. 808)
motor program (p. 808)
sensitive phase, or critical period (p. 810)
latent learning (p. 810)
habituation (p. 811)
reinforcement (p. 811)
classical conditioning (p. 811)
operant conditioning (p. 812)
trial-and-error learning (p. 812)
insight learning (p. 812)
parental imprinting (p. 812)
sexual imprinting (p. 812)
releaser pheromone (p. 815)
primer pheromone (p. 815)
agonistic behavior (p. 818)
threat display (p. 818)
appeasement display (p. 819)
round dance (p. 820)
waggle dance (p. 821)
circadian rhythm (p. 822)
map sense (p. 823)
compass sense (p. 824)
sociobiology (p. 827)
individual distance (p. 827)
territory (p. 827)
home range (p. 828)
altruistic behavior (p. 829)
kin selection (p. 830)
reciprocal altruism (p. 830)
monogamy (p. 830)
polygamy (p. 830)
polygyny (p. 830)
polyandry (p. 831)
promiscuous (p. 831)
parental investment (p. 831)
dominance hierarchy (p. 833)

SUMMARY

The behavior of an animal—what it does and how it does it—is the product of the functions and interactions of its various control and effector mechanisms. The analysis of behavior poses a problem because it is difficult to avoid anthropomorphic interpretations.

Behavior that depends largely on inherited mechanisms is *innate* behavior whereas behavior that is acquired as a result of experience is *learned*. *Instinct* is the heritable, genetically specified neural circuitry that organizes and guides behavior. Behavior cannot generally be classified as strictly innate or strictly learned; even the most rigidly automatic behavior depends on the environmental conditions for which it evolved, whereas most learning appears to be guided by innate mechanisms.

Fundamental components of behavior *Taxes, kineses,* and *reflexes* involve simple responses to stimuli. In a taxis, the response is oriented relative to the stimulus, but in a kinesis it is not. More complicated behavior patterns predominate in higher animals.

A signal that elicits a specific behavioral response is known as a *sign stimulus*. The animal must possess neural mechanisms that are selectively sensitive to sign stimuli. Certain sign stimuli are known as *releasers* because they "release" (trigger) specific innate behavior patterns. Depending on the sensory world of the organisms, behavior may be released by cues from many sensory modalities. *Pheromones* (odors), sounds, and colors may act as releasers. Releasers can be advantageous; they can initiate certain critical behavior responses automatically, and thus bypass the time-consuming and error-prone process of learning. However, they may be triggered by inappropriate stimuli. Exaggerated events are often superior to the natural stimulus in releasing a response; such stimuli are known as *supernormal stimuli*.

Motor programs Some behaviors appear to be all-or-none responses to releasers; once begun, they proceed to completion with little or no need for further feedback. Such behaviors were called *motor programs* or *fixed-action patterns*.

It is often difficult to be sure whether or not a motor behavior has been learned if it is not exhibited at birth. In some cases, particularly among invertebrates, opportunities for learning are so slight that many behaviors seen only in adults must be regarded as innate. Even in vertebrates many behaviors that appear to be learned are actually innate. By contrast, many behaviors we know to be learned can become stereotyped and largely automatic and take on the appearance of fully innate motor programs.

Learning and behavior Inheritance and learning are fundamental in determining the behavior of higher animals. In general, the inherited limits within which behavior patterns can be modified by learning are much narrower in the invertebrates than in the vertebrates, and narrower in the lower vertebrates than in the mammals.

Studying learning is difficult because one must determine whether the behavior results from learning or from maturation, whether an animal can learn one thing in one context but not another, and whether a sensitive phase or a latent period is involved.

Habituation is a simple type of learning in which the organism shows a gradual decline in response to repeated "insignificant" stimuli that bring no *reinforcement*. *Classical conditioning* is the association of a response with a stimulus with which it was not previously associated. In *operant conditioning* (trial-and-error learning), the animal learns to associate a certain activity with a reward or punishment.

In *imprinting* an animal learns to make a strong, long-lasting association with another organism (or object). Imprinting occurs only during a short sensitive phase. Various types of imprinting are known. *Parental imprinting* involves the recognition of the parent by the young. The critical period for parental imprinting is early and brief, and the learning is not reversible. During *sexual imprinting*, an animal learns to recognize its species, especially with regard to mate choices. Various other types of imprinting are now known.

Insight is the organism's ability to respond correctly the first time it encounters a novel situation; it is most prevalent in higher primates.

Motivation Animals may be motivated to do one thing now and another later. Current priorities can help the animal select among many behavioral possibilities.

Communication Communication most probably evolved first to help animals find appropriate mates. The wide variety of signals that now communicate mood, intention, and other useful information probably evolved from the mechanisms of sexual communication.

Chemical communication is the most primitive form of communication. Many animals secrete substances (pheromones) that influence the behavior of other organisms. Most pheromones can be classified into two groups: *releasers,* which trigger a rather immediate and reversible behavioral change in the recipient (e.g., sex attractants in insects), and *primers,* which initiate more profound long-term physiological changes in the recipient. Primers may not produce an immediate change.

Many animals can communicate information by sound. In many insects, sound functions as a mating call and a species recognition signal. Bird song functions as a species recognition signal, as a display that attracts females to the male and helps synchronize their reproductive drives, and as a display to defend territory.

Agonistic encounters between individuals of the same species may often be resolved by vocal and/or visual displays without physical contact. Through displays the combatants convey their attack motivation; the individual showing the higher attact motivation usually wins. The loser often responds with *appeasement displays;* these tend to inhibit further attacks by the antagonist.

Visual displays are frequently important in reproductive behavior. (A display is a behavior that has evolved specifically as a signal.) They function both in bringing together and synchronizing the two sexes in the mating act and in avoiding mating errors.

Scout honeybees are able to communicate to workers in the hive the quality of a food source as well as its distance and direction from the hive. When food is close to the hive, the bees dance a *round dance*, which means nearby food. When food is farther sway, the bees perform a *waggle dance*; the length and orientation of the waggle run conveys information on the distance of the food source and its direction relative to the sun's position. The honeybee dance is entirely instinctive.

Biological clocks All living things, whether individual cells or whole multicellular organisms, appear to have an internal timer—a "biological clock." Many organisms show *circadian rhythms,* activity rhythms that vary within a period of approximately 24 hours. Although the basic period of the clock is innate, it is constantly being reset by the environmental cycle.

Orientation and navigation Many animals can find their way over great distances through unfamiliar territory. Some birds are simply preprogramed to fly a certain course (the compass-and-timer strategy). Others have a special *map sense*, which makes them behave as though they are aware of longitude and latitude. Migrating birds and homing pigeons have remarkable navigational abilities. Birds (and other animals) can tell compass directions by observing the sun and stars. Pigeons can use magnetic cues in addition to sun cues for orientation. Other cues such as olfactory cues, meteorological parameters, and special sensory capabilities (e.g., barometric pressure detection, infrasound detection) may play some role in avian orientation.

Social behavior Many animal species show some degree of intraspecific cooperation. The cooperation varies from relatively simple forms to highly evolved and long-lasting societies in which almost every aspect of each individual's life depends on the activities of others. Effective communication between individuals ia a prerequisite for such cooperation.

The spatial arrangement of individuals influences the species' social behavior. In many species each individual is surrounded by a small volume of space that is its own—the *individual distance*. Moderately social animals maintain a relatively inviolate individual distance; it tends to be less rigid in highly social animals. A *territory* is an area an animal or group of animals actively defends from members of the same species. By spacing individuals, territoriality minimizes competition and reduces social conflict. The *home range* is the largest spacing unit, the total area in which an animal (or group of animals) normally travels.

Many social animals exhibit *altruistic behavior,* behavior that may reduce the personal reproductive success of the individual exhibiting the behavior while increasing the reproductive success of others. Behavior that benefits the altruist's kin may improve the chances that some of the altruist's genes will be perpetuated, since close relatives share some of the same genes *(kin selection).* In some instances *reciprocity* may be an explanation; in other cases an act that seems altruistic may actually serve *selfish* ends.

Several different mating systems are found in vertebrates. *Monogamy,* in which one male mates with one female, is common in birds but rare in mammals. There are two types of *polygamy: polyandry,* in which one female mates with many males, and the more common *polygyny,* in which one male mates with many females. Many species form no pair bonds; their mating is often *promiscuous*.

One of the principal determining factors in the evolution of mating systems is *parental investment*—the cost to the parent of behavior enhancing the likelihood that the offspring will survive and reproduce. In birds, monogamy is adaptive because the young are born helpless and it may require a large parental investment from both parents to raise them. In mammals the male and female have very unequal parental investments and polygyny may be an effective reproductive strategy.

In the social insects the unit of organization is the family, which typically consists of a single reproductive female—the queen—and her daughters (sometimes her sons). Division of labor is based on biologically determined castes. The queen's role is reproductive; the sterile workers build the hive or nest, nurse the young, gather the food, and care for the queen.

A vertebrate social group typically consists of a leader (usually a male) together with his mates and offspring. Such a grouping benefits both from reduced aggression and from the greater tendency for cooperativity and altruism to evolve among individuals with shared genes. There is no biologically determined caste system, and all adults are potential reproducers. An important aspect of vertebrate social organization is the *hierarchy,* a series of dominance-subordination relationships that tend to give order and stability to the group.

The evolution of behavior Behavior is adaptive; natural selection brings about an increase in well-adapted and a decrease in poorly adapted behavior patterns in the population. Behavior patterns evolve and can be studied in terms of the selection pressures that produce them. By comparison of the behavior patterns of a whole group of related species, the evolutionary derivation of the patterns can be clarified.

CONCEPT MAP

Map 35.1

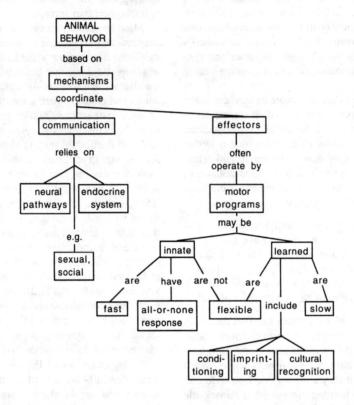

QUESTIONS

Testing recall

Select the correct term or terms to complete each statement.

1 Behavior that depends largely on inherited mechanisms is usually referred to as (innate, learned, instinctive, releasing) behavior. (p. 803)

2 An animal's simple movement toward or away from a stimulus is called a (tropism, taxis, kinesis, reflex). (pp. 805–807)

3 In humans, walking is an example of a(n) (innate motor program, learned motor program). (pp. 807–809)

4 The process by which newly hatched goslings follow their parents and recognize them is known as (parental imprinting, classical conditioning, trial-and-error learning, cultural learning). (p. 810)

5 A form of learning that ordinarily occurs quickly but only during a limited phase in the animal's life and that is difficult to unlearn is called (conditioning, insight, habituation, imprinting). (p. 811)

6 The area of the brain that acts as the center for learned motor programs in higher vertebrates is the (cerebral cortex, hypothalamus, cerebellum, midbrain). (p. 808)

7 (Visual, sound, chemical) signals represent the most primitive channel for communication among animals. (pp. 809–10)

8 Chemical substances, released by animals, that produce long-term physiological changes in other animals of the same species are called (primer pheromones, releaser pheromones, hormones.) (p. 815)

9 Bird songs may function (to indicate happiness, in territorial defense, in species recognition, as a display). (p. 818)

10 A dog approaches another with head up, hackles raised, and teeth bared. This is a(n) (greeting, appeasement, threat, mating) display. (p. 819)

11 Honeybees communicate (distance of the food source, direction of the food source relative to the sun) through the orientation of the waggle run on the (vertical, horizontal) comb in the hive. (p. 820)

12 Circadian rhythms have a period of about a (day, week, month, year). (p. 822)

13 Homing pigeons can use (the sun compass, the star compass, magnetic cues) as orientational cues. (pp. 823–26)

14 The total area in which an animal normally travels is known as its (individual distance, territory, home range). (p. 827)

15 In vertebrates, social hierarchies (give order and stability, decrease aggression). (pp. 831–34)

16 A mating system in which one female mates with many males is called (monogamy, polygyny, polyandry). (pp. 830–31)

17 Behavior patterns (do, do not) evolve. (pp. 834–35)

Testing knowledge and understanding

Choose the one best answer.

18 Behavior can be modified by

  a the nature of the sign stimulus.
  b the season of the year and the time of day.
  c the level of circulating hormones.
  d previous experience.
  e all the above.   (pp. 803–804)

19 In an experiment, a single frog tadpole was raised in an isolated, soundproof box. As soon as the tadpole transformed into a frog, several adult males of the same species were put into the box for several days; they called continuously during this period and were then removed. Except for these males, the original frog was never exposed to another frog. At maturity, it produced a call typical of its species. From the information provided here, what can you say about the basis of this call?

  a The call is entirely learned.
  b The call is entirely innate.
  c The call is partly learned, partly innate.
  d The call is neither learned nor innate.
  e None of the above can be concluded with certainty. (p. 804)

20 When a blowfly maggot is exposed to light it moves directly away from he light source. This is a(n)

  a reflex.   *c* kinesis.
  b taxis.   *d* instinct.   (p. 805)

21 When ready to reproduce, the adult male stickleback fish develops a bright red belly, which he displays aggressively to other males when staking out his nesting territory. Cardboard models of the fish placed in a nesting male's territory also elicit an aggressive display if the lower half of the model is painted red. The red belly is an example of a(n)

  a reflex.   *c* instinct.
  b sign stimulus. *d* learned pattern. (p. 805)

22 The courting male stickleback fish responds most vigorously to an extra-large model that looks like the drawing. This preferred but unnatural stimulus is sometimes referred to as a

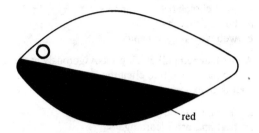

  a releaser.
  b supernormal stimulus.
  c kinesis.
  d displacement activity.
  e misdirected effort.   (p. 807)

23 Dung beetles of certain species dig burrows under a pile of dung and take the dung into the burrows to feed their larvae. When a female dung beetle finishes digging and stops bringing up soil from an underground borrow, the male seizes a piece of dung with his front legs and backs down into the burrow. When the tip of his abdomen strikes the female's head, he drops his load and returns to the surface for more. In this case, the striking of the female's head by the male's abdomen is probably an example of a(n)

  a sign stimulus.
  b supernormal stimulus.
  c sexual selection.
  d sexual imprinting.
  e operant conditioning.   (pp. 805–807)

24 The type of behavior seen in the above example would most likely be characterized as

  a innate.   *b* learned.   (p. 804)

25 A male dung beetle making a nuptial ball for a female to eat in relative safety underground exposes himself to considerable risk while reducing the risk to a potential mate. This is an example of

  a agonistic behavior.   *d* insight learning.
  b altruistic behavior. *e* conditioning
  c nuptial love.   (p. 829)

26 In human beings, certain perceptions of tactile stimuli release specific responses; for example, creeping things on the back of the hand release a shaking movement of the hand. This response is probably due to

  a innate or instinctive behavior.
  b imprinting.
  c habituation.
  d conditioning.
  e insight learning.   (p. 804)

27 Which one of the following best exemplifies an innate motor program?

 a a young child playing the piano
 b a baby learning to walk
 c food-choice behavior in rats
 d human speech
 e web weaving by a spider (p. 804)

28 A just-hatched gull chick pecks vigorously at the red spot on the parent's bill to elicit the feeding response. This red spot is an example of

 a imprinting.
 b trial-and-error learning.
 c classical conditioning.
 d a releaser.
 e a supernormal stimulus. (p. 815)

29 In many situations a behavior that is positively reinforced (for example by a reward) becomes more frequently practiced. This is known as

 a instinct.
 b classical conditioning.
 c operant conditioning.
 d imprinting.
 e habituation. (p. 812)

Questions 30 to 34 describe experiments or observations that can be classified as examples of one of the forms of learning listed below. For each example, choose the corresponding form of learning. Answers may be used once, more than once, or not at all.

 a insight
 b habituation
 c classical conditioning
 d trial and error
 e imprinting

30 A duckling will follow any large moving object presented to it right after hatching. (p. 812)

31 Flatworms respond to an electric shock by contracting their body muscles. If the worms are exposed to 50 or 100 electric shocks and at the same time to a beam of light, the worms will learn to contract immediately on presentation of the light beam even if no shock is present. (p. 811)

32 A hungry brown rat was placed in a closed metal box with a food slot and a food-releasing lever. The rat was allowed to poke around randomly until it accidentally tripped the lever and released a food pellet. The rat soon learned to trip the lever whenever it wanted food. (p. 812)

33 A chimpanzee is put in a cage with several boxes scattered across the floor and a clump of bananas hung from the top of the cage, out of reach. The ape piles the boxes on top of each other, climbs up, and eats the bananas. (p. 812)

34 Adult white-crowned sparrow males "defend" their territorial boundaries by vigorously singing their characteristic songs. If a tape-recorded sparrow song is played nearby, the male will first search frantically for the "other bird" and then sing his song at the speaker. When the tape is played repeatedly, the sparrow eventually gives up singing and resumes his normal activities. (p. 811)

35 Ivan Pavlov discovered that dogs would learn to salivate on hearing a sound if the sound was paired with the presentation of food over many trials. This type of learning is called

 a habituation.
 b classical conditioning.
 c operant conditioning.
 d cultural learning.
 e imprinting. (p. 811)

36 A bird is hatched and raised indoors in complete isolation from other members of its own species. It becomes imprinted on the student who raised it. When the bird is released to the wild as an adult, which one of the following behaviors would you *not* expect it to have difficulty performing correctly?

 a singing its species-typical song
 b using star patterns in migratory orientation
 c finding a mate
 d flying (pp. 811–13)

37 Which one of the following statements about imprinting is *true*?

 a Responsiveness to the releasing object is lost after the sensitive period is over.
 b It occurs only in mammals.
 c It is typically reversible, in both lab and natural situations.
 d It occurs early in the animal's life. (p. 812)

38 The chaffinch "learns" to sing when it is still a very young bird. Which of the following statements is correct?

 a Instinct does not play any role.
 b The chaffinch will sing its song without ever hearing another bird.
 c The chaffinch will not sing its song unless it hears the song early in life.
 d The chaffinch will sing the song of a number of other bird species if it hears the other song, rather than its own, during a critical phase early in life.
 e The chaffinch will begin to sing its song if, early in life, it hears the song of another bird species.
(pp. 817–18)

Questions 39 to 41 refer to the following diagram showing locations of some bees' food sources (A through C) and their hive, as well as the position of the sun. Assume that the sun remains stationary.

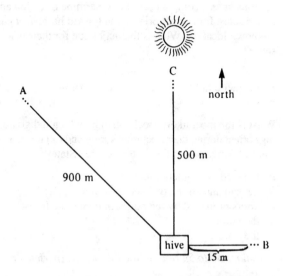

39 If a scout has returned to the hive from a successful flight to site A and performed its dance on a darkened vertical comb, what would you expect its dance to look like? (p. 820)

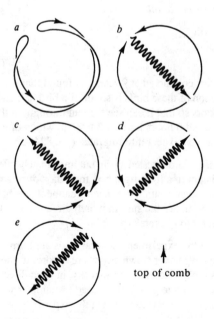

40 If another bee returned to the same hive from site B and danced on a vertical comb, what would you expect its dance to look like? (p. 820)

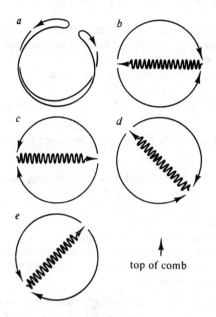

41 If another bee returned to the same hive from site C and danced on a vertical comb, what would you expect its dance to look like? (p. 820)

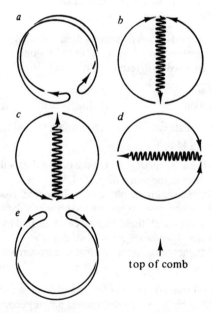

280 • CHAPTER 35

42 Which of the following statements best describes circadian rhythms?

　a　They are of exactly 24-hour duration, in the absence of external cues.
　b　Once the clock is set in motion, it is not possible to reset it.
　c　They are not exactly 24 hours in duration, but are generally reset each day.
　d　All biological rhythms are circadian.
　e　Circadian rhythms that are set by light/dark cycles stop when the organism is kept in complete darkness. (p. 822)

43 Which one of the following is *false* concerning circadian rhythms?

　a　The rhythmic activity pattern usually dies out quickly if the environment is kept constant.
　b　The rhythmic pattern can be reset to an altered pattern of light and dark cycles.
　c　The rhythm appears to be innate.
　d　The clock is not set precisely to a 24-hour day; it must be reset each day to keep it in synchrony with true time. (p. 822)

44 On which of the following flights would you expect passengers to experience the most severe case of jet lag?

　a　New York to San Francisco
　b　Cleveland to Miami
　c　Chicago to San Francisco
　d　San Francisco to New Orleans
　e　San Francisco to Chicago (p. 822)

45 Many animals can determine direction; that is, they have a compass sense. Which one of the following cues is *not* usually used to determine direction?

　a　sun　　　　c　stars
　b　moon　　　d　magnetism (pp. 823–25)

46 Which statement is *false* concerning the navigation of homing pigeons?

　a　Homing pigeons are not the only animals that can return home when displaced.
　b　Pigeons with the phase of their internal clock shifted by 6 hours orient homeward on sunny days.
　c　Pigeons with their internal clock shifted by 6 hours orient homeward on overcast days.
　d　Magnets can disorient pigeons on overcast days. (pp. 824–26)

47 A well-trained group of homing pigeons wearing magnets on their backs is released (one at a time) under clear, sunny skies. Which of the following results would you predict for the behavior of the group as a whole?

　a　They would be oriented in an approximately homeward direction.
　b　They would be randomly oriented.
　c　They would be oriented 90 degrees counterclockwise from the homeward direction.
　d　They would be oriented 90 degrees clockwise from the homeward direction.
　e　They would be oriented 180 degrees from the homeward direction. (pp. 824–26)

48 Pigeons have a compass sense to determine direction and a map sense for correct orientation toward home from an unknown location. What is the major cue for their map sense?

　a　sun　　　　　　d　moon
　b　magnetic field　e　We do not know.
　c　stars　　　　　　 (pp. 824–26)

49 What is the most likely mechanism by which indigo buntings orient during their nocturnal spring and fall migrations Central America and the eastern United States?

　a　the earth's magnetic field
　b　the sun and an internal sense of time
　c　odors produced by the northern temperate forest
　d　the moon
　e　the stars (pp. 824–26)

Questions 50 to 52 refer to the following experiment.

Several birds of the same species are kept in an outdoor cage and released under various circumstances.

For each set of circumstances described in a question, choose the direction in which the released bird would be expected to fly. Answers may be used once, more than once, or not at all.

　a　north　　　d　west
　b　south　　　e　northeast
　c　east　　　　　(pp. 824–26)

50 In a test conducted at 9 A.M. it is found that the migratory direction of these birds is south. If a bird is then kept outdoors so that it can see the sun throughout the day and is retested 6 hours later, at 3 P.M., in which direction would you expect it to migrate? (pp. 824–26)

51 The same bird is tested again, but immediately after its 9 A.M. test is put in a light-tight box for a day; during this time it cannot see the sun. After such treatment, the bird is tested at 3 P.M. under the sun. In which direction would you expect it to migrate? (pp. 824–26)

52 During the experiment, several birds are born in the cage; these have never flown anywhere before. If one of these birds is exposed to the normal sky, in what direction would you expect it to go when released? Assume that these birds are native to the area in which the experiment is conducted. (pp. 824–26)

53 A German fable goes: "One very cold night a group of porcupines were huddled together for warmth. However, their spines made proximity uncomfortable, so they moved apart again and got cold. After shuffling repeatedly in and out, they eventually found a distance at which they could still be comfortably warm without getting pricked. This

distance they then called decency and good manners." We would call it the

a individual distance.   c individual territory.
b home range.   (pp. 827–28)

54 The male Alaskan fur seal secures an area of beach and vigorously defends it from all other males. Mating occurs within the area. The bull guards his family but does not assist in feeding or caring for the young. The area defended by one male seal is called the

a home range.   d nursery.
b individual distance. e home turf.
c territory.   (pp. 827–28)

55 The mating system employed by the fur seal is most likely

a monogamy.   c polygyny.
b polyandry.   d promiscuity.   (p. 830)

56 All the following would be classified as illustrating social behavior *except*

a a redwing blackbird male with three females living within his territory.
b a monogamous male and female pair of Canada geese.
c a flock of tens laying their eggs within a restricted rocky area on a small island.
d a group of fish congregating in the last vestige of a drying pond.
e a male wood thrush defending his territory.   (p. 827)

57 Wolves urinate to mark the edges of their territories. This behavior is an example of

a a reflex.   d mating behavior.
b using pheromones.   e marking home range.
c marking individual distance.   (p. 827)

58 Two dogs have had an agonistic encounter and one of them shows appeasement displays. Which one of the following actions would you *not* expect to e a part of such appeasement displays?

a The tail is tucked between the legs.
b The hair is sleeked.
c The stance of the body is lowered.
d The head is turned away from the opponent.
e The ears are held erect.   (p. 819)

59 In a particular animal species the young are precocial; they can help care for themselves almost immediately. The care and feeding of the young is done by the female parent, though the male does guard the female and young. The most likely mating system for this species is

a monogamy.   c polygyny.
b polyandry.   d promiscuity.   (p. 830)

60 Which one of the following is true of *both* insect and vertebrate societies?

a The unit of organization is usually the family.
b Division of labor is based on biologically determined castes.
c All individuals in the society are capable of reproduction.
d Individual recognition between members is crucial to the organization.
e The organization is based on a rigid hierarchical system.
                                             (p. 833)

For further thought

1 A bird in Ithaca, New York, is trained to fly south and uses the sun as a reference point; normal sunrise is at 6 A.M. The bird then spends several weeks in a lab chamber where artificial lights come on 6 hours *later* than sunrise (though the number of "daylight" hours remains the same). In which compass direction would you expect this phase-shifted bird to fly if released at noon in Ithaca? Assume that the sun is in view.

2 Another bird, of a different species, normally migrates south from Ithaca at night, using star patterns to maintain its bearing. Normal sunrise is at 6 A.M. This bird spends several weeks in a lab chamber where artificial lights come on 6 hours *earlier* than sunrise (though the number of "daylight" hours remains the same). In which compass direction would you expect this phase-shifted bird to fly if released at midnight in Ithaca? Assume that the stars are in view.

3 Kramer showed the importance of the sun as a reference point for bird orientation with a simple experiment in which he blocked off the view of the sun and used mirrors to make it appear in other positions relative to a bird in an experimental arena. If the true position of the sun with respect to the bird's migratory direction (indicated by the heavy arrow) is shown on top, in which direction (*a* through *e*) would you expect the bird to attempt to fly if the sun's position is artificially altered as depicted at the bottom?

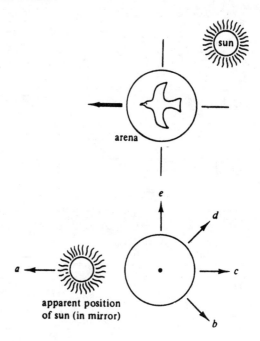

4. Over the years there has been a good deal of controversy concerning innate versus learned behavior patterns. Is it possible to distinguish between them? Explain.

5. When a young kitten plays, it frequently pounces on moving objects. As an adult it will use this same method to catch mice. Has the cat "learned" to catch mice or is catching mice a result of maturation? Design an experiment that would help you answer this question.

ANSWERS

Testing recall

1. innate, instinctive
2. taxis
3. learned motor program
4. parental imprinting
5. imprinting
6. cerebral cortex, cerebellum
7. chemical
8. primer pheromones
9. all except happiness
10. threat
11. distance, direction relative to the sun, vertical
12. day
13. sun position, magnetic cues
14. home range
15. both
16. polyandry
17. do

Testing knowledge and understanding

18	*e*	29	*c*	40	*a*	51	*b*
19	*e*	30	*e*	41	*c*	52	*b*
20	*b*	31	*c*	42	*c*	53	*a*
21	*b*	32	*d*	43	*a*	54	*c*
22	*b*	33	*a*	44	*a*	55	*c*
23	*a*	34	*b*	45	*b*	56	*d*
24	*e*	35	*b*	46	*b*	57	*b*
25	*b*	36	*d*	47	*a*	58	*e*
26	*a*	37	*d*	48	*e*	59	*c*
27	*e*	38	*c*	49	*e*	60	*a*
28	*d*	39	*c*	50	*b*		

Chapter 36

THE ORIGIN AND EVOLUTION OF EARLY LIFE

A GENERAL GUIDE TO THE READING

The diversity of life described in Part V is based on the evolutionary processes described in Part IV. Chapter 36 describes the theory of the origin of life widely held by most biologists, and the evidence that supports it. As you read Chapter 36 in your text, you will want to develop a sense of evolutionary time (millions-of-years order of magnitude) as you concentrate on the following topics.

1. Formation of the earth and its atmosphere. Carefully read the description of the proposed model of the early atmosphere (pp. 842–43), and compare them with earth's present atmosphere.

2. Formation of small organic molecules. In going over the material on this topic (pp. 843–45), you will want to pay special attention to Stanley Miller's experiment and its significance; study Figure 36.3 (p. 844).

3. Formation of polymers. You need to understand how polymers are formed from building-block molecules (p. 845). A review of Figures 3.8 (p. 55), 3.15 (p. 61), and 3.20 (p. 65) will refresh your memory about the condensation reactions that result in the formation of polymers.

4. Formation of molecular aggregates and primitive cells. As the discussion of this topic indicates (pp. 845–46), coacervate droplets and proteinoid microspheres are suggestive of what prebionts may have been like. Look ahead to Figure 37.16 (p. 871) and study the similarities between the bacteria shown there and the proteinoid microspheres shown in Figure 36.7 (p. 847).

5. Evolution of autotrophy. The evolution of photosynthetic pathways, particularly noncyclic photophosphorylation, was an immensely important event in the evolution of life (pp. 849–50). Be sure you understand what the oxygen revolution was and why it was significant.

6. The geologic time scale. It is important to have some knowledge of approximately when some evolutionary events occurred. Figure 36.18 is effective in summarizing the emergence of different life forms. (p. 855)

7. The kingdoms of life. Figure 36.23 (p. 858) is a schematic view of the emergence of kingdoms according to the six-kingdom system that is used by the text. This diagram serves as an advance organizer for Chapters 37 to 43, which describe the diversity of living things, and proceed in their organization from the simple to the more complex.

KEY CONCEPTS

1. Life arose spontaneously from nonliving matter under the conditions prevailing on the early earth; from these beginnings all present life on earth has descended. (p. 839)

2. All the events now hypothesized in the origin of life and all the known characteristics of life seem to fall well within the general laws of the universe. Energy could be provided by solar radiation, lightning, heat, or other sources. No supernatural event was necessary to the origin of life on earth. (p. 840)

3 There was no abrupt transition from "nonliving" prebionts to "living" cells; the attributes associated with life were acquired gradually. (pp. 843–49)

4 Living organisms, once they arose, changed their environment and so destroyed the conditions that had made possible the origin of life. Particularly important was the oxygen revolution. (p. 850)

5 The first cells were probably procaryotic; eucaryotic cells may have originated from a symbiotic union of ancient procaryotic cells of several types. (pp. 850–54)

6 The classification of living organisms into kingdoms is to some degree arbitrary. Depending on the criteria chosen, a number of different systems of classification are possible. (pp. 854–58)

CROSSLINKING CONCEPTS

1 The conditions under which life arose are compared with the results of Pasteur's experiment, which showed that spontaneous generation does not occur (Chapter 5).

2 Understanding the chemistry of life (Chapters 3 and 4) is very helpful in understanding the origin of life. Particularly important concepts include oxidation/reduction (especially significant for understanding how the early atmosphere of the earth differs from our current atmosphere), polymerization from condensation reactions, and the role of competition in evolution (applied to the chemical level in this chapter).

3 A general conception of heterotrophs, autotrophs, respiration (anaerobic and aerobic), and photosynthesis (cyclic and noncyclic), is crucial to understanding the evolutionary development of niches and the processes through which they were formed (Chapters 7 and 8).

4 The endosymbiotic hypothesis, first introduced when you studied chloroplasts and mitochondria (Chapter 6), takes on new significance as a step in the transition from procaryotes to eucaryotes.

5 The idea that living things alter the environment, and that these alterations can improve living conditions for other forms was introduced when you studied ecological succession in Chapter 33.

OBJECTIVES

1 Describe the formation of the earth according to the most widely held hypothesis. Describe the most widely accepted model for the composition of the early atmosphere; then discuss the differences between this proposed atmosphere and the atmosphere today. (pp. 838–43)

2 Explain how small organic molecules are thought to have formed in the early oceans, and specify what sources of energy were available to cause molecules to react together to form other compounds. (p. 843)

3 Describe Stanley Miller's experiment, and explain how its results support Oparin's hypothesis of the origin of life. (pp. 843–45)

4 Describe one method by which polymers may have been formed from the building-block molecules in the ancient oceans. (p. 845)

5 Contrast Oparin's coacervate droplets and Fox's proteinoid microspheres, and specify ways in which they resemble living cells. Indicate whether these prebionts have a genetic control system. (p. 846)

6 Explain why the earliest organisms were most likely heterotrophic procaryotes, and specify the role competition is thought to have played in the evolution of various chemical pathways in the early cells. (pp. 848–49)

7 Discuss reasons why the evolution of photosynthesis was necessary for life to continue, and indicate which pathway—cyclic or noncyclic phosphorylation—is believed to have evolved first. Explain what is meant by the term oxygen revolution. Give the number of ATP molecules synthesized from one mole of glucose under anaerobic conditions versus the number of ATPs that can be synthesized under aerobic conditions. (pp. 849–50)

8 Discuss possible reasons why the evolution of photosynthesis destroyed the conditions that made possible the origin of life. (p. 850)

9 Discuss the evidence for the endosymbiotic model for the origin of the eucaryotic cell. (pp. 851–54)

10 Name the eight kingdoms of organisms used in the classification scheme in your text, and list the distinguishing characteristics of each. (pp. 854–58)

KEY TERMS

first atmosphere (p. 842)
second atmosphere (p. 842)
abiotic synthesis (p. 845)
polymer (p. 845)
coacervate droplet (p. 846)
proteinoid microsphere (p. 846)
prebiont (p. 846)
ribozyme (p. 848)
oxygen revolution (p. 850)
stromatolite (p. 851)
endosymbiotic hypothesis (p. 852)
Precambrian (p. 854)
Cambrian (p. 854)

Archaebacteria (p. 857)
Eubacteria (p. 857)
Protista (p. 858)
Plantae (p. 858)
Fungi (p. 858)
Animalia (p. 858)

SUMMARY

Origin of life Scientists today believe that life could and did arise spontaneously from nonliving matter under the conditions that prevailed on the early earth, and that all present earthly life has descended from these beginnings. The basis for the current theory of the origin of life was stated by A. I. Oparin in 1936.

The solar system was probably formed between 4.5 and 5 billion years ago from a cloud of cosmic dust and gas. As the earth condensed, a stratification took place; heavier materials moved toward the center and lighter substances were concentrated at the surface. Eventually the lighter gases of the earth's first atmosphere escaped into space. The intense heat in the interior of the earth drove out various gases through volcanic action; these formed a second atmosphere.

The atmosphere of the early earth was quite different from today's oxidizing atmosphere. The most widely accepted model of the atmosphere assumes that it was made up primarily of gases that occur in present-day volcano outgassings: H_2O, N_2, H_2S, CO, CO_2, H_2. In this model, the early atmosphere contained almost no molecular oxygen but an abundance of hydrogen.

As the earth's crust cooled, the water vapor condensed into rain and began to form the oceans, in which gases from the atmosphere and salts and minerals from the land dissolved. Ultraviolet radiation, lightning, and heat could have provided substantial amounts of energy for reactions that produced organic building-block molecules from the combination of these substances. Experimental evidence shows that such abiotic synthesis of organic compounds can occur. These organic compounds could have accumulated slowly in the seas over millions of years since they would not have been destroyed by oxidation or decay.

The building-block molecules accumulating in the hot soup could have joined together by chance bonding or through some concentrating mechanisms (such as adsorption on clay minerals or evaporation from beach puddles) to form complex organic molecules such as polypeptides, nucleic acids, carbohydrates, and lipids. Short RNA sequences may have formed abiotically; these might have been able to reproduce themselves exactly if nucleotides and appropriate catalysts were present. How did the orderliness characteristic of living things emerge from this soup? Oparin speculated that *coacervate droplets* then formed; these are clusters of macromolecules surrounded by an orderly shell of water molecules. Such droplets have a definite internal structure and can absorb substances selectively. Fox proposes, instead, the formation of *proteinoid microspheres*, which exhibit many properties of living cells. Vast numbers of such prebiological systems (*prebionts*) may have arisen in the seas. Some may have contained favorable combinations of materials and grown in size; new droplets could have formed on fragmentation.

The first genes were probably composed of RNA rather than DNA. Some RNAs have catalytic activity and cut RNA chains at certain sites; others are self-splicing. Primitively, RNA may have served as both genes and catalysts, and only later given up those roles to DNA and enzymes. Eventually, the nucleotide sequences in the nucleic acids these prebionts contained came to code for the sequence of amino acids in protein, and transcription and translation evolved.

The earliest organisms were anaerobic, procaryotic heterotrophs that obtained energy from the nutrients available in the early ocean. As the nutrients disappeared, competition between the organisms must have increased. Natural selection favored any new mutation that enhanced an organism's ability to obtain or process food. Over time, various biochemical pathways evolved that enabled organisms to utilize different nutrients. The first form of metabolism was probably glycolysis and fermentation.

As the free nutrients were used up, some organisms evolved the ability to use the energy stored in inorganic molecules, and chemosynthetic organisms evolved. Later, some organisms evolved the ability to use the sun's energy to make ATP. Cyclic photophosphorylation probably evolved first. Later, the more complex pathways of noncyclic photophosphorylation and carbon dioxide fixation evolved. From this time onward, the continuation of life on earth depended on the activity of photosynthetic autotrophs.

An importnat by-product of noncyclic photophosphorylation is molecular oxygen. The oxygen released by photosynthesis converted the atmosphere into an oxidizing atmosphere in the *oxygen revolution*. The oxygen also gave rise to a layer of ozone in the upper atmosphere that shielded the earth's surface from intense ultraviolet radiation. In other words, living organisms, once they arose, changed their environment in a way that destroyed the conditions that had made possible the origin of life.

Once molecular oxygen became a major component of the atmosphere, both heterotrophic and autotrophic organisms could evolve the biochemical pathways of aerobic respiration, by which far more energy can be extracted from nutrient molecules than is obtainable by glycolysis alone.

Evolution of the eucaryotic cell The oldest cellular fossils are of procaryotic cells and are about 3.5 billion years old. Most authorities think the ancestors of the cyanobacteria evolved from bacteria about 3 billion years ago, perhaps even earlier. Eucaryotic cells probably arose about 1.8 billion years ago. The first fossils of multicellular animals are about 600 million years old.

Chloroplasts and mitochondria may be modern descendants of ancient procaryotic cells that became permanent residents within other cells and have evolved in concert with their hosts ever since. A growing list of characteristics indicates that mitochondria and chloroplasts have many features in common with free-living procaryotic organisms. If this model is correct, the chloroplasts are probably derived from cyanobacteria or prochlorophytes and the mitochondria from aerobic bacteria. There is little evidence on the origin of the nuclear membrane and the endoplasmic reticulum. The evolution of the eucaryotic cell represents an important landmark in the evolution of life; it is this cell that gave rise to all the eucaryotic lineages: protists, plants, animals, and fungi.

Few fossils of multicellular forms of life are found from the Precambrian, but they are abundant in the Cambrian, which began about 570 million years ago. By the end of the Cambrian, the ozone layer was well developed, and so permitted organisms to live on land. Plants invaded the land about 450 million years ago, and invertebrate animals followed.

The kingdoms of life Depending on the criteria chosen a number of different systems of classification are possible. The classification system used in your text recognizes six kingdoms. The *Archaebacteria* and *Eubacteria*, *Protista*, *Fungi*, and *Animalia*. The archaebacteria and eubacteria are procaryotic unicellular organisms. The kindom Protista consists of unicellular, colonial, and simple multicellular eucaryotic organisms. The protistan kingdom includes a variety of groups, many of which are not closely related evolutionarily. Three major lineages reached the higher multicellular level. Each exploits a different mode of nutrition; photosynthetic autotrophism is used by plants (kingdom Plantae), absorptive heterotrophism by fungi (kingdom Fungi), and ingestive heterotrophism by animals (kingdom Animalia).

CONCEPT MAPS

Map 36.1

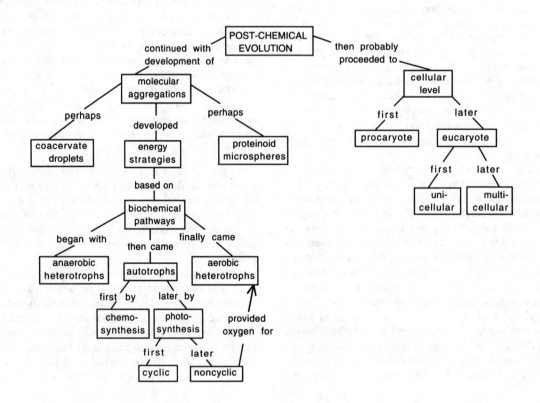

Map 36.2

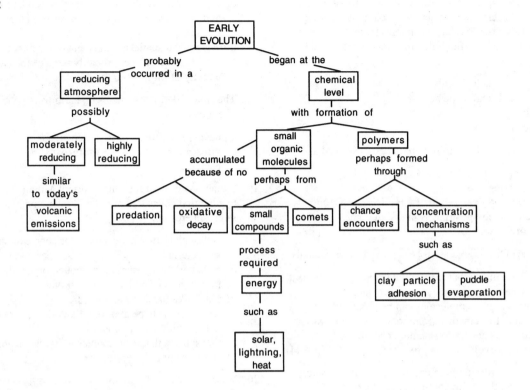

QUESTIONS

Testing recall

1. Arrange the following in the order of their appearance in the evolution of life. (pp. 839–54)

    a competition
    b formation of the earth
    c prebionts
    d hydrogen-rich atmosphere
    e oxidizing atmosphere
    f primitive procaryotic cell
    g abiotic synthesis of building-block molecules
    h complex biochemical pathways
    i O_2 revolution
    j great rains
    k polymerization
    l eucaryotic cell
    m photosynthesis

For questions 2 to 9, match the kingdoms with their descriptions.

2. Protista   a multicellular absorptive heterotrophs
3. Animalia   b multicellular photosynthetic autotrophs
4. Fungi   c multicellular ingestive heterotrophs
5. Eubacteria   d younger procaryotes
6. Archaebacteria e older procaryotes
7. Plantae   f unicellular eucaryotes

Testing knowledge and understanding

8. How old is the earth?

    a 4.5 million years
    b 15 million years
    c 4.5 billion years
    d 15 billion years
    e 4.5 trillion years   (p. 842)

9. Life would probably never have arisen on earth if the early atmosphere had contained

    a O_2.
    b H_2.
    c N_2.
    d NH_3.
    e CH_4.   (p. 842)

10. In 1953 Stanley Miller conducted a famous experiment in which he simulated conditions of the early earth in an enclosed apparatus. His experiment demonstrated that

    a life originated in the early oceans.
    b proteins and lipids will self-assemble into membranes.
    c amino acids could have been formed from the gases of the early atmosphere.
    d amino acids are necessary for life.
    e lightning was the source of energy for the synthesis of the first organic molecules.
    (p. 844)

11 In the Miller experiment, gases that were probably present in the early atmosphere were circulated through an apparatus past electrical discharges from tungsten electrodes. Which of the following were synthesized?

   a proteins
   b long-chain nucleic acids
   c amino acids and lactic acid
   d cellulose
   e all the above (p. 844)

12 The earth's primitive atmosphere is *best* described as

   a oxidizing.
   b nonoxidizing.
   c predominantly composed of CH_3, NH_3, H_2O, and O_2.
   d the same in composition as interstellar gas.
   e none of the above. (p. 844)

13 Which one of the following has been proved experimentally?

   a Life on earth originated by spontaneous generation.
   b Life on earth originated by special creation.
   c Complex organic molecules can be produced abiotically from methane, ammonia, and water vapor.
   d Spontaneous generation cannot occur today.
   e Eucaryotes evolved from procaryotes. (p. 844)

14 Energy for the formation of organic molecules possibly came from all the following sources *except*

   a solar radiation
   b electric discharge
   c volcanic eruptions
   d heat from the earth's core
   e magnetic fields (p. 844)

15 Which of the following is *not* a possible polymer formation process described in your text?

   a The concentrartion of organic material in the seas may have been sufficiently high for chance bonding to give rise to a considerable quantity of polymers.
   b Building-block molecules might have been concentrated by adsorption to clay particles before polymer formation.
   c Building-block molecules might have been oriented by enzymes to facilitate polymer formation.
   d Building-block molecules might have been concentrated by evaporation from beach puddles before polymer formation. (p. 845)

16 Nutrients could accumulate in the early organic soup because

   a heterotrophs had not yet evolved to eat them.
   b the atmosphere was a nonoxidizing one.
   c the atmosphere was an oxidizing one.
   d *a* and *b*.
   e *a* and *c*. (p. 846)

17 Which one of the following statements is *false*?

   a Coacervate droplets have a strong tendency toward formation of definite internal structure.
   b There is a definite interface between a coacervate droplet and the liquid in which it floats.
   c Coacervate droplets have a marked tendency to adsorb and incorporate various substances from the surrounding medium.
   d The regular spatial arrangement of the molecules within a coacervate droplet probably reduces the catalytic activity of proteins in the droplet. (p. 846)

18 The first *cellular* organisms on earth were probably most like today's

   a eucaryotic cells.
   b viruses.
   c bacteria.
   d green algae.
   e protozoans. (p. 848)

19 Sidney W. Fox, of the University of Miami, has suggested that the first cells were proteinoid microspheres. Fox's microspheres are

   a cells with approximately the same degree of organization as bacteria.
   b cells with many membrane-bound organelles, the membranes of which contain protein but not lipid.
   c much like present-day viroids.
   d spherical in shape and look like primitive cells, but do not exhibit osmotic properties.
   e tiny spheres that form spontaneously when hot aqueous solutions of polypeptides are cooled. (p. 845)

20 Which one of the following statements is *false* concerning Fox's proteinoid microspheres?

   a They have a double-layered outer boundary.
   b They can grow in size and complexity.
   c They exhibit electrical differences across their boundary.
   d They will shrink in hypotonic media and swell in hypertonic media. (p. 845)

21 Competition probably became important at an early stage in the origin of life because

   a nucleic acids cannot duplicate themselves.
   b available organic materials were being consumed faster than new materials were being synthesized.
   c cells arose early in the sequence as a result of the colloidal characteristic of proteins.
   d the early proteins probably acted as catalysts and thus accelerated the various chemical reactions that were taking place.
   e the ability to carry on photosynthesis had become widespread. (p. 844)

22 Which of the following events led most directly to the decrease in the amount of ultraviolet radiation reaching the surface of the earth?

   a evolution of cyclic photophosphorylation
   b evolution of noncyclic photophosphorylation
   c development of a nonoxidizing atmosphere
   d evolution of heterotrophy
   e evolution of fermentation (p. 850)

23 Early in the evolution of life on earth, metabolic breakdown of energy-rich compounds must have

been only by fermentation, because aerobic respiration

   a occurs only in animals, but the earliest organisms were plants.
   b could not occur before the evolution of noncyclic photophosphorylation.
   c could not occur after the so-called oxygen revolution.
   d could not occur before the first vascular plants evolved.
   e can only oxidize carbohydrates, but the early nutrients were mostly of other types. (p. 850)

24 Mode of nutrition is a very important characteristic of organisms. Which evolved first?

   a parasitism
   b herbivorous feeding
   c carnivorous feeding
   d heterotrophy
   e autotrophy (p. 849)

25 As the first cellular organisms evolved, they began to develop complex biochemical pathways; this development is thought to be due to the

   a depletion of abiotically produced nutrients.
   b presence of carbon dioxide in the atmosphere.
   c prior evolution of autotrophy.
   d abundance of abiotically produced nutrients.
   e increase in oxygen levels in the atmosphere. (p. 849)

26 Which of the following events in the history of the earth happened first?

   a evolution of eucaryotes
   b accumulation of oxygen in the atmosphere
   c evolution of *Homo sapiens*
   d evolution of photosynthetic flowering plants
   e evolution of procaryotes (p. 855)

27 Life first appeared on earth approximately

   a 6 billion years ago.
   b 3.5 billion years ago.
   c 2.3 billion years ago.
   d 500 million years ago.
   e 3.5 million years ago. (p. 855)

28 Which one of the following statements is *false* according to current scientific thinking on the origin of life?

   a The first autotrophs were probably eucaryotic cells.
   b Given great expanses of time, raw materials, and energy, the basic building-block organic compounds could be synthesized in the absence of living cells.
   c The first living organisms must have depended on glycolysis and fermentation to obtain usable energy.
   d As living organisms evolved, they changed the environment so much that the conditions that made possible the origin of life no longer exist.
   e Coacervate droplets have some attributes of living cells; they are orderly arrangements of molecules with a membrane that can selectively absorb substances from the surrounding medium. (p. 849)

29 According to the endosymbiotic theory of the origin of the eucaryotic cells,

   a mitochondria evolved from chloroplasts.
   b fungi, plants, and protists obtained their mitochondria from animals.
   c present-day procaryotes evolved from primitive eucaryotes.
   d mitochondria are remnants of ancient procaryotes that invaded other very primitive procaryotic cells.
   e mitochondria are remnants of ancient eucaryotes that invaded very primitive, evolving eucaryotic cells. (p. 853)

30 It has been proposed that mitochondria and chloroplasts are modern descendants of primitive forms of procaryotic cells that took up residence within primitive cells and evolved independently there. Which one of the following is *not* evidence of this view?

   a Like procaryotic cells, mitochondria and chloroplasts have DNA that is not closely associated with protein.
   b The ribosomes in procaryotic cells, mitochondria, and chloroplasts are very similar chemically, and differ from the cytoplasmic ribosomes of eucaryotes.
   c The genes in procaryotic cells, mitochondria, and chloroplasts lack the introns that are characterisitc of eucaryotic genes.
   d The single chromosome of chloroplasts and mitochondria is generally circular, like the bacterial chromosome.
   e Mitochondria and chloroplasts can be cultured outside cells. (p. 853)

31 At present, we are not sure how chloroplasts and mitochondria evolved. However, the most widely accepted theory states that

   a they are descendants of ancient procaryotic cells that became permanent residents inside other cells.
   b they arose when membranes surrounded existing structures inside cells.
   c they were first produced by budding from the plasma membrane and later DNA from the nucleus migrated to the organelles.
   d mitochondria evolved from chloroplasts.
   e chloroplasts evolved from mitochondria. (p. 853)

For further thought

1 Another hypothesis for the origin of life on earth suggests that life did not originate here; it was brought here by some extraterrestrial object, such as a meteor. The earth is constantly bombarded by meteorites. Analyses indicate that some meteorites contain molecules (some hydrocarbons) characteristic of living systems. Comment on this hypothesis.

2 Suppose you were given the following organisms to classify:
   a A unicellular organism that has chlorophyll and carries on photosynthesis. It lacks a cell wall and is highly motile.
   b A large multinucleated amoeboid heterotrophic organism that carries on phagocytosis. Its reproduction is plantlike; it forms groups of thin-walled spore cases similar to those of fungi.

In which kingdom would you classify each if you were using the six-kingdom system?

3 In Miller's first experiments with passing electrical charges through a mixture of gases, he found that a number of organic compounds had been synthesized. What possible complication was suggested that might have caused these results? How did he rule out this possibility as the cause of his results? What conclusions can you draw about the process of science from his research?

4 Describe the probable role of selection in the development of complex biochemical pathways.

5 Describe the origin of the oxygen revolution. How did the oxygen revolution affect the course of evolution of terrestrial organisms? Why should we be concerned about the prospect of a decrease in the ozone concentration in the earth's atmosphere?

6 Do you think there might be life on other planets? What is the basis for your belief?

ANSWERS

Testing recall

1 *b, d, j, g, k, c, f, a, h, m i, e, l* or *b, d, j, g, k, c, a, h, f, m, i, e, l*
2 *f*   4 *a*   6 *d*
3 *c*   5 *e*   7 *b*

Testing knowledge and understanding

8	*c*	14	*d*	20	*d*	26	*e*
9	*a*	15	*c*	21	*b*	27	*b*
10	*c*	16	*d*	22	*b*	28	*a*
11	*c*	17	*d*	23	*b*	29	*d*
12	*b*	18	*c*	24	*d*	30	*e*
13	*c*	19	*e*	25	*a*	31	*a*

Chapter 37

VIRUSES AND BACTERIA

A GENERAL GUIDE TO THE READING

This chapter focuses on the viruses, which occupy the border between life and nonlife, and on the primitive cellular organisms belonging to the two kingdoms Archaebacteria and Eubacteria. As you read Chapter 37 in your text, you will want to concentrate on the following topics.

1. The structure of viruses. In studying this topic (pp. 861–63), you will find it helpful to review Figure 10.2 (p. 204), which shows the structure of the bacteriophage, and to look carefully at Figures 37.1 (p. 862) and 37.2 (p. 863), which show the structure of plant and animal viruses, as well as bacteriophage.

2. The reproduction of viruses. As the discussion on page 864 indicates, the reproduction of plant and animal viruses is in some ways quite different from that of the bacteriophage. You will want to study the life cycle of the bacteriophage carefully (Fig. 37.3, p. 864). Make sure, too, that you understand what a provirus is; study carefully the material on RNA viruses (pp. 864–66) carefully. Figures 37.6 (p. 865) and 37.7 (p. 866) are helpful.

3. Archaebacteria. The structural differences among the archaebacteria, eubacteria, and eucaryotic cells are summarized in Table 37.3 (p. 869). Take particular note of the environments that the archaebacteria inhabit. (p. 869).

4. Bacterial anatomy. To prepare for your study of this topic, review the section on procaryotic cells in Chapter 6 (p. 134). Though bacterial cells (pp. 870–71) are relatively simple in structure, there is much to learn about them. Figures 37.14 to 37.18 (pp. 870–71) will help you appreciate their various forms.

5. Bacterial reproduction. Bacteria divide by binary fission (Fig. 9.2, p. 180) rather than by mitosis or meiosis.

6. Bacterial photosynthesis. Try to appreciate the fact that although many of the archaebacteria and eubacteria are photosynthetic, the manner in which photosynthesis is carried out may be quite different from the reactions we learned back in Chapter 7 (pp. 139–55). Only the cyanobacteria and prochlorophytes possess chlorophyll a, carry out noncyclic photophosphorylation, and release O_2.

7. Cyanobacteria. You will want to compare the cyanobacteria (pp. 875–77) with the other bacteria and try to understand why the cyanobacteria are such a successful group of organisms.

8. Beneficial bacteria. Try to remember that most bacteria are beneficial (p. 877); for example, recall the vital role that they play in decomposing organic material and in producing vitamins in the human body.

9. Bacteria as agents of disease. The text (pp. 878–79) describes how bacteria cause disease symptoms. You should review how the body defends against disease (Chapter 21). You will want to compare the treatments of bacterial diseases and of viral diseases.

KEY CONCEPTS

1. Viruses are on the borderline between living and nonliving; they lack the metabolic machinery to make ATP and proteins, and they cannot reproduce themselves in the

absence of a host, yet they have nucleic acid genes that encode information for their reproduction. (pp. 861–63)

2 The cells of archaebacteria and eubacteria are procaryotic; they lack a nuclear membrane and most other membranous organelles and thus are fundamentally different from all other living organisms. (pp. 869–70)

3 The archaebacteria differ in structure from the eubacteria. They are evolutionarily younger and typically occupy very challenging habitats. (pp. 870–72)

4 The eubacteria are an extraordinarily successful group of organisms; their extreme metabolic versatility and enormous reproductive potential have enabled them to survive in a wide variety of habitats. (p. 874)

5 The cyanobacteria are procaryotic cells that produce O_2 as a by-product of their photosynthesis; they may have initiated the oxygen revolution some 2.3 billion years ago. (p. 876)

6 Many bacteria are beneficial; all other organisms depend directly or indirectly on the activities of the bacteria. (p. 877)

CROSSLINKING CONCEPTS

1 This chapter deepens your understanding of the structure and reproduction of viruses, which were introduced in Chapter 10.

2 Bacteria are procaryotes and are among the simplest of living things. Refresh your memory with comparisons of procaryotic and eucaryotic cells (Chapter 6).

3 Review the differences between binary fission, the simpler method of cell reproduction used by bacteria, and mitosis, the more complex method of cell reproduction used by eucaryotes (Chapter 9).

4 The ecological importance of both autotrophic and heterotrophic bacteria was described in Chapter 34. This chapter discusses in detail the characteristics of bacteria—particularly their diversity of habitats and metabolic strategies—that enable them to occupy so many ecological niches.

OBJECTIVES

1 Explain how scientists showed in 1935 that viruses are distinct from bacteria. (p. 862)

2 Using diagrams and micrographs as in Figures 37.1 (p. 862) and 37.2 (p. 863), describe the basic structure of a virus. Indicate whether DNA is the genetic material in all viruses. Describe the diversity found in the viral genetic material, including the types of nucleic acid and the number of the nucleic acid strands. (p. 863)

3 Using Figures 10.4 (p. 205) and 37.3 to 37.5 (pp. 864–65), compare the reproductive cycle of a bacteriophage with that of a virus that infects the cells of higher organisms, pointing out both similarities and differences. Be sure to include in your answer the terms lytic cycle and provirus.

4 Compare and contrast the mode of reproduction of a DNA bacteriophage virus with that of a typical RNA virus and a retrovirus. Indicate whether eucaryotic cells normally possess RNA replicase or reverse transcriptase. See Figures 37.3 (p. 864), 37.6 (p. 865), and 37.7 (p. 866). (pp. 864–66)

5 Explain why most viral infections do not respond to treatment with antibiotics. Describe the role of interferon in the recovery from viral infection. (p. 866)

6 List three ways that archaebacteria differ from eubacteria in composition. Describe the environments inhabited by methanogens, halophiles, and thermoacidophilic bacteria. (pp. 868–69)

7 Explain how bacteria differ from viruses. Then list five differences between procaryotic and eucaryotic cells (Tables 6.1, p. 135, and 37.3, p. 864, may be helpful). (pp. 866–67)

8 Draw a diagram of a typical bacterial cell and label the cell wall, cell membrane, nucleoid, ribosomes, and flagellum. Then list the three different bacterial shapes and explain what the prefixes diplo-, staphylo-, and strepto- mean when applied to bacterial aggregations. (pp. 870–72)

9 Contrast the cell walls of bacteria with those of plants and fungi, and explain the significance of the cell wall in the life of bacteria. Explain what a capsule is and what role it plays. Distinguish between gram-positive and gram-negative bacteria. (p. 872)

10 Describe the structure and function of a bacterial endospore. (pp. 872–73)

11 Indicate how a bacterial flagellum differs from a eucaryotic flagellum in structure and function. (pp. 873–74)

12 Using Figure 9.2 (p. 180), describe the reproduction of a bacterial cell and show how it differs from that of a eucaryotic cell. Explain how genetic recombination can occur in some bacteria. (p. 872)

13 Differentiate among aerobes, obligate anaerobes, and facultative anaerobes. (p. 857)

14 Describe the three nutritive modes found in bacteria, and indicate which mode is the most common. For each division of photosynthetic bacteria, indicate which types of pigments are present, whether the process is anaerobic or aerobic, and whether molecular oxygen is produced as a by-product.

Indicate the possible evolutionary significance of the cyanobacteria and prochlorophytes. (pp. 875–78)

15 Discuss the role of bacteria as agents of disease, indicating how pathogenic bacteria may harm the body of their host. (pp. 878–79)

16 List five ways in which bacteria are beneficial. (p. 877)

KEY TERMS

capsid (p. 862)
envelope (p. 862)
obligate parasites (p. 863)
lytic cycle (p. 863)
provirus (p. 863)
vector (p. 864)
RNA replicase (p. 864)
retrovirus (p. 865)
reverse transcriptase (p. 865)
interferon (p. 867)
archaebacteria (p. 868)
eubacteria (p. 868)
coccus (p. 870)
bacillus (p. 870)
spirillum (p. 870)
diplococci (p. 871)
streptococci (p. 871)
staphylococci (p. 871)
capsule (p. 872)
endospore (p. 872)
nucleoid (p. 874)
binary fission (p. 874)
conjugation (p. 874)
obligate anaerobe (p. 875)
facultative anaerobe (p. 875)
heterocyst (p. 875)
toxin (p. 878)

SUMMARY

Viruses In the late nineteenth century, infectious agents so small that they could pass through an ultrafine filter were discovered. In 1935 the tobacco mosaic virus was crystallized, and the filterable viruses were recognized as different from bacteria.

Viruses are not cells; though some viruses possess a few enzymes, they lack the metabolic machinery for energy generation, and they do not have the ribosomes required for protein synthesis. The virus consists of a nucleic acid core covered with a protein coat (*capsid*). The nucleic acid can be DNA or RNA, double-stranded or single-stranded.

Viruses cannot reproduce themselves; the host cell manufactures new viruses using genetic instructions provided by the old virus. Once new viral components have been synthesized by the host cell, they assemble into new viruses that are released by lysis of the host cell or by its budding.

Sometimes viral DNA is integrated into the host chromosome as a *provirus*, which behaves as an additional part of the chromosome and is replicated with it. The virus remains there until the host cell is exposed to some stress; then the virus leaves the host chromosome and resumes normal replication.

DNA viruses use the enzymes of the host cell for their replication, but RNA viruses require special enzymes. *RNA replicase* catalyzes RNA-to-RNA replication. In retroviruses, *reverse transcriptase* transcribes RNA into DNA, which then becomes a provirus. Later it is transcribed into RNA for new viruses.

Viruses cause many diseases in various organisms, and have been associated with some kinds of cancer. Most viral infections do not respond to antibiotic treatment. A protein called *interferon* promotes recovery from viral infections. Produced by the host's cells in response to an invading virus, it is released from infected cells and interacts with uninfected cells and gives them a resistance to infection.

The procaryotic kingdoms The procaryotes are placed in two kingdoms, the Archaebacteria, or "ancient" bacteria, and the Eubacteria, the "true" bacteria. In contrast to the viruses, bacteria are cellular, have the metabolic machinery to generate ATP and to reproduce themselves, and they have ribosomes for protein synthesis. As procaryotes, they have no nuclear membrane, mitochondria, endoplasmic reticulum, Golgi apparatus, or lysosomes.

The archaebacteria live in very harsh environments: high temperature, high salinity, and high acidity. There are three main groups of archaebacteria: the methanogens, (methane-producers), the extreme halophiles ("salt-loving"), and the thermoacidophiles ("heat- and acid-loving").

The cells of most euacteria are classified according to their shape: spherical (*cocci*), rod-shaped (*bacilli*), or helical (*spirilla*). Some spherical bacteria remain together after cell division and may form pairs (*diplococci*), chains (*streptococci*), or clusters (*staphylococci*).

The cell walls of the eubacteria differ from those of plants and fungi; they are made of murein, a polymer of polysaccharide chains crosslinked by amino acids. The cell wall protects the bacteria from physical and osmotic assault. Penicillin is toxic to growing bacteria because it inhibits murein formation. Differences in the cell wall chemistry of different bacteria are associated with differences in their staining properties; these can be used for identification. An envelope or *capsule* may surround the cell wall.

Some eubacteria can form special resting cells called *endospores*, which enable them to withstand adverse conditions. When conditions improve, the endospores produce new bacterial cells.

Many eubacteria are motile. Some have flagella, which have a structure entirely different from that of eucaryotic flagella and cilia. The flagella have a rotary, propellerlike motion rather than the whiplike beating of eucaryotic flagella.

Bacteria have enormous reproductive potential. The single circular chromosome called a plasmid, is found in the *nucleoid* area of the cell; it replicates before division. Bacteria reproduce by *binary fission*, a type of nonmitotic cell division that produces two daughter cells exactly like the parent. Although this process is asexual, many eubacteria obtain new genetic material by conjugation.

Most eubacteria are heterotrophic, and either saprophytic or parasitic. The oxygen requirement of bacteria varies; most are *aerobic* (require oxygen), but some are *facultative anaerobes* (can live with or without oxygen) and others are *obligate anaerobes* (poisoned by oxygen).

Some bacteria are *chemosynthetic autotrophs* and others are *photosynthetic autotrophs*. The chemoautotrophs oxidize certain inorganic compounds and trap the released energy. The photoautotrophs get energy from light and fix carbon. The green and purple bacteria, unlike higher plants, lack chlorophyll *a* or *b*; instead they have their own types of bacteriochlorophyll. They do not use water as an electron source and do not produce O_2.

The major group of aerobic photosynthetic bacteria are the cyanobacteria. All cyanobacteria possess chlorophyll *a*, the pigment found in higher plants, and generate O_2 as a by-product of photosynthesis. The pigments are located in membranous disks (thylakoids), but are not contained within chloroplasts. The cyanobacteria or prochlorophyta (which also have chlorophyll *a*) may have been the ancestors of the chloroplasts of eucaryotes.

Many cyanobacteria can fix atmospheric nitrogen; this allows them to survive in minimal environments. Certain cells, the *heterocysts*, are specialized for nitrogen fixation. High levels of O_2 inhibit the process of nitrogen fixation.

Beneficial bacteria far outnumber harmful ones. They play a vital role in the cycling of materials in the ecosystem, are important in many food and industrial processes, and are used to produce antibiotics.

Eubacteria cause disease symptoms by interfering with normal function, by destroying cells and tissues, or by producing poisons, called *toxins*. Antibiotics are useful against eubacteria, but unfortunately many bacteria have evolved resistance to antibiotics.

CONCEPT MAPS

Map 37.1

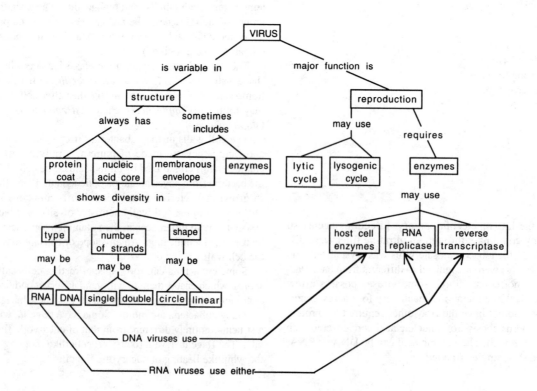

Map 37.2

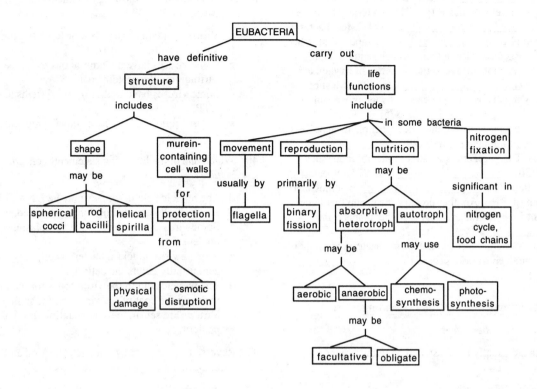

QUESTIONS

Testing recall

Fill in the blanks.

A virus consists of an inner core of a single molecule of nucleic acid which may be (1) _____ or _____ (p. 862). Surrounding the nucleic acid is a coat of protein called the (2) _____ (p. 862). The viruses that attack bacterial cells are called (3) _____ (p. 863). They attach by their tail to the bacterial cell wall and the (4) _____ (p. 863) is injected into the host cell. The phage nucleic acid provides the genetic information for the synthesis of new viral (5) _____ and _____ (p. 863). The new viruses escape by (6 _____ of the cell. (p. 864)

In plant and animal viruses, the (7) _____ enters the host cell (p. 864). If the virus is an RNA virus, the RNA molecule can serve as a template for the synthesis of new RNA, provided the enzyme (8) _____ is present to catalyze the reaction (p. 864). In the retroviruses, however, the RNA serves as a template for the synthesis of (9) _____ (p. 865). The enzyme (10) _____ is required for this reaction (p. 865). This nucleic acid becomes integrated into the host's chromosome and is called a (11) _____ (p. 865). Once the new virus particles have been assembled they may be released by (12) _____ of the cell or by (13) _____ through the cell membrane. (p. 865)

The procaryotic kingdoms include the (14) _____ and _____ (p. 868). The cells of both these groups lack a (15) _____ membrane and most other membranous organelles (p. 868). The three basic shapes of bacteria are (16) _____, _____, and _____. (p. 870)

The kingdom Archaebacteria appears to be somewhat (17) _____ than the kingdom Eubacteria (p. 868). The archaebacteria occupy challenging habitats. Methanogens are found mostly in bogs and marshes and get energy from (18) _____ (p. 868). Halophiles live in very (19) _____ environments (p. 868). Thermoacidophilic bacteria utilize (20) _____ in their metabolism. (p. 868)

The bacterial cell is surrounded by a cell wall made of (21) _____ (p. 872). Under certain conditions some bacteria form (22) _____, which are resistant to destruction (p. 872). Some bacteria have (23) _____, which move the cell by their rotary motion (p. 873). Bacteria have

(24) _____ within the cytoplasm to carry out protein synthesis (p. 871). Bacteria reproduce by (25) _____ (p. 874). Most bacteria show the (26) _____ mode of nutrition (p. 875). Bacteria vary in their oxygen requirements; those that cannot survive in the presence of oxygen are called (27) _____ (p. 875). Most of these organisms obtain their energy by the process of (28) _____. (p. 875)

The cyanobacteria resemble the higher plants in possessing the light-trapping pigment (29) _____ (p. 875). This pigment is located in membranous sacs called (30) _____ (p. 875). Special cells called (31) _____ fix atmospheric nitrogen (p. 876). Another procaryotic group that has a pigment system similar to that of higher plants is the (32) _____. (p. 875)

Many bacteria cause disease. Some bacteria harm their host by producing poisons called (33) _____. (p. 878)

Testing knowledge and understanding

Choose the one best answer.

34 Viruses are usually not classified as living organisms because they do not

   a contain DNA or RNA.
   b self-replicate.
   c possess chlorophyll.
   d possess nuclear membranes.
   e possess enzymes.   (p. 862)

35 The coat of a virus is generally made of

   a nucleic acid.   d murein.
   b protein.   e glucose.
   c cellulose.   (p. 862)

36 Which one of the following statements concerning viruses is *false*?

   a Viruses have no metabolic machinery of their own.
   b Viral genes are composed only of DNA.
   c Viruses cannot be cultured on artificial media.
   d Viruses will form crystals under certain conditions.
                                              (p. 862)

37 Viruses are

   a naked, circular pieces of RNA.
   b procaryotic cells without membranes.
   c infectious particles composed of nucleic acid cores and protein coats.
   d fragments of dead procaryotic or eucaryotic cells.
   e infectious proteins.   (p. 862)

38 Which entity consists only of DNA?

   a a bacteriophage   d a bacterial spore
   b a viral capsid   e none of the above
   c a human chromosome   (pp. 862–66)

39 Which one of the following statements concerning viruses is *false*?

   a Viruses contain either DNA or RNA but not both.
   b Viruses can be crystallized.
   c Some viruses have a membranous envelope surrounding the protein coat.
   d Some viruses have multienzyme systems to generate ATP.
   e Viruses require living host cells for replication.
                                              (p. 862)

40 Which one of the following statements concerning viruses is *false*?

   a The nucleic acid in viruses may be DNA or RNA, either single-stranded or double-stranded.
   b Some viruses reproduce in the cytoplasm, others in the nucleus.
   c When a virus attacks a bacterial cell, only the nucleic acid usually enters the cell.
   d The nucleic acid in the virus codes for the proteins that make up the coat of the virus and for some enzymes.
   e Viruses are sensitive to most antibiotics, particularly penicillin.   (p. 863)

41 Reverse transcriptase synthesizes DNA on an RNA template. This enzyme is characteristic of

   a animal cells.
   b bacterial viruses.
   c fungal cells.
   d certain DNA viruses.
   e certain RNA viruses.   (p. 865)

42 In which *one* of the following ways do viruses resemble cellular organisms? Viruses

   a divide by mitosis.
   b can undergo mutation.
   c exhibit aerobic respiration.
   d have an extensive endoplasmic reticulum.
   e have a membrane composed largely of phospholipids.
                                           (pp. 863–65)

43 Animal cells produce an antiviral protein known as

   a interferon.   c lysozyme.
   b viralase.   d penicillin.   (p. 867)

44 The cells of procaryotess differ from those of eucaryotic organisms in many ways. Which of the following is *not* one of those ways?

   a Cells of procaryotess lack a nuclear membrane.
   b When chlorophyll *a* is present in moneran cells, it is located in thylakoids, but these are not contained in chloroplasts.
   c Cells of procaryotes lack mitochondria and an endoplasmic reticulum.
   d Cells of procaryotess lack ribosomes.
   e When the cells of procaryotes possess flagella, those flagella lack a 9 + 2 microtubular structure.
                                           (pp. 868–69)

45 A bacterial cell does *not* possess

   a a cell wall.   d mitochondria.
   b DNA.   e ribosomes.
   c a plasma membrane.   (p. 872)

46 The cell walls of bacteria

   a enable the bacteria to survive in a hypotonic medium.
   b contain cellulose.
   c are called capsules.
   d are composed of chitin.
   e are composed of protein.   (p. 872)

47 The antibiotic penicillin inhibits the ability of bacteria to

   a synthesize proteins.
   b replicate DNA.
   c synthesize cell-wall material.
   d perform respiration.   (p. 872)

48 Bacteria generally reproduce by

   a conjugation.   d mitosis.
   b transformation.   e meiosis.
   c binary fission.   (p. 874)

49 Which one of the following statements is *true*?

   a All bacteria are heterotrophs.
   b The flagella of bacteria move by rotary motion.
   c The genetic material of bacteria is RNA.
   d The cell walls of bacteria contain cellulose.
   e All bacteria are aerobic.   (p. 873)

50 Bacteria that are poisoned by oxygen are called

   a aerobic.
   b obligate anaerobes.
   c facultative anaerobes.
   d obligate parasites.   (p. 875)

51 Most bacteria are

   a ingestive heterotrophs.
   b absorptive heterotrophs.
   c photosynthetic heterotrophs.
   d photosynthetic autotrophs.
   e chemosynthetic autotrophs.   (p. 875)

52 Which one of the following groups possesses chlorophyll *a* and releases molecular oxygen as a by-product of photosynthesis?

   a cyanobacteria
   b halophilic archaebacteria
   c purple bacteria
   d green bacteria   (p. 875)

53 Cyanobacteria are

   a heterotrophic procaryotes.
   b heterotrophic eucaryotes.
   c autotrophic procaryotes.
   d autotrophic eucaryotes.
   e chemosynthetic procaryotes.   (p. 875)

54 Cyanobacteria differ from all other bacteria but the prochlorophytes in

   a having a nuclear membrane.
   b possessing chlorophyll *a*.
   c having amino acids in their cell walls.
   d lacking lysosomes.
   e exhibiting no mitosis.   (p. 876)

55 Cyanobacteria do *not* have

   a chlorophyll *a*.
   b chlorophyll *b*.
   c photosynthetic membranes.
   d autotrophic nutrition.
   e ribosomes.   (p. 876)

56 Penicillin would be *least* effective against the organisms causing

   a scarlet fever.   d tuberculosis.
   b gonorrhea.   e strep throat.
   c AIDS.   (p. 867)

57 Which one of the following is *never* exhibited by any kind of bacteria?

   a conjugation   d spore formation
   b photosynthesis   e chemosynthesis
   c meiosis   (p. 874)

58 Which one of the following groups of bacteria belongs to the kingdom Archaebacteria?

   a prochlorophytes
   b cyanobacteria
   c eubacteria
   d green photosynthetic bacteria
   e methanogens   (p. 868)

For further thought

1 Would you classify viruses as living or nonliving? Justify your answer. (pp. 861–67)

2 Antibiotics are effective against many bacterial diseases, but most viral infections do not respond to treatment with them. However, a few antibiotics are somewhat effective against certain viruses. Below are listed four antibiotics and their mode of action. For each, determine whether it would be effective against bacteria and/or viruses and explain your answer. (pp. 866–67)

   a penicillin—affects procaryotic cell-wall synthesis
   b streptomycin—inhibits protein synthesis and causes misreadings of the mRNA code at the procaryotic ribosome
   c tetracycline—inhibits protein synthesis on the procaryotic ribosome
   d actinomycin—inhibits mRNA synthesis

3 When penicillin was first introduced to treat gonorrhea, it was extraordinarily effective, but certain physicians predicted that eventually it would become useless. In time,

a strain of *Gonococcus* (the organism causing gonorrhea) began to show resistance. Why have these organisms become resistant? (Chapter 31, pp. 878–79)

4 Food poisoning can develop very rapidly after eating contaminated foods. Explain the speed with which this malady develops. (p. 874)

5 In the late 1940s it was discovered that feeding antibiotics to farm animals not only controlled infection but also stimulated growth. Today, half the antibiotics produced in the United States are used for animal feeds. Some scientists view this practice with alarm; in their opinion the widespread use of antibiotics is dangerous, and we may eventually return to the pre-antibiotic era in the treatment of bacterial infections. Can you think of reasons for the scientists' concern? (p. 878)

6 Certain bacteria are essential components of the scheme of life. Discuss the role of bacteria in the cycling of inorganic nutrients. Could life survive if all bacteria were eliminated? (Chapter 34)

ANSWERS

Testizng recall

1 DNA, RNA *or* single/double-stranded
2 capsid
3 bacteriophages
4 nucleic acid
5 nucleic acid and protein
6 lysis
7 whole virus particle
8 RNA transcriptase
9 DNA
10 reverse transcriptase
11 provirus
12 lysis
13 extrusion
14 Archaebacteria, Eubacteria
15 nuclear
16 coccus, spirillum, bacillus
17 younger
18 anaerobic chemosynthesis
19 salty
20 sulfur
21 murein
22 endospores
23 flagella
24 ribosomes
25 binary fission
26 heterotrophic
27 obligate anaerobes
28 fermentation
29 chlorophyll *a*
30 thylakoids
31 heterocysts
32 Prochlorophyta
33 toxins

Testing knowledge and understanding

34	*b*	41	*e*	47	*c*	53	*c*
35	*b*	42	*b*	48	*c*	54	*b*
36	*b*	43	*a*	49	*b*	55	*b*
37	*c*	44	*d*	50	*b*	56	*c*
38	*e*	45	*d*	51	*b*	57	*c*
39	*d*	46	*a*	52	*a*	58	*e*
40	*e*						

Chapter 38

THE PROTISTAN KINGDOM

A GENERAL GUIDE TO THE READING

You will remember from Chapter 36 that the unicellular eucaryotic organisms are frequently difficult to classify as either plants or animals, and consequently are placed in a kingdom of their own, the kingdom Protista. Also included in this kingdom are two divisions of multicellular algae, the brown and red algae, whose cells are relatively unspecialized. As you read about this kingdoms in Chapter 38 in your text, you will want to focus on the following topics.

1. Animal-like protists. Be sure you understand why these organisms, discussed on pages 881 to 884, are referred to as acellular. Also, for each of the phyla or subphyla described, learn the form of locomotion characteristic of its members.

2. Funguslike protists. You need to understand which part of the life cycle of slime molds is plantlike and which part is animal-like. (pp. 884–85).

3. Plantlike protists: the unicellular algae. In studying the material on these organisms (pp. 886–89), be sure to distinguish the plantlike and animal-like qualities of the euglenoids. Also note the important role of the diatons and dinoflagellates in aquatic food webs.

4. Algal life cycles. Be sure you understand the differences among the three types of life cycles depicted in Figure 38.17 (p. 890). Alternation of generations is a particularly important concept to grasp because this type of life cycle is found widely among multicellular algae and in all the land plants discussed in Chapter 39.

5. Plantlike protists: the multicellular algae. Note that the brown algae (pp. 889–92) are mostly large and complex in structure. In reading about the red algae (p. 892), pay special attention to the role of their accessory pigments.

KEY CONCEPTS

1. Protists include primarily unicellular or colonial eucaryotic organisms and plantlike multicellular ogranisms whose bodies show relatively little distinction between tissues. (p. 881)

2. Protists can be separated into three groups: animal-like (protozoans), funguslike, and plantlike. However, there are also many similarities among them. (p. 881)

3. Each protozoan should be regarded not as equivalent to a cell of a more complex animal but as a complete organism with the same properties and characteristics as cellular animals. Protozoans are generally distinguished from each other by their locomotion. (p. 881)

4 The members of the subphylum Mastigophora are protozoans that possess flagella as their principal locomotor organelles. They appear to be the most primitive of all the protozoans and may have given rise to the other protozoan groups. (p. 882)

5 The members of the subphylum Sarcodina are the amoeboid protozoans. This subphylum is important for two subgroups that have played major roles in the geologic history of the earth. (pp. 882–83)

6 The slime molds are funguslike protists that are animal-like in some stages of their life cycle and plantlike in others. (pp. 884–85)

7 The unicellular plantlike protists show a combination of plantlike and animal-like characteristics. (pp. 886–89)

8 The multicellular plantlike protists include the brown and red algae. The aquatic environment provides water, oxygen, nutrients, and even mechanical support; consequently algal tissues are not very differentiated. (pp. 889–92)

CROSSLINKING CONCEPTS

1 Because most groups or multicellular plants and all multicellular animals have flagellated stages in their life cycle, many biologists believe that the flagellate protozoans may be the ancestors of all the multicellular plants and animals discussed in Chapters 39, 41, 42, and 43.

2 This chapter describes the important roles of some protistans in the marine communities discussed in Chapter 34. In particular, dinoflagellates and diatoms are major components of the phytoplankton, the principal producers in marine food webs. In addition, some dinoflagellates live symbiotically in coral organisms, and thus are largely responsible for the great productivity of the coral reef community.

3 The different types of life cycles, treated briefly in Chapter 9, are described in more detail in this chapter. Alternation of generations is found in all the terrestrial plants discussed in Chapter 39.

OBJECTIVES

1 List the three major groups of protists, explain how the members generally obtain the high-energy organic compounds necessary to carry out their life processes. (p. 881)

2 Describe how protozoans generally carry out the life processes of nutrition, gas exchange, regulation, locomotion, and reproduction. (pp. 881–82)

3 List the three protozoan groups discussed in your text, give two distinguishing characteristics for each, and explain why many biologists prefer to consider the protozoans acellular rather than unicellular organisms. (p. 882)

4 Name the protozoan phylum that contains organisms believed to be ancestral to the others. (p. 882)

5 Briefly describe the geological significance of the Sarcodina subphylum. (p. 882)

6 Using Figure 38.7 (p. 885), describe the life cycle of the organism that causes malaria. (pp. 884–85)

7 Contrast the life cycle of the true slime mold with that of the cellular slime mold, discussing both the similarities and differences. (pp. 884–85)

8 Give ways in which the euglenoids resemble plants and ways in which they resemble animals. (pp. 887–88)

9 Specify two distinguishing characteristics shared by the Chrysophyta and the dinoflagellates, and indicate the role of these organisms in aquatic food webs. (p. 889)

10 Using Figure 38.17 (p. 890), describe and compare the three types of life cycle exhibited by the algae. Explain what is meant by alternation of generations, and define the terms sporophyte, gametophyte, sporangium, and spore. (pp. 889–90)

11 Contrast the brown algae and the red algae with respect to where they grow and to their size, type of cell wall, storage product, and types of chlorophyll and other pigments. (pp. 890–92)

KEY TERMS

protist (p. 881)
protozoa (p. 881)
acellular (p. 881)
Kinetoplastida (p. 882)
Mastigophora (p. 882)
Sarcodina (p. 882)
Ciliata (p. 882)
sporozoan (p. 884)
malaria (p. 884)
cellular slime mold (p. 884)
true slime mold (p. 884)
zoospore (p. 885)
euglenoid (p. 887)
dinoflagellate (p. 888)
Chrysophyta (p. 889)
diatom (p. 889)
plankton (p. 889)
thallus (p. 889)

gametophyte (p. 890)
spore (p. 890)
sporophyte (p. 890)
sporangium (p. 890)
alternation of generations (p. 890)
Phaeophyta (p. 890)
fucoxanthin (p. 891)
holdfast (p. 891)
stipe (p. 891)
blade (p. 891)
Rhodophyta (p. 891)

SUMMARY

The kingdom Protista includes primarily unicellular or colonial eucaryotic organisms and the plantlike multicellular organisms whose bodies show relatively little distinction between tissues. Though some of the protists tend to be plantlike, others animal-like, and still others funguslike, they share many characteristics.

Animal-like protists (protozoans) Although protozoans are usually said to be unicellular, many biologists consider them *acellular*, since they are far more complex than other individual cells. They live in a variety of aquatic and moist habitats, and exhibit great diversity of form. They are heterotrophic, and their reproduction method is either asexual or sexual.

The protozoans have been divided into two groups, Kinetoplastida and Ciliata, which are heterogeneous groups of structurally similar, but probably not closely related, organisms. Kinetoplastida includes two subphyla, Mastigophora and Sarcodina.

The mastigophorans, or flagellates, move by flagella and appear to be the most primitive of the protozoans. They may have given rise to other protozoan groups and possibly to the multicellular plants and animals. Most mastigophorans live as symbionts in the bodies of higher plants and animals.

The sarcodines are the amoeboid protozoans; they move by means of pseudopods. They are thought to be closely related to the flagellates because some flagellates undergo amoeboid phases, and some sarcodines have flagellated stages. Several groups of sarcodines secrete hard calcareous or siliceous shells. The shells of two of these groups, the foraminiferans and radiolarians, have formed much of the limestone and some of the siliceous rocks on the earth's surface.

The members of the phylum Ciliata possess numerous cilia for movement and feeding. Their other organelles are greatly elaborated. The ciliates differ from other protozoans in having a macronucleus and one or more micronuclei. The macronucleus controls normal cell metabolism; the micronuclei are concerned with reproduction of the cell and production of the macronucleus.

Funguslike protists The Sporozoa are distinguished by a spore-like infective cyst stage in their life cycles. All are nonmotile internal parasites. Malaria is caused by species of the genus *Plasmodium*; the pathogen is transmitted from host by female *Anopheles* mosquitoes.

Two groups of slime molds are recognized: the *cellular slime molds* and the *true slime molds*. Though funguslike in many ways, they are not closely related to the fungi or to one another. Their life cycle includes an animal-like amoeboid stage and a funguslike sedentary stage in which a fruiting body develops and produces spores.

The water molds are funguslike in their possession of hyphae and heterotrophic nutrition, but animal-like in their production of flagellated zoospores.

Plantlike Protista: the unicellular algae Several groups of unicellular organisms show a combination of plantlike and animal-like characteristics. They are plantlike because many have chlorophyll and often a cell wall; they are animal-like because they are highly motile. The various algae divisions are probably not closely related; they evolved from different eucaryotic cell lineages and acquired different bacterial lineages as chloroplasts.

The euglenoids are highly motile unicellular flagellates that lack cell walls. Like higher plants, most have chlorophylls *a* and *b*. They probably evolved from flagellates that acquired photosynthetic procaryotes and chloroplasts.

The dinoflagellates are small, usually unicellular organisms that typically possess two unequal flagella. The photosynthetic species possess chlorophylls *a* and *c*; they are second only to the cyanobacteria and diatoms as primary producers of organic matter in the marine environment. Some species are luminescent, and some are poisonous. The so-called red tides that sometimes kill millions of fish are caused by species of red-pigmented dinoflagellates.

The yellow-green algae, the golden-brown algae, and the diatoms belong to the division Chrysophyta. These predominantly unicellular or colonial algae are shades of yellow or brown and possess chlorophylls *a* and *c* (no *b*). The walls of many contain silica. The shells of diatoms form deposits of what is known as diatomaceous earth. The diatoms are an extremely important link in most freshwater and marine food chains; they are the most abundant component of the marine *plankton*. The phytoplankton are the principal photosynthetic producers in marine communities.

Life cycles Many algae show a primitive life cycle in which thehaploid gametophyte stage of the cycle is dominant and the only diploid stage—the zygote—is very tansitory. Dominance of the haploid stages is characteristics of most primitive plantlike protists and fungi; this was the ancestral condition.

Many of the multicellular algae and most plants show a life cylce that is more complex in that it exhibits *alternation fo generations*: a haploid multicellular stage (gametophyte) alternates with a diploid multicellular stage (sporophyte).

Animals and a very few multicellular algae have a life cycle in which meiosis produces gametes directly; the haploid gametes then fuse to form a diploid zygote, which develops into a new diploid organism.

Platnlike Protista: the multicellular algae The kingdom Protista includes two division of algae whose members are predominantly multicellular, the brown and red algae. Because algae are aquatic, they need litle in the way of tissue differentiation. The entire plant is considered to consist of a single tissue, called a *thallus*.

The brown algae are large seaweeds. The thallus may be a filament or a large, compelx three-dimensional structure. They possess chlorophull *a* and *c* and the brownish pigment fucoxanthin. Some show tissue differentiation. they are complex plants that have convergently evolved many similarities to the vascular plants. The sexual life cycle usually exhibits alternation of generations.

The red algae are mostly marine. They possess chlorophylls *a* and *d*, and blue and red pigments. The latter are important in absorbing the light that epentrates into deep water and transferring the energy to chlorophyll *a* for photosynthesis. The life cycles of the red algae are complex; alternation of generations is usual. Flagellated cells never occur.

CONCEPT MAPS

Map 38.1

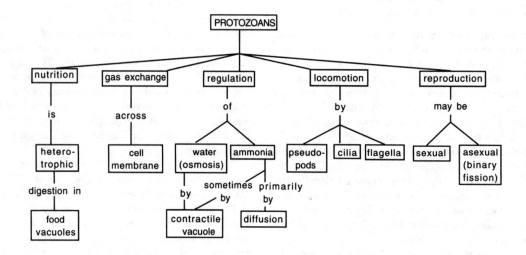

QUESTIONS

Testing recall

1 The chart on the next page lists the various groups of Protista and some important characteristics. After studying this chapter, you should be able to fill in the chart using the directions below. The correctly completed chart will be a good study aid.

    Column 1: Indicate whether the phylum or division contains organisms that are autotrophic, heterotrophic, or both.
    Column 2: Indicate whether flagella are typically present or absent.
    Column 3: List any mechanism of locomotion other than flagella.
    Column 4: Indicate whether chlorophyll is present or absent. (If present, give types.)
    Column 5: List any other important or distinguishing characteristics.

THE PROTISTAN KINGDOM • 303

Group	1 Autrotrophic or heterotrophic	2 Flagella	3 Other means of movement	4 Chlorophyll pigments	5 Other distinguishing characteristics
Flagellates (Mastigophora)					
Ameboid protozoans (Sarcodina)					
Ciliates (Ciliata)					
Sporozoans					
Slime molds (Mycetozoa)					
Water molds (Oømycota)					
Euglenoids (Euglenophyta)					
Dinoflagellates (Dinoflagellata)					
Yellow-green algae, golden-brown algae, and diatoms (Chrysophyta)					
Brown algae (Phaeaphyta)					
Red Algae (Rhodophyta)					

Match each statement below with the division or divisions to which it most clearly applies. Answers may be used once, more than once, or not at all.

  a Euglenophyta (euglenoids)
  b Dinoflagellata (dinoflagellates)
  c Chrysophyta (golden algae)
  d Phaeophyta (brown algae)
  e Rhodophyta (red algae)

2 The division members possess the pigment fucoxanthin. (p. 891)

3 The division members possess the pigments chlorophyll *a* and chlorophyll *b*. (p. 887)

4 Some members of the division, having conductive tissue similar to phloem, show a primitive type of tissue differentiation. (pp. 890–91)

5 Some division members have a life cycle in which gametes are produced directly by the diploid plant (p. 891)

6 The division members have chlorophyll *a* and *c* (pp. 888–91)

7 Members of this division are an important source of commercial colloids used in agar, ice cream, some cheeses, and salad dressing. (p. 892)

8 A member of this division forms diatomaceous earth. (p. 889)

9 Members of this division typically possess red and blue pigments in addition to chlorophylls *a* and *d* (p. 892)

10 The members of this division lack flagellated cells at all stages of their life cycle. (p. 892)

11 The members of this division lack a cell wall. (p. 887)

Testing knowledge and understanding

Choose the one best answer.

12 Which one of the following statements is *false* concerning protozoans?

  a Their food is usually digested in food vacuoles.
  b Movement is by pseudopods or by cilia or flagella that beat.
  c Reproduction is primarily by binary fission.
  d They release ammonia as their primary nitrogenous waste.
  e Most freshwater forms have contractile vacuoles.
  (p. 881)

13 The most "primitive" protozoans are thought to be

  a zflagellates.
  b amoeboid protozoans.
  c ciliates.
  d sporozoans.   (p. 882)

14 The organism that causes malaria is a(n)

  a ciliate.
  b zooflagellate.
  c dinoflagellate.
  d sporozoan.
  e amoeboid protozoan.   (p. 884)

15 Which one of the following groups contains organisms with no flagellated cells?

  a mastigophorans   d dinoflagellates
  b euglenoids   e yellow-green algae
  c sporozoans   (p. 884)

16 A piece of chalk is made of the shells of

  a ciliates.   d zooflagellates.
  b sarcodines.   e dinoflagellates.
  c diatoms.   (p. 882)

17 The protozoans showing the most complex and well-developed organelles belong to the phylum or subphylum

  a Mastigophora.   c Ciliata.
  b Sarcodina.   (p. 883)

18 Suppose you are given an unknown organism to identify. You find that it is unicellular and lacks a cell wall and locomotor organelles. It is an internal parasite. You conclude that it is most likely to be a(n)

  a ciliate.   d flagellate.
  b sporozoan.   e amoeboid protozoan
  c slime mold.   (p. 884)

19 Which one of the following statements is *false* concerning the organism that causes malaria?

  a It is a flagellate that is transmitted to humans by the bite of a termite.
  b It completes part of its life cycle in the *anopheles* mosquito and part in the liver and blood of humans.
  c It is a sporozoan with a complex life cycle.
  d Efforts at control have been directed at eradication of the insect vectors by pesticides.   (p. 884)

20 The movement of slime molds most closely resembles that of

  a ciliates.
  b amoeboid protozoans.
  c golden-brown algae.
  d flagellates.
  e euglenoids.   (p. 884)

21 All the animal-like and funguslike protists

  a are parasitic.
  b are heterotrophic.
  c have cell walls.
  d move by pseudopods.
  e reproduce by spores.   (pp. 884–85)

22 Which one of the following groups includes both autotrophic and heterotrophic forms?

   a euglenoids
   b flagellates
   c sporozoans
   d amoeboid protozoans
   e slime molds
   (pp. 886–87)

23 Which one of the following characteristics is unique to the euglenoids?

   a flagella
   b absence of a cell wall
   c chloroplasts with three membranes
   d chlorophylls *a* and *b*
   e photosynthetic and nonphotosynthetic forms
   (pp. 886–87)

24 Which one of the following statements is *false* concerning the dinoflagellates?

   a Some can produce light.
   b Some cause the so-called red tides.
   c They have chlorophyll and are photosynthetic.
   d They are an important component of the marine food chain.
   e Their boxlike shells contain silica.
   (p. 888)

25 Red tides, which sometimes kill fish, result from blooms of

   a euglenoids.
   b dinoflagellates.
   c cyanobacteria.
   d slime molds.
   e ciliates.
   (p. 888)

26 A unicellular organism has a siliceous (glasslike) shell and is autotrophic. It is most likely a

   a dinoflagellate.
   b Euglena.
   c diatom.
   d red algae.
   e brown algae.
   (p. 889)

27 In the ocean, which algae would normally be found in the deepest water?

   a Dinoflagellata
   b Chrysophyta (yellow-green, golden-brown)
   c Phaeophyta (brown)
   d Rhodophyta (red)
   (p. 892)

28 The algal division that has evolved the most complex multicellular thallus is

   a Euglenophyta.
   b Chrysophyta (yellow-green, golden-brown)
   c Phaeophyta (brown)
   d Rhodophyta (red)
   (p. 890)

For further thought

1 Some experimental evidence suggests that persistent pesticides such as DDT may affect the activity of the important marine unicellular plants. What ecological problems might result?

2 The World Health Organization lists malaria as one of the most serious infectious diseases. Malarial control programs center around the control of *Anopheles* mosquitoes. What stages of the life cycle of *Plasmodium* take place in the mosquito? If all mosquito populations were exterminated, would malaria be eliminated?

3 In what ways do the plantlike protists resemble higher plants? How are they different?

ANSWERS

Testing recall

1 Information for the chart can be found as follows: flagellates, pages 881–82; amoeboid protozoans, pages 882–83; ciliates, pages 883–84; sporozoans, page 884; slime molds, pages 884–85; water molds, page 885; euglenoids, page 887; dinoflagellates, page 888; yellow-green algae, golden-brown algae, and diatoms, pages 889–91; brown algae, pages 890–91; and red algae pages 891–92.

2 *d*
3 *a*
4 *d*
5 *d*
6 *b, c, d*
7 *e*
8 *c*
9 *e*
10 *e*
11 *a*

Testing knowledge and undersanding

12 *c*	17 *c*	21 *b*	25 *b*
13 *a*	18 *b*	22 *a*	26 *c*
14 *d*	19 *a*	23 *c*	27 *d*
15 *c*	20 *b*	24 *e*	28 *c*
16 *b*			

Chapter 39

THE PLANT KINGDOM

A GENERAL GUIDE TO THE READING

This chapter describes the diverse organisms within the plant kingdom and the evolutionary relationships among them. The terms in this chapter are easy to master if you learn them one at a time, as you encounter them in your reading. Some hints that may help you learn the life cycle of plants are presented in the discussion of this topic just below. As you read Chapter 39 in your text, you will want to concentrate on the following topics.

1. Plant life cycles. Life cycles were introduced in Chapter 9 and the three major types were described in more detail in Chapter 38 (pp. 889–90). Most multicellular green algae have the type shown in Figure 39.5 (p. 898). In this type there is no sporophyte stage; the zygote is the only diploid stage. This is believed to be the ancestral or primitive condition. All other plant life cycles exhibit alternation of generations, in which a diploid sporophyte (spore-producing) stage alternates with a haploid gametophyte (gamete-producing) stage. Careful study of Figure 39.10 (p. 901), which depicts the cycle in generalized form, will make it easier for you to grasp the variations exhibited by the bryophytes, ferns, conifers, and flowering plants. You will notice a major evolutionary trend in the life cycles of plant groups—a progressive decrease in the size and importance of the gametophyte and a corresponding increase in the size and importance of the sporophyte.

2. The volvocine series. This line of green algae, discussed on pages 897, exemplifies a number of evolutionary trends and provides a clue as to how multicellularity may have arisen in green plants.

3. The movement onto land. Seven basic problems faced by plants on land are listed (p. 900); you need to understand these problems, since much of the evolution of the terrestrial plants can best be understood in terms of adaptations that helped to solve these problems.

4. The nonvascular plants. The bryophytes (pp. 901–904) are the simplest in structure of the land plants. You will want to know why they have never become totally free of their dependence on an aquatic environment.

5. Evolutionary innovations of the vascular plants. These are listed on page 905. Be sure you understand how these adaptations are related to the basic problems of life on land.

6. Psilopphyta, club mosses, and horsetails. These groups, described on page 905, are the earliest vascular plants; from them we can infer the evolutionary history of the higher plants. Notice in particular the evolutionary advances exhibited by the club mosses and horsetails over the psilophytes.

7. Ferns. The characteristics of ferns (pp. 905–908) and their life cycle (Fig. 39.24, p. 908) are important.

8. Evolutionary advances in the seed plants. Pollen and seeds, the two most evolutionary innovations of the seed plants are described briefly on page 908. They are discussed in more detail in the subsequent specific descriptions of the conifers and the flowering plants.

9. Conifers. You need to learn the life cycle of the pine (pp. 909–11 and Fig. 39.29, p. 911) and understand the advances in the life cycle of a pine with reference to that of a typical fern.

10 Angiosperms. Pay special attention to the flowering plants (pp. 911-13) since they are the most abundant plants and hence the dominant group in the plant kingdom. You may wish to review pages 494 to 497 in Chapter 24, which describe the details of angiosperm reproducition. You need to understand how the life cycle of the angiosperm differs from that of the conifer. Be sure you know the differences between monocots and dicots (Fig. 39.33, p. 913).

11 Summary of evolutionary trends. This important section (pp. 913-14) provides an excellent summary of much that was discussed in the chapter. Figure 39.34 (p. 914) is a helpful pictorial representation of the evolutionary trends among plant groups.

KEY CONCEPTS

1 Terrestrial plants have evolved specialized tissues and organs that provide support, protection from desiccation, and efficient transport of nutrients and water. (p. 900)

2 The green algae (Chlorophyta) are regarded as the group from which the land plants arose. (pp. 899-900)

3 In the life cycle of the most primitive plants, the haploid stages are dominant; this apparently was the ancestral condition. (p. 896)

4 The evolution of most plant groups shows a tendency toward reduction of the gametophyte (multicellular haploid stage) and increasing importance of the sporophyte (multicellular diploid stage). (pp. 900-901)

5 The evolutionary move from an aquatic existence to a terrestrial one was not simple, for the terrestrial environment is in many ways hostile to life. Plants evolved a number of characteristics that enabled them to survive on land. (p. 900)

6 The bryophytes represent the most conspicuous exception to the evolutionary trend toward reduction of the gametophyte and increasing importance of the sporophyte. (pp. 902-904)

7 The evolution of the tracheophytes is best understood in terms of adaptations for life in a terrestrial environment. (pp. 904-905)

8 The angiosperms are the most successful land plants because they have evolved the most efficient adaptations for living and reproducing on land. (pp. 911-12)

9 The key evolutionary trends among the plant groups are the progressive decrease in the size and importance of the gametophyte; the evolution of vascular tissue for support and transport; and, in the angiosperms and conifers, the evolution of pollen and seeds. (pp. 913-14)

CROSSLINKING CONCEPTS

1 Alternation of generations was first introduced in Chapter 9. This chapter emphasizes the significance of alternation of generations in providing options that help organisms survive adverse conditions, disperse their offspring, and benefit from genetic recombination.

2 There is a very strong linkage between terrestrial reproductive styles in plants and animals. The offspring of terrestrial organisms need more protection than those of aquatic organisms. Terrestrial plants evolved a multicellular embryo that develops within the female archegonium. Terrestrial placental mammals evolved a multicellular embryo that develops within the uterus.

3 It is useful to look at the diversity of plants with an evolutionarily aware eye. Plants have evolved from more simple, less differentiated organisms to more complex, more differentiated organisms as they invaded and adapted to terrestrial environments.

OBJECTIVES

1 List two characteristics of the chlorophytes (green algae); then list three evolutionary trends exemplified by this group. (pp. 895-96)

2 Using Figure 39.2 (p. 896), describe the life cycle of *Chlamydomonas*, a representative unicellular green alga. Indicate whether the dominant stage is haploid or diploid. Explain what the term isogamy means and why *Chlamydomonas* is regarded as isogamous. (p. 897)

3 Trace the evolutionary changes in the volvocine series from *Chlamydomonas* through *Gonium* and *Pandorina* to *Volvox*. In doing so, specify the evolutionary advances at each stage and use the terms: heterogamy and oogamy. (p. 897)

4 Using Figure 39.2 (p. 896), describe the life cycle typical of very primitive plants, and compare it to the life cycle of more advanced alga, as shown in Figure 39.7 (p. 899). Explain what is meant by alternation of generations, and define the terms sporophyte, gametophyte, and sporangium. (pp. 896-99)

5 Discuss seven problems plants face in a terrestrial environment, and indicate how bryophytes and vascular plants have solved these problems. Define the terms antheridium and archegonium. (p. 900)

6 List the major characteristics of the bryophytes; then, using Figure 39.12 (p. 902), describe their life cycle and name the dominant stage. On a drawing of a bryophyte such as Figure 39.11 (p. 902), point out the sporophyte stage and the gametophyte stage. (pp. 902-904)

7 List six evolutionary innovations in the ancestral vascular plants, and trace the evolutionary advances of the vascular plants from the psilophytes through the club mosses, horsetails, and ferns to the seed plants. In doing so, indicate the order in which these groups appear in the fossil record. (pp. 905–12)

8 Define and use correctly each of the terms: sporophyll, megaspore, and microspore. (pp. 905–909)

9 Using Figure 39.24 (p. 908), describe the life cycle of a typical fern and name the dominant stage. Explain why the ferns are in some respects no better adapted than the bryophytes for life on land. (p. 908)

10 Describe two adaptations that probably contributed to the great evolutionary success of the seed plants. (p. 908)

11 Using Figure 39.29 (p. 911), describe the life cycle of a pine tree and discuss the ways in which this life cycle is more advanced than that of a typical fern. Then indicate whether the pine produces one or two types of spores. (p. 909)

12 Using Figure 39.30 (p. 911), identify the parts of a flower and describe the process of reproduction (Figs. 39.31, p. 911, and 24.8, p. 496, may be helpful). Summarize the main ways in which the life cycle of an angiosperm differs from that of the pine, a representative conifer. (pp. 911–12)

13 List at least five differences between monocots and dicots. (p. 913)

14 Explain how the flowering plants are adapted to meet the seven problems of a terrestrial environment referred to in Objective 5. (pp. 911–12)

15 Using Figure 39.34 (p. 914) and the material on pages 913 to 914, summarize the evolutionary trends among plant groups. (pp. 913–14)

16 For each of the following pairs, indicate which of the specified conditions is primitive and which is advanced:

   isogamy—heterogamy
   one type of spore—two types of spores (megaspores and microspores)
   diploid stage dominant—haploid stage dominant
   no protection for gametes—protection for gametes
   multicellularity—unicellularity
   water needed for fertilization—no water needed for fertilization
   tissue differentiation into roots, stems, leaves—little tissue differentiation
   gametophyte independent—gametophyte dependent on sporophyte
   reproductive structure multicellular—reproductive structure unicellular   (pp. 897–909)

KEY TERMS

isogamous (p. 897)
volvocine series (p. 897)
heterogamous (p. 897)
oogamous (p. 897)
antheridia (p. 900)
archegonia (p. 900)
embryophyta (p. 900)
Bryophyta (p. 901)
foot (p. 902)
stalk (p. 902)
capsule (p. 902)
protonema (p. 902)
gemmae (p. 903)
Psilophyta (p. 904)
Lycopphyta (p. 905)
Sphenophyta (p. 905)
Pterophyta (p. 905)
sporangium (p. 905)
sporophyll (p. 905)
megaspore (p. 908)
microspore (p. 908)
pollen grain (p. 908)
seed (p. 908)
Coniferophyta (p. 908)
integument (p. 909)
ovule (p. 909)
Angiophyta, or angiosperms (p. 911)
carpel (p. 911)
embryo sac (p. 911)
polar nuclei (p. 911)
double fertilization (p. 911)
endosperm (p. 912)
fruit (p. 912)
dicot (p. 913)
monocot (p. 913)

SUMMARY

Green algae The green algae (*Chlorophyta*) are probably the group from which the land plants arose. Like land plants, the green algae possess chlorophylls *a* and *b* and carotenoids, their storage product is starch, and most have cellulose in their walls.

Many divergent evolutionary tendencies can be perceived among the chlorophytes, including the evolution of colonies, both nonmotile and motile, the evolution of multicellular filaments and the evolution of three-dimensional leaflike thalluses.

The *volvocine series* is a series of genera showing a gradual progression from the unicellular condition to an elaborate colonial organization. A number of evolutionary changes are seen in this series: a change from unicellular to colonial life, increased coordination of activity and increased interdependence among the cells, increased division of labor, and a gradual change from *isogamy* to *heterogamy* to *oogamy* (motile male gamete, nonmotile female gamete).

The life cycles of many green algae include a multicellular stage. Often this stage is a filamentous thallus, which may be branching or nonbranching, depending on the species. Sexual reproduction is usually isogamous. The multicellular stage is haploid; the zygote is the only diploid stage. *Ulva* is a multicellular green alga with an expanded leaflike thallus two cells thick. It undergoes an *alternation of generations* where its haploid multicellular phase (*gametophyte*) alternates with a diploid multicellular phase (*sporophyte*).

Terrestrial plant adaptations Life probably arose in the water, and the evolutionary move to land was not a simple one. A terrestrial environment poses many problems for plants, including a novel need to obtain enough water and transport it and dissolved materials from one part of the plant to another, to prevent excessive evaporation, to maintain a moist surface for gas exchange, to support the plant against gravity, to carry out reproduction, and to withstand the extreme environmental fluctuations.

Much of the evolution of the land plants can best be understood in terms of adaptations to help solve these problems. The sterile jacket cells around the multicellular sex organs help protect the enclosed gametes from desiccation. Such male and female sex organs are known as *antheridia* and *archegonia*. All terrestrial plants are oogamous, and fertilization and early embryonic development take place within the moist environment of the archegonium. The aerial surfaces of the plant are often covered by a waxy cuticle, which prevents excessive water loss.

All land plants have a life cycle with an alternation of generations. The haploid gametophyte stage is dominant in the bryophytes whereas the sporophyte is dominant in the vascular plants.

The nonvascular plants (bryophytes) Two lineages of land plants evolved from green algae, the *bryophytes* and the *vascular plants*. The bryophytes (liverworts, hornworts, and mosses) are small plants that grow only in moist places. Moisture must be available for the flagellated sperm to swim to the egg. Bryophytes lack vascular tissues for transport and support. The bryophytes show alternation of generations; the haploid stage is dominant. The sporophyte is attached to the gametophyte; it consists of a foot embedded in the gametophyte, a stalk, and a capsule. Meiosis occurs within the capsule and produces haploid spores that fall to the ground, germinate, and eventually develop into mature gametophytes.

The seedless vascular plants The vascular plants evolved many adaptations enabling them to live successfully in almost all land habitats. The first vascular plants were simple plants that lacked true roots and leaves. Their naked stems showed a Y-shaped branching pattern. A *cuticle* and tracheids were present in the stems. The evolution of tracheids was a crucial adaptation because they permitted transport of water and nutrients and provided support against gravity. The most primitive vascular plants living today belong to the division *Psilopshyta* (the psilophytes); they also lack leaves and roots.

The club mosses (*Lycophyta*) and horsetails (*Sphenophyta*) show many advances over the psilopsids. They have true roots, stems, and needlelike leaves. Some of the leaves have become specialized for reproduction and bear *sporangia* on their surfaces; these leaves are called *sporophylls*. Although the living members of these divisions are small, many of the ancient forms were large trees; much of today's coal was formed from these plants.

The ferns (*Pterophyta*) have well-developed vascular systems and true roots, stems, and leaves. The leaves of ferns are large and flat and are thought to have evolved from branched and webbed stems. The large, leafy fern plant is the diploid sporophyte phase. The spores are produced in sporangia on the fertile leaves (sporophylls). On germination the spores develop into small, independent, nonvascular gametophytes bearing antheridia and archegonia. The flagellated sperm must swim to the archegonium. The gametophytes therefore require a moist habitat; thus ferns are restricted to a moist environment. A life cycle with a small nonvascular gametophyte, flagellated sperm, and dominant sporophyte is characteristic of all seedless vascular plants.

The seed plants The dominant and best-adapted land plants are the seed plants. These plants produce two types of spores (*microspores* and *megaspores*), which give rise to very samll male and female gametophytes. The male gametophyte (*pollen grain*) is transported to the female gametophyte so water is not necessary for the transport of sperm. The seed consists of the embryo and stored food, surrounded by a seed coat.

The conifers (*Coniferophyta*) are a large and important division of seed plants. The leaves of most of these plants are small evergreen needles or scales. The life cycle of the pine provides a good example of the seed method of reproduction in conifers. The pine tree is the diploid sporophyte stage. It bears two kinds of cones: large female cones, which produce spores that develop into female gametophytes, and small male cones, which produce spores that develop into pollen grains (the male gametophyte). The pollen grains are released into the air; some may land on the scales of a female cone. The pollen grain develops a *pollen tube*, which grows through the tissues of the female sporangium into one of the archegonia. There it discharges its two sperm nuclei, one of which fertilizes the egg. The zygote develops into an embryo, and a tough seed coat surrounds it and its stored food material. The seeds are borne exposed on the surface of the scale of the female cone. Finally, the seed is shed from the cone.

The flowering plants (*Anthophyta*, or angiosperms) are the dominant land plants. Their reproductive structures are the flowers. In these plants the ovules are enclosed within modified leaves called *carpels*, which unite to form the ovary of the flower. Meiosis occurs in each ovule and produces one functional megaspore, which divides mitotically to produce the *embryo sac*—the female gametophyte.

Each anther has sporangia in which meiosis occurs and produces haploid microspores. These develop into two-nucleate pollen grains—the male gametophytes. When a pollen grain lands on the female part of a flower, a pollen tube grows downward, enters the ovary, and discharges two sperm nuclei into the female gametophyte. *Double fertilization* occurs: one sperm nucleus fertilizes the egg; the other combines with the two *polar nuclei* to form a triploid nucleus, which will give rise to the *endosperm*. After fertilization, the ovule matures into a seed, which consists of a seed coat, stored food (endosperm), and embryo. The seeds are enclosed in *fruits* that develop from the ovaries and associated structures.

Angiosperms differ from other seed plants in many reproductive characteristics: their reproductive structure is the flower, they have double fertilization, their endosperm is triploid, and their seeds are enclosed within fruits.

For these reasons the angiosperms are the most successful and dominant land plants. They are divided into two classes: the *monocots*, whose embryos each have one cotyledon, and the *dicots*, whose embryos each have two cotyledons. There are many basic differences between the two groups.

QUESTIONS

Testing recall

Mark each characteristic below A if it is found in an aquatic environment and T if it is found in a terrestrial environment. (Some characteristics may be found in both groups.)

1 tissue differentiation (p. 902)

2 isogamy (p. 897)

3 multicellular sex organs (p. 900)

	Obtaining enough water	Transport of materials	Supporting plant against pull of gravity	Preventing excessive water loss	Carrying out reproduction with little water	Protection for the embryo
Bryophyta						
Seedless vascular plants						
Seed plants						

4 sterile jacket cells around reproductive organs (p. 900)

5 dominant gametophyte. (pp. 896, 902–904)

6 zygote development outside female reproductive organs (pp. 896–97)

7 cuticle on aerial parts (p. 904)

8 embryo development within archegonium (p. 900)

9 The chart below lists various groups of land plants and some of the problems faced by land plants. Complete it by filling in the features evolved by each group to solve these problems. The completed chart will be a useful study aid. (pp. 901–12)

Testing knowledge and understanding

Choose the one best answer.

10 The evolutionary trend toward multicellularity is best demonstrated by comparing various members of the

  a Phaeophyta (brown algae).
  b Bryophyta.
  c Cyanobacteria.
  d Chlorophyta (green algae).
  e Rhodophyta (red algae).   (p. 897)

11. In which one of the following divisions would expect to find some members that lack a multicellular sporophyte stage in the life cycle?

    a Chlorophyta
    b Pterophyta
    c Coniferophyta
    d Bryophyta
    e Psilophyta
                                           (pp. 895–97)

12. Land plants have probably evolved from which algal line?

    a Chlorophyta (green)
    b Rhodophyta (red)
    c Phaeophyta (brown)
    d Chrysophyta (yellow-green, golden-brown)   (p. 899)

13. Which one of the following best describes the type of sexual reproduction in which the gametes are all motile but are of two sizes, one type being smaller than the other?

    a homospory
    b heterospory
    c heterogamy
    d oogamy
    e isogamy   (p. 897)

Questions 14 to 16 refer to the following diagram of a moss plant.

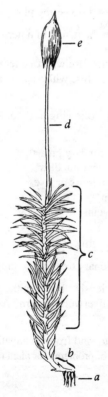

14. Which of the labeled structures (*a, b, c, d, e*) are composed of diploid cells? (p. 902)

15. In which parts of the plant does photosynthesis primarily occur? (p. 902)

16. In which part of the plant are the spores produced? (p. 902)

17. The evolution of plants completely adapted to the terrestrial habitat and able to grow to large size was made possible by the appearance of

    a collenchyma tissue.
    b parenchyma tissue.
    c tracheids.
    d vessel elements.
    e sclereids.
                                                 (p. 904)

18. As plants occupied drier habitats they could no longer rely on the transport of sperm in water. The gymnosperms overcame this problem by evolving

    a pollen.
    b cones.
    c motile sperm.
    d eggs and sperm in the same plant.
    e mature sperm produced only in the rainy season.
                                                 (p. 908)

19. In plants, the gametophyte generation is produced by

    a fertilization of an egg cell by a sperm cell.
    b the formation of four gametes by meiosis.
    c mitosis of sporophyte cells.
    d haploid spores that divide by mitosis.
    e fusion of male and female sporophyte cells.   (p. 900)

20. In plants, the sporophyte generation is produced by

    a fusion of two diploid cells.
    b growth of a sperm cell.
    c mitosis of haploid cells.
    d mitosis of diploid cells.
    e meiosis.   (p. 901)

21. Which one of the following does *not* function specifically as an adaptation to the land environment?

    a multicellular sex organ with jacket cells
    b structures to attach the plant to the substrate
    c cuticle
    d embryonic development within the archegonium
    e vascular tissue   (p. 901)

22. You are given an unknown plant to study in the laboratory. You find that it has chlorophyll but no xylem. Its multicellular sex organs are enclosed in a layer of jacket cells. Its gametophyte stage is free-living. The plant probably belongs in the

    a Chlorophyta (green algae).
    b Phaeophyta (brown algae).
    c Bryophyta.
    d Rhodophyta (red algae).
    e Psiloophyta.   (pp. 901–903)

23 The sporophylls of ferns produce spores by

  a mitosis. *b* meiosis. (p. 907)

24 A plant has vascular tissue, scalelike leaves, and a reproductive cone but no seeds. The plant probably should be classified as a

  a moss. *c* horsetail.
  b conifer. *d* fern. (pp. 905–908)

25 The seedless vascular plants are regarded as less highly adapted to life on land than the seed plants. Which one of the following traits of seedless plants illustrates this condition?

  a distinctive diploid and haploid stages
  b presence of vascular tissue
  c dominant sporophyte generation
  d sperm cells with flagella
  e sporophyte attached to gametophyte (p. 908)

26 Along with the seed, the seed plants have evolved several additional adaptations to the land environment. Which one of the following is *not* such an adaptation?

  a Flagellated gametes are not required for seed formation.
  b The female gametophyte is protected from desiccation by the surrounding tissues of the sporophyte.
  c The seed and/or associated structures serve as a means of dispersal.
  d The method of seed formation introduces a new type of genetic recombination.
  e The seed has its own food supply for the enclosed embryo. (p. 908)

27 Which one of the following statements is *true*?

  a Gymnosperms have a free-living gametophyte stage in their life cycle.
  b Most gymnosperms have flagellated sperm cells.
  c In a spruce seed, the tissue that serves as stored food is haploid.
  d The spruce seed is protected by a fruit. (p. 909)

28 Place the following groups of plants in order, beginning with those that first appeared on the earth and progressing toward those that appeared most recently in time.

  a conifers, angiosperms, ferns, bryophytes, algae
  b algae, bryophytes, ferns, conifers, angiosperms
  c algae, ferns, bryophytes, conifers, angiosperms
  d algae, bryophytes, ferns, angiosperms, conifers
  e bryophytes, algae, angiosperms, conifers, ferns (p. 905)

29 Which one of the following traits is characteristic of *both* the gymnosperms and the angiosperms?

  a double fertilization to form embryo and endosperm
  b haploid endosperm
  c fruit derived from ovaries
  d sporophylls modified into stamens and pistils
  e heterospory (p. 908)

30 The female gametophyte of a flowering plant

  a contains a single haploid nucleus.
  b gives rise to the ovary.
  c fuses with a pollen grain during pollination.
  d is produced in the ovule.
  e is fertilized by the filament. (p. 911)

31 A flower has six petals, three sepals, nine stamens, three ovules, and one ovary. How many seeds will this flower produce?

  a 1 *d* 9
  b 3 *e* 27
  c 6 (p. 911)

32 The plant in question is a

  a monocot. *b* dicot. (p. 913)

33 The endosperm of a flowering plant is triploid because

  a it results from the fusion of a normal gamete and a diploid gamete.
  b two male nuclei fuse with one female nucleus.
  c two polar nuclei fuse with a sperm nucleus.
  d none of the above. (p. 912)

34 Which one of the following is *not* a characteristic of monocots?

  a absence of secondary growth
  b scattered vascular bundles
  c flower parts in fours or fives
  d one cotyledon
  e parallel leaf veins (p. 913)

For further thought

1 Draw an evolutionary tree for the kingdom Plantae.

2 What is the relationship between the gametophyte and sporophyte of a liverwort? A fern? A pine tree? A flowering plant?

3 The movement to land from an aquatic environment presented a number of problems for plant reproduction. Describe

these problems and explain how the angiosperms have adapted to survive and reproduce on land. In your answer, diagram the life cycle of the angiosperms and describe in some detail the process of sexual reproduction.

ANSWERS

Testing recall

1	T	3	T	5	T, A	7	T
2	A	4	T	6	A	8	T

9 Chart: Information can be found on the following pages: Bryophyta, pages 901 to 904; seedless vascular plants, pages 904 to 908; and seed plants pages 908 to 912.

Testing knowledge and understanding

10	d	17	c	23	b	29	e
11	a	18	a	24	d	30	d
12	a	19	d	25	d	31	b
13	c	20	d	26	d	32	a
14	d, e	21	b	27	c	33	c
15	c	22	c	28	b	34	c
16	e						

Chapter 40

THE FUNGAL KINGDOM

A GENERAL GUIDE TO THE READING

This chapter provides a brief overview of the fungal kingdom. Fungi used to be included in the plant kingdom because they are predominantly sedentary and because their cells have walls. However, they differ from plants in so many ways that they are now placed in a kingdom of their own. Your understanding of fungi will be enhanced if you concentrate on the following topics.

1. General characteristics of fungi. The main characteristics of fungi are described on page 919.

2. Forms of sexual reproduction. Since the characteristics of sexual reproduction (or lack thereof) are important in distinguishing among the fungi, you will want to learn these characteristics for each division. Figures 40.4 (p. 920), 40.7 (p. 921), and 40.11 (p. 924) are the key to understanding the various forms of reproduction in fungi.

KEY CONCEPTS

1. The fungi are a diverse group of sedentary, absorptive heterotrophic organisms; they have great economic importance and play a vital role as decomposers. (p. 917)

2. The cells of fungi are eucaryotic. They have walls, but are usually organized into threadlike hyphae in which the cellular partitions may be absent or incomplete. (p. 917)

3. The characteristics of sexual reproduction are significant in distinguishing among the several divisions of fungi. (p. 918)

CROSSLINKING CONCEPTS

1. As with primitive plants, the haploid stage is usually dominant in the life cycle of the fungi.

2. Bacteria and fungi play a crucial role in decomposition. The importance of this role was particularly emphasized in the study of ecosystems in Chapter 40.

3. This chapter describes the fungi that take part in the important mutualistic relationship with plant roots known as mycorrhizae, first discussed in Chapter 16.

OBJECTIVES

1. Give the main characteristics of the fungi. (p. 919)

2. Compare the four major groups with respect to occurrence, presence or absence of cellular partitions in their hyphae, mode of reproduction, and economic importance. (pp. 918–25)

3. Using Figure 40.4 (p. 920), describe the life cycle of *Rhizopus*. Explain why fungi like *Rhizopus* are referred to as conjugation fungi. (pp. 918–20)

4. Using Figure 40.7 (p. 921), describe the processes of asexual and sexual reproduction in a cup fungus. (pp. 920–22)

5. Using Figure 40.11 (p. 924), describe the process of sexual reproduction in a club fungus. (pp. 922–24)

6. Indicate the distinguishing characteristic of the imperfect fungi. Explain how they are able to reproduce. (p. 925)

KEY TERMS

hypha (p. 917)
mycelium (p. 917)
chitin (p. 918)
fragmentation (p. 918)
budding (p. 918)
spores (p. 918)
sporangium (p. 918)
conidia (p. 918)
Zygomycota (p. 918)
zygospore (p. 918)
rhizoid (p. 919)
Ascomycota (p. 920)
ascus (p. 921)
ascospores (p. 922)
mycorrhizae (p. 922)
lichen (p. 922)
Basidiomycota (p. 922)
basidium (p. 922)
basiospores (p. 922)
Deuteromycota (p. 925)

SUMMARY

Fungi are eucaryotic organisms that are absorptive heterotrophs. They reproduce asexually by means of spores, and are parasitic or saprophytic. Fungi live on or in their food supply; most carry out extracellular digestion and then absorb the products. The fungi are important decomposers and many are of great economic importance.

The fungal body usually consists of threadlike *hyphae*, a mass of which is called a *mycelium*. Cell walls, which are often composed of chitin, are present and the hyphae are usually divided into cells by cross walls. The cross walls are often incomplete or absent, so the cytoplasm may be continuous. The individual cells often have more than one nucleus.

Most fungi carry out both sexual and asexual reproduction, but the vast majority reproduce more often by asexual means. The haploid stage is the dominant stage in fungi. Asexual reproduction may occur in the fungi by fragmentzation, budding, conidia, or spore formation. Sexual reproduction involves conjugation between two cells to form a two-nucleate cell, zygote formation, and then meiosis to restore the haploid condition. The text describes four divisions of fungi: Zygomycota, Ascomycota, Basidiomycota, and Deuteromycota. Sexual reproductive structures are important in defining the first three groups, whereas the Deutreromycota have no known form of sexual reproduction.

Members of the Zygomycota (conjugation fungi) are widespread as saprophytes in soil and dung. Their hyphae characteristically lack crosswalls, so the multinuclear cytoplasm is continuous. Reproduction may be asexual (by spores) or sexual. Sexual reproduction occurs by *conjugation* between two cells from hyphae of two different mycelia. These gamete cells fuse to form a zygote, which develops a thick wall and becomes dormant. At germination, the nucleus undergoes meiosis and a hypha develops. It releases asexual spores that grow into new mycelia. *Rhizopus* (black bread mold) is an example of this phylum.

The members of Ascomycota (sac fungi) are a diverse group, varying from unicellular yeasts through powdery mildews to cup fungi. The hyphae have cross walls, but they usually are incomplete and have large holes. All produce a reproductive structure called an ascus during their sexual cycle. An ascus is a sac within which haploid spores, usually eight, are produced. These fungi also reproduce asexually by means of special spores called *conidia*. Sac fungi form the fungal component of some *mycorrhizae* and of *lichens*. They are also important as decomposers and as the cause of many plant and animal diseases.

The members of Basidiomycota (club fungi) include the puffballs, mushrooms, toadstools, and bracket fungi. The large fruiting bodies of these fungi are composed of compacted hyphae with cross walls. During sexual reproduction, hyphae from two different mycelia with uninucleate cells unite and give rise to hyphae with binucleate cells. Certain terminal cells become zygotes when their two nuclei fuse. The zygote then becomes a *basidium*, a club-shaped reproductive structure. Meiosis occurs within the basidium, and four haploid spores are produced. Each spore may give rise to a new mycelium.

CONCEPT MAP

Map 40.1

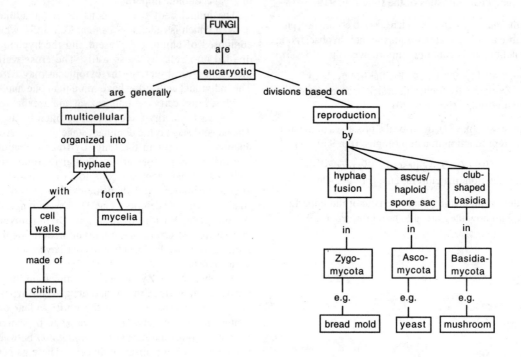

QUESTIONS

<u>Testing recall</u>

1 Complete the following chart. It will be a useful study aid.

	Cross walls in hyphae (present, absent or incomplete	Sexual reproductive structure (if any)	Representative members	Economic importance
Conjugation fungi (Zygomycota)				
Sac fungi (Ascomycota)				
Club fungi (Basidiomycota)				
Imperfect fungi (Deuteromycota)				

Testing knowledge and understanding

Choose the one best answer.

2 In most of the fungi, the dominant stage is

   a haploid.　　　*b* diploid.　　　(p. 918)

3 Which one of the following pairs is *incorrectly* matched?

   a Deuteroomycota—exclusively asexual reproduction
   b Zygomycota—conjugation
   c Zygomycota—hyphae with cross walls
   d Ascomycota—conidia　　　(p. 918)

4 Which one of the following is *not* a characteristic of fungi?

   a absorptive heterotrophic forms
   b body made up of threadlike hyphae
   c reproduction by spores
   d both saprophytic and autotrophic forms
   e cell wall containing chitin　　　(p. 917)

5 Fungi are

   a heterotrophic.
   b autotrophic.
   c photosynthetic.
   d more than one of the above　　　(p. 917)

6 Zygotes of fungi are

   a haploid.　　　*b* diploid.　　　(p. 918)

7 In the fungi, the spores are

   a haploid.　　　*b* diploid.　　　(p. 918)

8 During their sexual cycle, the members of Zygomycota (conjugation fungi) produce a(n)

   a conidium.
   b ascus.
   c basidium.
   d flagellated gamete.
   e thick-walled zygote.　　　(p. 918)

9 Yeast is classified as one of the

   a Protista.
   b Ascomycota (sac fungi).
   c Basidiomycota (club fungi).
   d Monera.
   e Zygomycota (conjugation fungi).　　　(p. 920)

10 In the stalk of a mushroom each of the cells

   a is haploid.
   b is diploid.
   c has two separate haploid nuclei.
   d has two separate diploid nuclei.
   e has four separate haploid nuclei.　　　(p. 922)

11 Which one of the following is *not* found in the Basidiomycota (club fungi)?

   a mycelium　　　*d* binucleate hyphae
   b basidium　　　*e* hyphae with cross walls
   c ascus　　　　　(pp. 920–21)

12 When we eat a mushroom we are eating

   a a mycelium.　　*d* hyphae.
   b the fruiting body.　*e* all the above
   c basidia.　　　　(pp. 922–24)

13 A student was given an unknown organism to identify. Though it looked like a solid mass of tissue, the student found it was made up of a mass of filaments, the cells of which lacked chlorophyll. Cross walls were present in the filaments, however, and two nuclei were found in many of the cells. The reproductive structures were club-shaped, and four spores were found on the tip of these structures. The student decided that this plant probably belonged in the

   a bacteria.
   b Ascomycota (sac fungi).
   c Deuteroomycota (imperfect fungi).
   d Basidiomycota (club fungi).
   e Zygomycota (conjugation fungi).　　　(pp. 922–24)

ANSWERS

Testing recall

1 Information of the chart can be found as follows: Zygomycota, pages 918 to 920; Ascomycota, pages 920 to 922; Basidiomycota, pages 922 to 925; and Deuteromycota, page 925.

Testing knowledge and understanding

2	*a*	5	*a*	8	*e*	11	*c*
3	*c*	6	*b*	9	*b*	12	*e*
4	*d*	7	*a*	10	*c*	13	*d*

Chapter 41

THE ANIMAL KINGDOM: THE RADIATES AND THE PROTOSTOMES

A GENERAL GUIDE TO THE READING

This chapter and the next two provide an overview of the diversity of animals and the evolutionary relationships among them. Much of the material should be familiar to you; the different organ systems and how they work were discussed in detail in Part III, "The Biology of Organisms," and your text includes cross references to this material. The emphasis in this chapter is on evolutionary relationships, so you will want to concentrate on them as you read. One interpretation of these relationships is shown in the phylogenetic tree in Figure 41.1 (p. 928). In addition, the following topics in Chapter 41 deserve special attention.

1. Sponges. The sponges (pp. 928–30) are believed to have evolved from the protists independently of other multicellular animals, as Figure 41.1 (p. 928) makes clear.

2. Radial symmetry versus bilateral symmetry. Since you will be studying both the radiate phyla and the Bilateria, you need to understand the differences between radial and bilateral symmetry.

3. Presence of a coelom. A coelom is a cavity between the digestive tract and the body wall. The absence or presence of a coelom, and the type of coelom, when present, are important in animal classification. Four different types of body are discussed: the acoelomate (pp. 932–36), the pseudocoelomate (pp. 939–40), and those having one or the other of the true coeloms—the coelom that arises as a split in an initially solid mass of mesoderm and the coelom that arises as cavities in mesodermal pouches (pp. 941–44). Figure 41.12 (p. 933) is a useful summary of body types.

4. Divergence of the protostomes and deuterostomes. The two main branches of the animal kingdom are the Protostomia and the Deuterostomia. Because the differences between the protostomes and the deuterostomes are fundamental to our classification scheme, you will want to make sure you understand the characteristics separating them. (pp. 937–38)

5. Molluscs. Learn the fundamental molluscan body plan, and then see how it is modified in the various classes. (pp. 941–44)

6. Segmented worms (Annelida). Look closely at the polychaetes, which are considered to show the more primitive features of this phylum, discussed on pages 945 to 946.

7. Arthropods. In studying the arthropods (pp. 946–54), pay particular attention to the insects, since they are an enormously successful group of organisms.

KEY CONCEPTS

1. Characteristics such as level of organization, type of symmetry, presence or absence of segmentation, nature of embryonic development, and form of larva, if any, have been used to establish hypothetical phylogenetic relationships among the various animal groups. (p. 927)

2. The members of the phylum Porifera (sponges) differ greatly from other multicellular animals, and are believed to have evolved on a line of their own. (pp. 929–30)

3. The two radiate phyla—Cnidaria and Ctenophora—comprise radially symmetrical animals with bodies that are relatively simple in structure. (pp. 930–32)

4. The members of the phyla Platyhelminthes (flatworms) and Nemertea (proboscis worms) are thought to be the

most primitive bilaterally symmetrical animals; in each, the body is composed of three well-developed tissue layers, and is acoelomate. (pp. 932–33)

5 A major split probably occurred in the animal kingdom soon after the emergence of a bilateral organism. One evolutionary line led to the phyla in which the blastopore becomes the mouth; these animals are called protostomes. The other line led to the phyla in which the blastopore becomes the anus and a new mouth is formed; these animals are called the deuterostomes. (pp. 937–38)

6 In several protostome phyla, the body cavity is not completely enclosed by mesoderm; it is a pseudocoelom. All the other protostome phyla have true coeloms (at least at some stage of development); in most groups these arise as a split in the initially solid mass of mesoderm. (p. 933)

7 The arthropods are generally regarded as the most highly evolved representatives of the protostome line. (p. 946)

CROSSLINKING CONCEPTS

1 Concepts from your general study of animal development (Chapter 13), particularly indeterminate and determinate cleavage and the three tissue layers of endoderm, ectoderm, and mesoderm, are absolutely necessary for understanding the development of invertebrate organisms in this chapter.

2 This chapter discusses the evolutionary development of many structures and organs found in invertebrates. The function of these structures and organs was more fully explained in Part III.

3 This chapter discusses the evolutionary development of many of the invertebrate structures and organs whose functions were described in Part III.

OBJECTIVES

1 Name the classes and a few of the members of the phylum Porifera (sponges), and explain why they are believed to have evolved independently of other multicellular organisms and hence not to be closely related to the other animal groups. Indicate a possible evolutionary relationship between the sponges and the animal-like protists. (pp. 928–30)

2 Describe the distinguishing characteristics of the cnidarians; then name the three cnidarian classes, and list the identifying features of each. (pp. 930–32)

3 Using Figure 41.7 (p. 931), describe the life cycle of a typical hydrozoan such as *Obelia*; then explain how the life cycles of the jellyfish (a scyphozoan) and the sea anemone (an anthozoan) would differ from that of *Obelia*. (pp. 930–31)

4 Describe the major characteristics of the phylum Platyhelminthes; then name the three classes in this phylum, and list the distinguishing features of each. Discuss the tapeworm's special adaptations for parasitism. (pp. 933–36)

5 Specify two important evolutionary advances found in the nemerteans. (p. 937)

6 Complete the following chart, comparing the protostomes and the deuterostomes. (pp. 937–38)

	Protostomia	Deuterostomia
Fate of the blastopore		
Type of early cleavage (determinate or indeterminate, radial or spiral)		
Origin of the mesoderm		
Formation of the coelom (as a split in a solid mass of mesoderm or as cavities in mesodermal pouches)		
Major phyla		

7 Using Figures 41.12 (p. 933) and 41.23 (p. 940), explain why the body cavity of the rotifers and nematodes is called a pseudocoelom rather than a true coelom. Then list the identifying features of the nematodes and discuss the economic importance of this phylum. (pp. 939–40)

8 Describe the body plan common to all molluscs, and indicate the modifications it exhibits in the classes Polyplacophora (chitons), Monoplacophora, Gastropoda (snails), Bivalvia (bivalves), and Cephalopoda (squids and octopuses). (pp. 941–44)

9 Describe the distinguishing characteristics of the annelids, and compare the polychaetes, the earthworms (Oligochaeta), and the leeches (Hirudinea) with respect to the presence or absence of a distinct head, the degree of segmentation, the presence or absence of parapodia, and the number of setae (bristles). (pp. 945–46)

10 Describe the ways in which the onychophorans resemble the annelids and the ways in which they resemble the arthropods. (pp. 946–47)

11 Describe the distinguishing characteristics of the arthropods, and indicate the major modifications of the ancestral body plan that have arisen among the various arthropod subphyla. Distinguish between chelicerae and mandibles. (pp. 946–50)

12 Describe the distinguishing characteristics of the Crustacea and Insecta, and list representative examples of each. Briefly describe the structure of the generalized insect body. (pp. 950–54)

KEY TERMS

Porifera (p. 928)
collar cell (p. 928)
filter feeder (p. 928)
spicule (p. 928)
Parazoa (p. 930)
Eumetazoa (p. 930)
Radiata (p. 930)
Bilateria (p. 930)
radial symmetry (p. 928)
Cnidaria (Coelenterata (p. 930)
gastrovascular cavity (p. 930)
cnidocyte (p. 930)
nematocyst (p. 930)
mesolamella (p. 930)
Hydrozoa (p. 931)
polyp stage (p. 930)
medusa stage (p. 930)
planula (p. 931)
Scyphozoa (p. 931)
Anthozoa (p. 931)

true coelom (p. 933)
acoelomate (p. 933)
pseudocoelom (p. 933)
bilateral symmetry (p. 928)
Platyhelminthes (p. 933)
Turbellaria (p. 934)
Trematoda (p. 934)
schistosomiasis (p. 935)
Cestoda (p. 935)
scolex (p. 935)
proglottid (p. 935)
Nemertea (p. 936)
complete digestive system (p. 937)
Protostomia (p. 937)
Deuterostomia (p. 937)
determinate cleavage (p. 937)
indeterminate cleavage (p. 937)
Rotifera (p. 939)
Nemata (p. 939)
trichinosis (p. 940)
Mollusca (p. 941)
foot (p. 941)
visceral mass (p. 941)
mantle (p. 941)
mantle cavity (p. 941)
Polyplacophora (p. 942)
radula (p. 941)
Monoplacophora (p. 942)
Gastropoda (p. 942)
torsion (p. 942)
Bivalvia (p. 944)
Cephalopoda (p. 943)
Annelida (p. 945)
segmentation (p. 945)
Polychaeta (p. 945)
parapodium (p. 945)
Oligochaeta (p. 946)
clitellum (p. 946)
Hirudinea (p. 946)
Onychophora (p. 947)
Arthropoda (p. 946)
biramous (p. 948)
uniramous (p. 948)
Trilobita (p. 948)
Chelicerata (p. 949)
chelicerae (p. 950)
cephalothorax (p. 950)
pedipalp (p. 950)
mandibles (p. 951)
maxillae (p. 951)
Crustacea (p. 950)
Uniramia (p. 952)
Insecta (p. 952)
labium (p. 953)
labrum (p. 953)

SUMMARY

The animal kingdom includes many diverse groups of organisms.

The members of the phylum *Porifera* (sponges) differ greatly from other multicellular animals and are believed to have evolved on a line of their own. They are multicellular animals at a primitive, pretissue level of organization. Sponges are filter feeders; their bodies are perforated sacs through which water flows. The flagellated collar cells lining the interior are remarkably similar to protozoan collared flagellates.

The radiate phyla These phyla—Cnidaria (Coelenterata) and Ctenophora—comprise radially symmetrical animals whose bodies have definite tissue layers but no distinct internal organs. They have a *gastrovascular cavity*. Their bodies consist of two well-developed tissue layers, an outer epidermis (ectoderm) and an inner gastrodermis (endoderm), with a third layer of mesolamella between them. All have cnidocytes with nematocysts.

The *cnidarians* include the hydras, jellyfishes, sea anemones, and corals. There is some division of labor among their tissues, but it is never as complete as in most bilateral multicellular animals, and most functions performed by mesodermal tissues in other animals are performed by ectodermal or endodermal cells in coelenterates. There are three classes discussed: Hydrozoa, Scyphozoa, and Anthozoa.

Many hydrozoans have a sedentary *polyp* stage that alternates with a free-swimming jellyfishlike *medusa* stage. Certain cells of the medusa produce gametes. The zygote develops into an elongate ciliated planula larva, which eventually settles and gives rise to the polyp. In the scyphozoans (jellyfishes), the medusa stage is dominant; in the anthozoans (sea anemones and corals), it is absent altogether. The anthozoans are the most advanced of the coelenterates.

Acoelomate bilateria The members of the phyla Platyhelminthes and Nemertea are regarded as the most primitive bilaterally symmetrical animals. Their bodies are composed of three well-developed tissue layers; there is no coelom.

The *Platyhelminthians* (flatworms) are dorsoventrally flattened, elongate animals with (usually) a gastrovascular cavity. The flatworms have advanced to the organ level of construction. The more extensive development of mesoderm probably made this advance possible. The phylum is divided into three classes: Turbellaria, Trematoda, and Cestoda.

Turbellarians, of which planarians are examples, are free-living. Members of the classes Trematoda (flukes) and Cestoda (tapeworms) are entirely parasitic. They have evolved many adaptations for parasitism: a resistant cuticle, loss or reduction of structures, extremely well-developed reproductive systems, and elaborate life cycles. The members of the phylum *Nemertea* probably evolved from the turbellarian flatworms. However, they have a *complete digestive tract* and a simple blood circulatory system.

Divergence of the protostomes and deuterostomes The fate of the embryonic blastopore differs in various animal groups. In many animals the blastopore becomes the mouth, and a new anus develops; in others the situation is reversed. This fundamental difference in development suggests that a major evolutionary split occurred soon after the origin of a bilateral ancestor. One evolutionary line led to all the phyla in which the blastopore becomes the mouth, the *Protostomia* (first mouth). The other line led to the phyla in which the blastopore forms the anus and a new mouth is formed, the *Deuterostomia* (second mouth). The protostomes and deuterostomes also differ in the pattern of the initial cleavages and whether they are determinate, the method of origin of the mesoderm and coelom, and the type of larva.

The pseudocoelomate protostomes A *true coelom* is a cavity enclosed entirely by mesoderm and located between the digestive tract and body wall. In several protostome phyla the body cavity is partly bounded by ectoderm and endoderm; it is called a *pseudocoelom*.

The members of two pseudocoelomate phyla are very abundant: the *Rotifera* (wheel animals) and the *Nemata* (nematodes). Both phyla comprise small, wormlike animals without a definite head but with a complete digestive tract. Nematodes have only longitudinal muscles in their body wall and lack cilia or flagella. Many nematodes are parasitic in plants and animals. *Trichinella*, *Ascaris*, hookworms, pinworms, and filaria worms parasitize humans.

Coelomate protostomes Sometime during their development, all the other protostome phyla possess true coeloms, which usually develop as a split in the mesoderm. All have a complete digestive tract, and most have well-developed circulatory, excretory, and nervous systems.

The phylum *Mollusca* is the second largest in the animal kingdom; it includes the snails and slugs, clams and oysters, and squids and octopuses. The molluscs are soft-bodied animals composed of a ventral muscular *foot*, a *visceral mass*, and a *mantle*, which covers the visceral mass and usually contains glands that secrete a shell. The mantle often overhangs the visceral mass, and thus encloses a *mantle cavity*, which frequently contains gills. Most molluscs have an open circulatory system.

The molluscs are usually divided into eight classes, of which we consider five: Polyplacophora (chitons), Monoplacophora, Gastropoda (snails), Bivalvia (bivalves), and Cephalopoda (squids, octopuses). The cephalopods are specialized for rapid movement and predation; they have convergently evolved many similarities to vertebrates.

The phylum *Annelida* (segmented worms) is usually divided into three classes: Polychaeta, Oligochaeta, and Hirudinea (leeches). The polychaetes are marine animals with a well-defined head bearing eyes and antennae. Each body segment bears a pair of *parapodia*, which function in movement and gas exchange. Oligochaetes include earthworms and freshwater species; they lack a well-developed head and parapodia, and have few setae. The hirudineans (leeches) are the most specialized of the annelids and show little internal segmentation.

The members of the phylum *Onychophora* are of special interest because they combine annelid and arthropod characters and are regarded as an early evolutionary offshoot from the line leading to the arthropods from an ancient annelidlike ancestor.

The phylum *Arthropoda* is the largest of the phyla. Arthropods are characterized by a jointed chitinous exoskeleton and jointed legs. They have elaborate musculature, a well-developed nervous system and sense organs, and an open circulatory system.

The arthropod body may be an elaboration and specialization of the segmented body of a presumed annelid ancestor. The first arthropods had long wormlike bodies composed of nearly identical segments, each bearing a pair of legs. Among evolutionary trends seen in the arthropods are reduction in the number of segments; groupings of segments into body regions; increasing cephalization; and specialization of some legs for functions other than movement, with loss of legs from other segments.

The arthropods can be divided into four subphyla: Trilobita, Chelicerata, Crustacea, and Uniramia. The *trilobites* are now extinct; they more closely resemble the hypothetical arthropod ancestor than any other known group. Fossils show that every segment bore a pair of nearly identical legs. In the Chelicerata, Crustacea, and Uniramia the tendency toward specialization of some appendages and loss of others is apparent.

The members of *Chelicerata* include the horseshoe crabs and arachnids (spiders, ticks, mites, scorpions). The body is usually divided into a cephalothorax and an abdomen. The first pair of postoral legs is modified as mouthparts called *chelicerae*. There are five other pairs of appendages on the cephalothorax; in some the first pair is modified as feeding devices called *pedipalps*.

The members of *Crustacea* and *Uniramia* have mandibles as their first pair of mouthparts, and two additional pairs of mouthparts called maxillae. The crustaceans include lobsters, crabs, shrimps, water fleas, sow bugs. They are characterized by two pairs of antennae. The uniramians (centipedes, millipedes, and insects) have only a single pair of antennae and unbranched appendages.

The insects are an enormous group of diverse organisms that occupy almost every conceivable habitat on land and in freshwater. There are more species of insects than all other animal groups combined. The insect body is divided into a head, thorax, and abdomen. The head bears many sensory receptors, usually including compound eyes; one pair of antennae; and three mouthparts (mandibles, maxillae, and *labium*) derived from ancestral legs. The upper lip, or *labrum*, may also be derived from ancestral legs. The thorax bears three pairs of walking legs and usually one or two pairs of wings. Insects were the first animals to evolve the ability to fly, which opened a new way of life and enabled them to occupy different niches.

CONCEPT MAP

Map 41.1 Arthropods This map is a detail map showing the structure and functional relationships of the phyla of arthropods. For the broad picture in this chapter, as with all the chapters of Part IV, it is a good idea to use the phylogenetic trees and charts, which compare key characteristics of different phyla. For a more detailed view, you may wish to map particularly important phyla, such as Mollusca or Annelida.

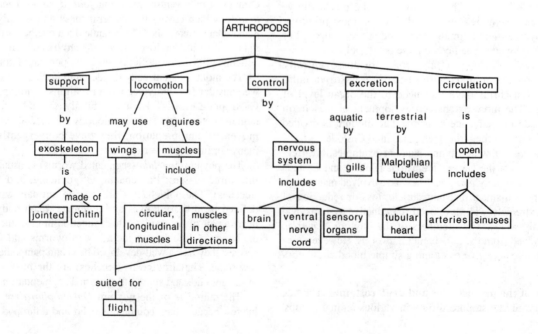

QUESTIONS

Testing recall

The following is a chart of the animal kingdom, set up as a phylogenetic tree. (It shows only one of many possibilities, since much is uncertain about the evolution of animal groups.) A phylogenetic tree can be useful in learning the various groups, particularly if you label the sites of various evolutionary advances. The dashed lines on the tree represent certain evolutionary advances. For example, above line a, all organisms are multicellular. Match each of the characteristics listed below with the letter of a line dividing the tree.

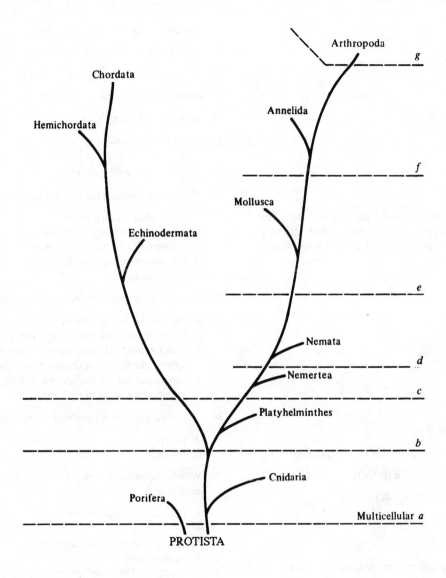

1. complete digestive tract (p. 937)
2. body cavity that is either a pseudocoelom or a true coelom (p. 939)
3. true coelom (p. 941)
4. jointed appendages and exoskeleton (p. 947)
5. segmentation in adult (p. 945)
6. three well-developed tissue layers (p. 933)
7. fundamental bilateral symmetry (p. 933)
8. Identify on the tree the deuterostome and the protostome branches. (pp. 937–38)

Match each organism below with the taxonomic category in which it belongs. Answers may be used once, more than once, or not at all.

a Radiata
b Parazoa
c Protostomia

9 jellyfish (p. 930)
10 sponge (p. 930)
11 earthworm (p. 937)
12 snail (p. 937)
13 hydra (p. 930)
14 nematode worm (p. 937)
15 cockroach (p. 937)
16 squid (p. 937)
17 coral (p. 932)

Match each of the important animal characteristics below with the phylum or phyla whose members possess that characteristic. Answers may be used once, more than once, or not at all.

a Annelida
b Arthropoda
c Nemata
d Cnidaria
e Mollusca
f Platyhelminthes
g Porifera
h Nemertea
i Rotifera

18 radial symmetry (p. 930)
19 complete digestive tract, no body cavity (pp. 936–37)
20 jointed legs and exoskeleton (p. 947)
21 segments with parapodia (p. 945)
22 mantle, foot, visceral mass (p. 941)
23 pseudocoelom (p. 939)
24 bilateral symmetry, gastrovascular cavity (p. 933)
25 saclike body lined with collar cells (p. 928)

Testing knowledge and understanding

The following key lists five animals, each representing a different phylum. The order in which they are listed is random and does not reflect the evolutionary placement of the phyla. Recalling the evolutionary relationships postulated in your text, match each characteristic below with the animal representing the most "primitive" phylum in which the characteristic appears. Answers may be used once, more than once, or not at all.

a sponge
b nematode worm
c hydra
d clam
e flatworm

26 true coelom (p. 941)
27 circulatory system (p. 941)
28 complete digestive tract (p. 939)
29 well-defined tissue organization (p. 930)
30 bilateral symmetry (p. 933)
31 organ level of development (p. 934)
32 nervous system (p. 934)
33 gastrovascular cavity (p. 930)

Choose the one best answer.

34 The evidence linking the sponges to the flagellated protists is

a the sponge's collar cells.
b similar photosynthetic pigments.
c a common reserve food, starch.
d the volvoxlike sponge blastula.
e nonexistent. (p. 929)

35 A major difference between a sponge and a hydra is that

a the sponge, but not the hydra, has a primitive nerve net.
b the hydra, but not the sponge, has three cell layers.
c water enters and exits through separate openings in the sponge, but through a single opening (mouth) in the hydra.
d the hydra, but not the sponge, is bounded on the outside by a layer of epidermal cells. (pp. 928–30)

36 Adult sea anemones have which cnidarian body form?

a polyp   *c* planula
b medusa   (p. 931)

37 Which one of the following is *not* a characteristic of the cnidarians?

a flame-cell excretory system
b nerve net
c digestive tract with one opening
d nematocysts
e tissue level of development (pp. 930–32)

38 Which one of the following animals does *not* have a well-developed mesoderm?

a housefly   *d* planarian
b jellyfish   *e* earthworm
c snail   (pp. 930–31)

39 A major evolutionary advance seen in Platyhelminthes (flatworms) but not in Cnidaria is

a a gastrovascular cavity.   *c* bilateral symmetry.
b a coelom.   *d* endoderm. (p. 933)

40 Which one of the following has a pseudocoelom?

   a roundworm (nematode)   d cockroach
   b earthworm   e planarian
   c clam   (p. 939)

41 Which one of the following has a planula larva?

   a annelids   d cnidarians
   b molluscs   e nematodes
   c arthropods   (p. 931)

42 All the following are adaptations of tapeworms *except*

   a resistant cuticle.
   b phagocytic proglottids.
   c no mouth.
   d no digestive tract.
   e diffusion of nutrients and gases across body surfaces.
                                              (pp. 935–36)

43 Filaria worms cause elephantiasis by

   a living in the digestive tract and absorbing nutrients.
   b boring into skeletal muscles and encysting.
   c boring through bare feet.
   d living in the lymphatic system and blocking the flow of lymph.   (p. 940)

44 Trichinosis, a disease human beings can get from eating insufficiently cooked pork, is caused by a

   a virus.   d tapeworm.
   b fungus.   e nematode worm.
   c fluke.   (p. 940)

45 Which one of the following is *not* a characteristic of molluscs?

   a open circulatory system
   b foot with visceral mass
   c shell secreted by the mantle
   d anus developed from the blastopore   (pp. 937–38)

46 *All* molluscs and annelids have

   a gills for gas exchange.
   b an open circulatory system.
   c segmentation.
   d a true coelom.   (p. 941)

47 *All* annelids have

   a segmentation.   c setae.
   b parapodia.   d suckers.   (p. 945)

48 Annelids differ from flatworms in that

   a all annelids live in the soil, whereas all flatworms live in freshwater.
   b annelids have a digestive tract, whereas flatworms do not.
   c annelids have a cavity between the digestive tract and the body wall, whereas flatworms do not.
   d annelids have a nervous system, whereas flatworms do not.   (p. 945)

49 Annelids differ from nematode worms in that

   a annelids have a complete digestive tract, whereas nematodes do not.
   b annelids have circular muscles, whereas nematodes have only longitudinal muscles.
   c annelids are exclusively free-living, whereas nematodes are exclusively parasitic.
   d annelids have a pseudocoelom, whereas nematodes are acoelomate.
   e annelids have a mouth derived from the blastopore, whereas nematodes have an anus derived from the blastopore.   (p. 939)

50 Which one of the following animal phyla is *least* closely related to all the others?

   a Cnidaria   d Arthropoda
   b Porifera (sponges) e Mollusca
   c Annelida   (p. 929)

51 Which one of the following statements is *true* of both annelids and arthropods?

   a Both exhibit indeterminate cleavage in early development.
   b The blastopore becomes the anus in both.
   c Both exhibit spiral cleavage.
   d The mesoderm arises as outpocketings of the endoderm in both.   (pp. 937–38)

52 Which one of the following *arthropod* characteristics was probably the most important adaptation for the land environment?

   a coelom   d exoskeleton
   b segmentation   e circulatory system
   c complete digestive tract   (p. 947)

53 Which one of the following is *not* typical of insects?

   a body divided into a head, thorax, and abdomen
   b compound eyes
   c pincerlike chelicerae
   d one pair of antennae
   e three pairs of legs   (p. 954)

For further thought

1 Explain and discuss the pattern of increasing complexity within the animal kingdom (at the phylum level).

2 Contrast the annelids and arthropods with respect to the degree of segmentation, development of the nervous system, adaptations for digestion, and the gas-exchange and circulatory systems.

3 Compare the molluscs and annelids with respect to the basic body plan, the gas-exchange systems, circulatory systems, and excretory systems.

4 Compare alternation of generations in plants and animals.

ANSWERS

Testing recall

1 *c* 3 *e* 5 *f* 7 *b*
2 *d* 4 *g* 6 *b*
8 The deuterostomes are shown on the branch at left (Echinodermata—Chordata); the protostomes are shown on the branch at right (Platyhelminthes—Arthropoda).
9 *a*   14 *c*   18 *d*   22 *e*
10 *b*   15 *c*   19 *h*   23 *c, i*
11 *c*   16 *c*   20 *b*   24 *f*
12 *c*   17 *a*   21 *a*   25 *g*
13 *a*

Testing knowledge and understanding

26 *d*   33 *c*   40 *a*   47 *a*
27 *d*   34 *a*   41 *d*   48 *c*
28 *b*   35 *c*   42 *b*   49 *b*
29 *c*   36 *a*   43 *d*   50 *b*
30 *e*   37 *a*   44 *e*   51 *c*
31 *e*   38 *b*   45 *d*   52 *d*
32 *c*   39 *c*   46 *d*   53 *c*

Chapter 42

THE ANIMAL KINGDOM: THE DEUTEROSTOMES

A GENERAL GUIDE TO THE READING

This last chapter of Part IV is devoted to the deuterostome phyla, including the Chordata, the phylum to which we belong. The emphasis is on the evolutionary relationships of the various groups. You will want to concentrate on the following topics.

1. The relationships between echinoderms, hemichordates, and chordates. You need to learn the evolutionary relationships linking these three diverse groups of animals. (p. 957)

2. Chordate characteristics. Begin by memorizing the three basic chordate characteristics on page 961.

2. Tunicates. In reading about the tunicates (pp. 961–62), notice that though the adult tunicate does not resemble a chordate in any way, the larva (Fig. 42.11, p. 962) bears some resemblance to what we might expect a primitive chordate to look like.

3. Vertebrate evolution. The vertebrate classes are discussed on pages 963 to 967. You will want to focus on the evolutionary relationships among them, shown in Figure 42.13 (p. 964), and on the key advances in each vertebrate class. Try to understand which of the advances were adaptations to overcome the difficulties of life on land.

KEY CONCEPTS

1. The deuterostomes form the second main branch of the animal kingdom. The coelom arises as cavities in mesodermal pouches. (p. 957)

2. The phylum Echinodermata, the most primitive of the major deuterostome phyla, is linked in important ways to the phyla Hemichordata and Chordata. (p. 957)

3. All chordates possess, at some time in their life cycle, a notochord, pharyngeal slits (or pouches), and a dorsal hollow nerve cord. (p. 961)

4. The members of the subphylum Vertebrata may be regarded as the most highly evolved representatives of the deuterostome line. (pp. 963–77)

5. The acquisition of hinged jaws was one of the most important events in the history of vertebrates, since it made possible the development of a variety of feeding methods and life styles. (p. 965)

6. Any fish that had appendages better suited for land locomotion than those of their fellows would have been able to exploit the ecological opportunities open to them on land more fully. Through selection pressure exerted over millions of years, the fins of these first vertebrates to walk on land have slowly evolved into legs. (p. 968)

7. The evolution of the amniotic egg, which provides a fluid-filled chamber in which the embryo may develop even when the egg itself is in a dry place, was an important evolutionary advance in the conquest of land. (p. 969)

8. Other important adaptations to a terrestrial environment include internal fertilization, no larval stage, relatively impermeable skin, and strong legs to prevent the body from dragging on the ground. (p. 969)

9 Many reptilian lines, including the dinosaurs, died out at the end of the Cretaceous era. The mass extinction may have resulted from an asteroid collision, comet impact, cycles of intense volcanic activity, or long-term climatic change. (p. 973)

10 Birds adapted to flight by the evolution of many characteristics, including feathers, wings, endothermy (warm-blooded body), a four-chambered heart, hollow bones, and good vision, hearing, and equilibrium. (pp. 973–75)

11 Mammals' adaptations to terrestrial life include endothermy, a four-chambered heart, a diaphragm, hair, and ventral limbs to keep the body high off the ground. (p. 975)

CROSSLINKING CONCEPTS

1 The chart of the geological eras (Chapter 19) is a useful tool to keep available as you study the emergence of each group of chordates. It will help you link the groups together instead of trying to learn each one separately.

2 Prior study of homologous and analogous structures (Chapter 32) gives you the ability to analyze the relationship of the hinged jaws of arthropods and vertebrates.

3 Your study of adaptive radiation in the Galapagos finches (Chapter 32) gives you a basis for appreciating the tremendous evolutionary radiation of the era of fishes and the era of terrestrial invasion.

4 A general understanding of preadaptations (Chapter 31) lays the basis for the appreciation of lungs and large fleshy bases of paired pectoral and pelvic fins as preadaptations for terrestrial life.

5 Some of the challenges faced by plants as they adapted to terrestrial life (such as desiccation and the need for support) are very similar to those faced by animals. Your prior study of plants (Chapter 39) gives you a basis for comparison.

6 Successful adaptations to terrestrial life made by arthropods (Chapter 41) can be compared with vertebrate adaptations in this chapter.

OBJECTIVES

1 Name the three phyla of the Deuteostomia that are discussed, and list representative examples of each. (p. 957)

2 Describe the distinguishing characteristics of the echinoderms, and list representative examples of the phylum. (pp. 957–59)

3 Discuss the evolutionary relationships linking the echinoderms, hemichordates, and chordates. (p. 957)

4 Give three distinguishing features of the chordates, list the three subphyla, and give an example of each. (p. 961)

5 Describe the larval and adult forms of the tunicates, and suggest an evolutionary relationship between the tunicates and the vertebrates. (pp. 961–62)

6 List the seven classes of modern vertebrates, and indicate the evolutionary relationships among them. State the order in which they appear in the fossil record. (pp. 963–76)

7 Give a possible explanation for the mass extinctions of the dinosaurs and other animals at the end of the Cretaceous. (p. 973)

8 Describe the key innovations of each vertebrate class, and specify which of these were adaptations to overcome the difficulties of life on land. (pp. 965–76)

KEY TERMS

Echinodermata (p. 957)
water-vascular system (p. 958)
tube feet (p. 959)
disk (p. 959)
arm (p. 959)
Hemichordata (p. 960)
pharyngeal slit (p. 960)
dipleurula larva (p. 960)
Chordata (p. 961)
notochord (p. 961)
Tunicata (p. 961)
Cephalochordata (p. 962)
lancelet (p. 962)
Agnatha (p. 964)
hinged jaws (p. 965)
paired fins (p. 965)
bony scales (p. 965)
Chondrichthyes (p. 966)
cartilaginous fish (p. 966)
Osteichthyes (p. 966)
bony fish (p. 966)
fleshy-finned fish (p. 967)
lobe-finned fish (p. 967)
Tetrapoda (p. 967)
Amphibia (p. 968)
Reptilia (p. 969)
amniotic egg (p. 969)
stem reptiles (p. 971)
Aves (p. 973)
Mammalia (p. 975)
monotreme (p. 976)
marsupial (p. 976)
placental (p. 976)

SUMMARY

The phyla Echinodermata, Hemichordata, and Chordata are in the Deuterostomia and share certain deuterostome characteristics, such as formation of the anus from the blastopore, radial and indeterminate cleavage, origin of the mesoderm in pouches, and formation of the coelom as the cavities in the mesodermal pouches.

Echinoderms Members of the phylum *Echinodermata* are exclusively marine, mostly bottom-dwelling animals. The adults are radially symmetrical; all show five-part symmetry. Because the larvae are bilaterally symmetrical, it is generally held that the echinoderm evolved from bilateral ancestors. All members have an internal skeleton made of bony plates, a well-developed coelom, and a unique *water-vascular system*. The sea stars have a body that consists of a central disc with five arms. Other members of the phylum include the brittle stars, sea urchins, sea cucumbers, and sea lilies.

Hemichordates Members of the phylum *Hemichordata* (acorn worms) show affinities with both the echinoderms and the chordates. They resemble the chordates in their pharyngeal gill slits and dorsal nerve cord However, their ciliated lava *(dipleurula)* is almost identical to that of the echinoderms. The hemichordates' ties to both these groups have helped clarify the relationships between these two major phyla. The chordates and echinoderms are believed to have evolved from a common ancestor at some remote time.

Chordates The phylum *Chordata* contains both invertebrate and vertebrate members. It is divided into three subphyla: Tunicata, Cephalochordata, and Chordata. At some time in their life cycle, all chordates have a *notochord*, pharyngeal slits, and a dorsal hollow nerve cord.

The adults of the subphylum *Tunicata* (tunicates) are sessile filter feeders. The larvae, however, are free-swimming, bilaterally symmetrical, tadpolelike organisms with pharyngeal slits, notochords, and dorsal hollow nerve cords.

The members of the subphylum *Cephalochordata* (lancelets) are small marine organisms with a permanent dorsal hollow nerve cord and notochord. They are filter feeders and show segmentation.

Vertebrate evolution The members of the subphylum *Vertebrata* are characterized by an endoskeleton that includes a segmented backbone composed of a series of vertebrae that develop around the embryonic notochord.

Fish The first vertebrates belonged to the class *Agnatha*; they were fishlike animals encased in an armor of bony plates. They lacked jaws and paired fins. The lamprey and hagfishes are the only jawless fishes found today and are very unlike their ancient ancestors.

A line of fishes with hinged jaws and paired fins probably arose from the ancient agnaths. The evolution of hinged jaws was an important innovation; it allowed alternative feeding methods and life styles to develop. The jaw was developed from a set of gill support bars. Four lines evolved from the ancestral jawed fish; two, the Chondrichthyes and Osteichthyes, are living today.

The members of the modern class *Chondrichthyes* (sharks, skates, rays) have a cartilaginous skeleton, whereas those of the class *Osteichthyes* have a bony skeleton. The primitive bony fishes had lungs in addition to gills. Early in their evolution the class split into two divergent groups. One group (ray-fins) underwent great radiation and gave rise to most of the bony fishes alive today; the other group (fleshy-fin), which is now mostly extinct, gave rise to the present-day lungfishes and lobe-finned fish. The lobe-fins and fleshy-fins had lungs and leglike fleshy paired fins that may have enabled them to crawl. They appear to be ancestral to the amphibians.

Amphibians The first members of the class *Amphibia* were quite fishlike, but they became a large and diverse group as they exploited the ecological opportunities on land. Most of the amphibians became extinct by the end of the Permian; the few that survived included the ancestors of the salamanders, the apodes, and the frogs and toads. Amphibians use their moist skin for gas exchange and so must live in a moist environment. Also, they must return to water to reproduce since they utilize external reproduction, lay fishlike eggs, and have an aquatic larva.

Reptiles The members of the class *Reptilia* are the first truly terrestrial group. They use internal fertilization, lay amniotic shelled eggs, have no larval stage, and have scaly, relatively impermeable skin. In addition, their legs are larger and stronger than those of amphibians, their lungs are better developed, and their heart is almost four-chambered.

The stem reptiles underwent a great deal of adaptive radiation and produced many different lineages of reptiles. One line, the archosaurs, gave rise to the crocodiles, flying reptiles, dinosaurs, and birds. Another line of reptiles (therapsids) gave rise to the mammals. At the end of the Cretaceous, there was a mass extinction of the dinosaurs, as well as many invertebrates. Many hypotheses have been proposed to explain this event, including an asteroid collision, comet impact, intense volcanic activity, and long-term climatic change.

Birds The members of the class *Aves* (birds) have evolved many adaptations for their active way of life. Birds are endothermic and homeothermic, their heart is four-chambered, they have feathers for insulation, their bones are light, and their forelimbs are adapted for flight.

Mammals The members of the class *Mammalia* also are endothermic and homeothermic and have a four-chambered heart. They have a diaphragm, hair for insulation, differentiated teeth, and an enlarged neopallium. Embryonic development occurs within the female; the young are born alive and are nourished by milk from the mammary glands. Early in their evolution the mammals split into two groups: the monotremes (egg-laying mammals) and the group that includes the two major mammalian lines, the marsupials and placentals. Marsupial embryos remain in the uterus for a short time and then undergo further development attached to a nipple in the mother's abdominal pouch. Placental embryos complete their development in the uterus.

CONCEPT MAP

Map 42.1

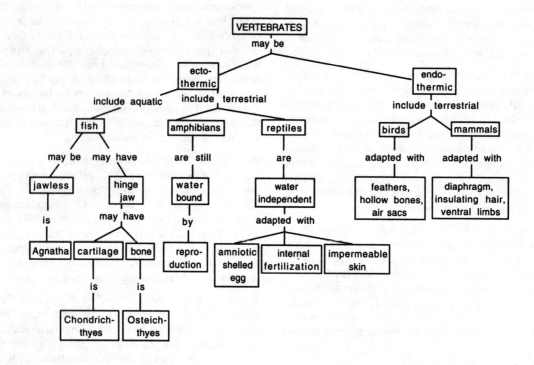

QUESTIONS

Testing recall

Match each item with the one best answer. Each answer may be used more than once.

- a Agnatha
- b Amphibia
- c Aves
- d Cephalochordata
- e Chondrichthyes
- f Echinodermata
- g Hemichordata
- h Mammalia
- i Osteichthyes
- j Reptilia
- k Tunicata

1 feathers a characteristic (p. 975)

2 jawless fish with cartilaginous skeletons and scaleless skin (p. 964)

3 saclike marine forms whose larval stage has well-developed notochord and pharyngeal slits (p. 961)

4 body hair a characteristic (p. 976)

5 primitive animals in which the adult form possesses pharyngeal slits but no notochord (p. 960)

6 first amniotic vertebrates (p. 969)

7 aquatic animals with bony skeletons and respiration by gills (p. 966)

8 slender elongate animals whose body form closely approaches that of early fishlike vertebrates; have fins, distinct tail, show segmentation, and are filter feeders. (p. 962)

9 endothermic animals with a muscular diaphragm (p. 975)

10 marine animals with hinged jaws and cartilaginous skeletons believed to have evolved from ancient jawed fishes (p. 965)

Testing knowledge and understanding

Choose the one best answer.

11 All the following statements are true of both chordates and echinoderms *except*

  a both exhibit indeterminate cleavage in early development.
  b the embryonic blastopore becomes the anus in both.
  c both have true coeloms.
  d both exhibit radial cleavage.
  e both develop pharyngeal slits and a dorsal hollow nerve cord. (pp. 957–61)

12 In humans, the blastopore becomes the anus, and cleavage is radial. An invertebrate phylum sharing these characteristics is

   a Porifera (sponges).
   b Arthropoda.
   c Hemichordata.
   d Mollusca.
   e none of the above. (p. 960)

13 In which pair do the organisms named have very similar larval stages?

   a starfish—vertebrate
   b starfish—annelid
   c hemichordate—arthropod
   d mollusc—starfish
   e starfish—hemichordate (pp. 958–60)

14 According to the classification in your text, the lowest level of classification shared by a frog and a goldfish is the

   a kingdom.
   b class.
   c phylum.
   d order.
   e subphylum. (pp. 961–68)

15 In both arthropods and vertebrates

   a the blastopore becomes the mouth.
   b cleavage is indeterminate.
   c a ciliated larval form is present.
   d the mesoderm originates as outpocketings of the endoderm.
   e a true coelom is present. (p. 961)

16 Both the living jawless fishes and the cartilaginous fishes lack true bone. The ancestors of both these groups

   a had cartilaginous skeletons rather than bony skeletons.
   b had bony skeletons.
   c had no skeleton of any kind.
   d lived on land.
   e were capable of flight. (pp. 965–66)

17 Segmentation is a characteristic found in all the following organisms *except*

   a fish.
   b lancelet.
   c tunicate.
   d frog.
   e snake. (p. 961)

18 Insects and vertebrates may both be regarded as highly evolved representatives of their evolutionary lines. They have convergently evolved many striking similarities, presumably in response to similar selection pressures. Which one of the following is *not* a true statement about a similarity between the two groups of animals?

   a Most members of both groups possess jointed appendages.
   b At least some members of both groups have evolved wings.
   c At least some members of both groups have a structurally distinct and highly specialized head region.
   d In both groups the circulatory system has become highly specialized for oxygen transport, as is necessary in such active animals.
   e Hinged jaws occur in both groups, though their evolutionary derivations are quite different. (p. 961)

19 Two fossil vertebrates, each representing a different class, are found in the undisturbed rock layers of a cliff. One is an early amphibian. The other fossil, found in an older rock layer below the amphibian, is most likely to be

   a a dinosaur.
   b a snake.
   c an insectivorous mammal.
   d a fish.
   e a bird. (p. 968)

20 The first vertebrate land colonizers were

   a amphibians.
   b reptiles.
   c mammals.
   d clams.
   e birds (p. 968)

21 At some stage of development, all animals in the phylum Chordata have

   a a notochord.
   b a backbone.
   c pharyngeal slits or pouches.
   d all the above.
   e *a* and *c*. (p. 963)

22 Which one of the following is the correct sequence of evolution in the subphylum Vertebrata?

   a cartilaginous fishes—jawless fishes—bony fishes—amphibians—reptiles
   b jawless fishes—cartilaginous fishes—bony fishes—birds—reptiles
   c jawless fishes—bony fishes—amphibians—reptiles—mammals
   d jawless fishes—cartilaginous fishes—bony fishes—reptiles—amphibians—mammals (pp. 964–75)

23 The jawless fishes

   a are now extinct.
   b were the direct ancestors of the lobe-finned fish and the lungfish.
   c gave rise to the placoderms.
   d were very large fish with paired fins and scales, much like our modern fish. (p. 965)

24 The lobe-finned fishes appear to have been the direct ancestors of present-day

   a freshwater bony fish.
   b stem reptiles.
   c amphibians.
   d cartilaginous fishes.
   e placoderms. (p. 967)

25 All members of three vertebrate classes *must* have internal fertilization. These three classes are

   a Agnatha, Aves, and Mammalia.
   b Amphibia, Reptilia, and Mammalia.
   c Reptilia, Aves, and Mammalia.
   d Osteichthyes, Amphibia, and Reptilia.
   e Amphibia, Reptilia, and Aves. (pp. 969–75)

26 In which one of the following animals is the embryo *not* protected by amniotic fluid?

   a robin   *d* human being
   b alligator   *e* lion
   c bullfrog   (p. 968)

27 A study of fossils suggests that mammals evolved directly from

   a therapsid reptiles.   *d* lobe-finned fishes.
   b dinosaurs.   *e* modern reptiles.
   c amphibians.   (p. 975)

28 Birds show all the following adaptations for flight *except*

   a endothermy (warm-bloodedness).
   b four-chambered heart.
   c good vision, hearing, and equilibrium.
   d solid bones for strength.
   e insulating feathers.   (p. 975)

29 Mammals share all the following characteristics *except*

   a endothermy (warm-bloodedness).
   b four-chambered heart.
   c hollow bones.
   d insulating hair.
   e diaphragm.   (p. 976)

30 A study of fossils suggests that the most immediate ancestors of birds were

   a amphibians.   *d* mammals.
   b placoderms.   *e* reptiles.
   c bony fishes.   (p. 973)

For further thought

1 Discuss the problems faced by animals in terrestrial life, and explain how the important adaptations in the insects and terrestrial vertebrates enable them to solve these problems.

2 Arthropods and vertebrates may both be regarded, in a sense, as the most highly evolved representatives of their respective evolutionary lines. They have independently evolved many similar solutions to the problems of survival and reproduction. Discuss ways in which the groups resemble each other and ways in which they differ.

3 Movement onto land presented reproduction problems for both plants and animals.

   a Describe the processes of reproduction in the primitive plants and in lower animals.
   b Compare and contrast the manner in which the terrestrial higher plants and animals "solved" the problem of no longer being able to fertilize eggs in water and the mechanisms that evolved to supply nutrition to the growing embryo and protect it against desiccation.

4 Draw a phylogenetic tree of the seven vertebrate classes with modern representatives. Explain in some detail the evolutionary significance of the agnatha, lobe-finned fishes, and therapsids.

ANSWERS

Testing recall

1 *c*   4 *h*   7 *i*   9 *h*
2 *a*   5 *g*   8 *d*   10 *e*
3 *k*   6 *j*

Testing knowledge and understanding

11 *e*   16 *b*   21 *e*   26 *c*
12 *c*   17 *c*   22 *c*   27 *a*
13 *e*   18 *d*   23 *c*   28 *d*
14 *e*   19 *d*   24 *c*   29 *c*
15 *e*   20 *a*   25 *c*   30 *e*

Chapter 43

THE EVOLUTION OF PRIMATES

A GENERAL GUIDE TO THE READING

This last chapter of the text describes the evolution of the primates and of our own species. As you read Chapter 43 in the text, you will want to focus on the following topics.

1 Evolution of the primates. Especially important in the discussion of this topic (pp. 979–81) are the basic primate characteristics listed on pages 979 to 983. You will want to learn them.

2 Evolution of human beings. Learn the anatomical changes that occurred in the evolution of human beings from an apelike ancestor (pp. 984–87). The relationships among the fossil hominids are summarized in Figure 43.15 (p. 988).

3 The interaction of cultural and biological evolution. You will want to understand how human beings are able to influence not only the evolution of other species but also our own biological evolution. (pp. 993–95)

KEY CONCEPTS

1 Fossil evidence indicates that the members of the mammalian order Primata arose from a tree-dwelling stock of small, shrewlike insectivores. (p. 479)

2 Many of the traits most important to us as human beings first evolved because our distant ancestors lived in trees. (pp. 979–81)

3 The upright posture and bipedal locomotion of human beings may have evolved as adaptations for freeing the hands to carry objects or to gather or manipulate food, or they may have made it easier to maintain surveillance. (pp. 985–86)

4 Cultural and biological evolution are interwoven, though cultural evolution can proceed mor rapidly than biological evolution. Humans can now directly affect their future biological evolution. (pp. 993–95)

CROSSLINKING CONCEPTS

1 Knowledge of the molecular techniques for determining phylogenetic relationships described in Chapter 32 enables you to understand how these tools have been used to establish the degree of relatedness of human beings to the living great apes.

2 Your study of preadaptation (Chapter 31) provides the basis for understanding the evolution of the human hand.

3 Your understanding of biological evolution gives you a basis for comparing it with cultural evolution.

OBJECTIVES

1 List 11 characteristics shared by all primates. (pp. 979–80)

2 Describe the characteristics of the two primate suborders and name some representatives of each group. Also describe the differences between Old World and New

World monkeys and indicate which group is more closely related to hominoids Figure 43.5 (p. 982) may be helpful.

3 Discuss current ideas of the evolutionary history of humans. In doing so, describe some of the anatomical changes that occurred in the course of evolution from ape ancestor to modern human, and discuss the roles that bipedalism, tool use, and increased brain size may have played in human evolution. (pp. 984–86)

4 State the order in which early hominids appear in the fossil record and describe the major morphological and cultural characteristics of each form. (pp. 986–90)

5 Discuss how biological and cultural factors have interacted in the course of human evolutionary history. (pp. 990–95)

KEY TERMS

Homo sapiens (p. 979)
Primata (p. 979)
prosimian (p. 981)
tarsier (p. 981)
anthropoid (p. 981)
New World monkey (p. 982)
Old World monkey (p. 982)
gibbon (p. 983)
hominoid (p. 981)
ape (p. 984)
Hominidae (p. 984)
bipedal locomotion (p. 985)
Australopithecus afarensis (p. 986)
A. africanus (p. 987)
A. robustus (p. 987)
A. boisei (p. 987)
Homo habilis (p. 987)
H. erectus (p. 989)
archaic *H. sapiens* (p. 989)
Neanderthal man (p. 989)
Cro-Magnon man (p. 989)
cultural evolution (p. 993)

SUMMARY

Evolution of the primates Fossil evidence indicates that the members of the mammalian order Primata arose from a tree-dwelling stock of small shrewlike insectivores. Important characteristics of the primates include grasping hands (with nails and opposable first digits), increased mobility of the shoulder joint (braced by the clavicle), rotational movement of the elbow, good stereoscopic vision, and expansion of the brain. These traits are associated with life in the trees.

The living primates are classified in two suborders: Prosimii and Haplorhini. The prosimians (lemurs, lorises, and others) are more primitive; they are small arboreal animals with opposable first digits. The Haplorhini include tarsiers, New World monkeys, Old World monkeys, gibbons, and the hominoids: apes and humans.

The New World and Old World monkeys differ from each other in many ways; for example, most New World monkeys have a prehensile tail, nostrils of New World monkeys are separated by a wide partition, and New World monkeys tend to be smaller than Old World monkeys. The gibbons and apes evolved from the line leading to the Old World Monkeys. Gibbons are smaller than the apes. Apes, including orangutans, gorillas, and chimpanzees, are all fairly large animals that have no tail, a relatively large skull and brain, very long arms, and a tendency to walk semierect on the ground.

Evolution of human beings The earliest members of the family Hominidae (humans) probably arose from the same stock that produced the gorillas and chimpanzees. In the course of human evolution from a man-ape ancestor, the jaw became shorter, the skull more balanced on top of the vertebral column, the braincase larger, the nose more prominent, the arms shorter, and the feet flatter and the big toe ceased being opposable. The ape and human lineages probably split between 4 and 6 million years ago.

The upright posture and *bipedal locomotion* of human beings may have evolved as adaptations for freeing the hands to carry objects or to gather or manipulate food, or they may have made it easier to maintain surveillance.

The first truly humanlike fossils belong to the genus *Australopithecus*, the earliest species being *A. afarensis* is. Fossil indicate that these early hominids were bipedal but rather apelike in their skeletal features. At least three later species of *Australopithecus* lived contemporaneously.

Another lineage of the early australopithecines diverged greatly and gave rise to the genus *Homo*. *Homo habilis*, the first clear representative of the *Homo* line, evolved from *australopithecus* about 2 million years ago. They had larger brains and used stone tools. *H. habilis* was later replaced by *Homo erecturs*. *Homo sapiens* probably evolved from this form about 300,000 years ago.

Two forms of *H. sapiens* arose: *H. sapiens neanderthalensis* in Europe and modern *H. sapiens sapiens* in Africa. Modern humans spread over Eurasia and replaced the Neanderthals about 30,000 years ago.

Cultural and biological evolution are interwoven, although cultural evolution can proceed more rapidly than biological evolution. Human beings profoundly influence the evolution of other species and now have the ability to influence and control their own biological evolution.

CONCEPT MAP

Construct a concept map relating the terms Hominidae, apes, bipedal locomotion, upright posture, australopithecines, *Australopithecus afarensis*, later australopithecines, *Homo*, *H. habilis*, *H. erectus*, *H. sapiens*, Neanderthals, modern humans, tool-making, fire, biological evolution, and cultural evolution.

QUESTIONS

Testing recall

Match each item with the one best answer. Answers may be used once, more than once, or not at all.

a Australopithecus afarensis
b Australopithecus africanus
c Australopithecus boisei
d Homo erectus
e Homo habilis
f Homo sapiens
g tree-dwelling insectivore

1 earliest tool-making hominid (p. 988)
2 probable ancestor of both the *Homo* and *Australopithecus* lines (p. 986)
3 robust species of australopithecine (p. 987)
4 first hominid to use fire and leave Africa (p. 989)
5 earliest fossils are about 300,000 years old (p. 989)
6 common ancestor of all the primates (p. 981)
7 earliest representative of our own genus (p. 988)
8 *first* true hominid to walk upright (p. 986)

Testing knowledge and understanding

Choose the one best answer.

9 Which of the following is *not* a characteristic associated with primates in general?

   a location of the foramen magnum, the hole in the skull through which the spine passes, in the bottom of the skull
   b mobile digits (fingers and toes)
   c eyes placed toward the front of the head
   d fingernails rather than claws
   e two mammary glands (pp. 979–80)

10 Which one of the following human characteristics evolved primarily as an adaptation for an arboreal (tree-dwelling) life?

   a hair
   b freely rotating shoulder joint
   c bipedalism
   d bowl-shaped pelvis
   e head balanced over the vertebral column (p. 979)

11 The evolution of which one of the following human characteristics probably *cannot* be explained in part, either directly or indirectly, by the fact that the ancestors of human beings were adapted for an arboreal existence?

   a binocular (3-D) vision
   b freely rotating shoulder joint
   c opposable thumb
   d great reduction of body hair
   e enhanced eye-hand coordination (p. 980)

12 According to the best available evidence, which one of the following was the most recent ancestor of *Homo sapiens*?

   a Homo habilis
   b Homo erectus
   c Australopithecus robustus
   d Australopithecus africanus
   e Australopithecus afarensis (p. 989)

13 The animal most closely related to *Homo sapiens* is the

   a gibbon.   d orangutan.
   b chimpanzee.   e monkey.
   c gorilla.   (p. 984)

14 What is the scientific name given to Neanderthal man?

   a Australopithecus africanus
   b Australopithecus robustus
   c Homo erectus
   d Homo habilis
   e Homo sapiens (p. 989)

For further thought

1 Do you think that the concept of race has any biological validity as applied to human geographical variation? Justify your answer.

2 Why is it reasonable to regard the highly developed brain of human beings as both a cause and a consequence of human culture?

ANSWERS

Testing recall

| 1 e | 3 c | 5 f | 7 e |
| 2 a | 4 d | 6 g | 8 a |

Testing knowledge and understanding

9 a	12 b
10 b	13 b
11 d	14 e